Cellular Networks - Positioning, Performance Analysis, Reliability

Cellular Networks - Positioning, Performance Analysis, Reliability

Editor

Fredrik Jonasen

scitus
academics

Cellular Networks - Positioning, Performance Analysis, Reliability
Edited by **Fredrik Jonasen**

ISBN: 978-1-68117-282-8
Library of Congress Control Number: 2016937454

SCITUS Academics LLC,
www.scitusacademics.com
Box No. 4766, 616 Corporate Way,
Suite 2, Valley Cottage,
NY 10989

Notice

Printed in United States of America on Acid Free Paper

Preface

Signals must be separated in time, space or frequency in order to prevent multiple transmission overlap and interference. FDMA and TDMA techniques aim at separating transmissions into different frequency bands and time frames. These systems allow multiple users to communicate in a single area without data collisions. Networks can also be physically separated by space in order to avoid data collisions. In such cases, users can simultaneously communicate on the same channel, provided that they are located in different networks. A mobile network or mobile network is the latest wireless communication network. Modern telecommunications systems comprise wireless cellular networks. Today it is hard to imagine our lives without these networks. Nevertheless, the development, implementation and operation of these networks require engineers and scientists to address a number of related issues. The network is distributed over land areas known as cells, each served by at least one fixed-site transceiver, known as a base station. This base station provides cell network coverage that can be used for voice, data and other data transmission. Each cell has different frequency characteristics. This means that frequencies can be re-used by non-adjacent cells without interference. User devices such as mobile phones can therefore communicate even if the device moves through cells during transmission. Cellular networks provide subscribers with advanced features on alternative solutions, including increased capacity, low battery power consumption, increased geographical coverage and reduced signal interference. The Global Mobile Communication System, the general packet radio service, 3GSM and the multiple access code division are popular cellular technologies.

This book will be of immense valuable for professionals in academia and industry as well as for telecommunications graduates.

Table of Contents

Chapter 1

Cluster-Based Design for Two-hop Cellular Networks

Hrishikesh VENKATARAMAN[1], Sinan SINANOVI'C[2], Harald HAAS[2]

[1]*School of Engineering and Science, Jacobs University Bremen, Germany*

[2]*Institute for Digital Communications, University of Edinburgh, UK*

ABSTRACT

Optimal resource allocation with an objective of maximizing the system capacity is an NP-hard problem in multihop cellular networks. Hence, different heuristic algorithms have been developed over the years that would improve the network system capacity. In this paper, a novel cluster-based architecture is proposed for a two-hop cellular network whereby the transmission distance between any communicating pair is restricted to half the cell radius. In this design, a given radio resource is used by two simultaneously communicating pairs in every hexagonal cell, but for only half the time slot period. The characteristic feature of this cluster-based design is that it enables a frequency reuse ratio of one. The proposed hierarchical system is analyzed and tested under realistic propagation conditions including lognormal shadowing. It has been observed that the system capacity of a cluster-based design is 2.5 times that obtained from the single-hop cellular system with no relaying. In addition, the cluster-based design achieves higher capacity compared to state-of-the-art two-hop algorithms. This is an important finding since the hierarchical cluster-based approach has fewer degrees of freedom in the selection of the routing path for the end-to-end connection. Practical routing algorithms should be able to benefit from this.

Keywords: Cluster-Based Design, Synchronized Resource Reuse, Interference Model, Spatial Protection Margin

1. INTRODUCTION

Some of today's key challenges in the design of wireless systems are to provide high peak data rates as well as to provide a network architecture that allows for an efficient utilization of the scarce spectrum resources while the power consumption of the network is minimized. Currently deployed single-hop cellular networks as a stand-alone technology are handicapped by numerous limitations, viz., inability to cover dead zones, high attenuation of signals, high shadowing, inefficient use of energy, etc. A direct evolution of the existing cellular network architecture are multihop hybrid cellular networks, where the communication between the mobile station (MS) and the base station (BS) takes place in multiple hops [1]. One can envisage a multihop hybrid cellular network as a means to enable sharing of information between possibly mobile sensor nodes or gathering of sensed information toward query points on a wireline network [2]. Alternatively, one can view multihop cellular network models as a method to extend the communication coverage and provide higher data rate for an infrastructure-based cellular network [3]. Such a hybrid network model aims at providing global connectivity. At the same time, it seeks to mitigate interference, and to maximize the system capacity of the network while achieving a frequency reuse of one. This is the target being aimed for the development of 4th generation, or IMT-Advanced, wireless networks. These systems are primarily based on orthogonal frequency division multiplexing (OFDM), time division multiple access (TDMA), frequency division multiple access (FDMA) and time division duplexing (TDD) [4].

In recent years, there has been extensive research work in the direction of capacity scaling for multihop hybrid wireless networks. It is shown in [5] that having the infrastructure based BS component over the multihop ad hoc network drastically increases the connectivity of the network. For a multihop hybrid cellular network with n nodes and m BSs, the results in [6] show that if m grows asymptotically slower than $\sqrt{n}$, the benefit of adding BSs on capacity is insignificant. However, if m grows faster than $\sqrt{n}$, the system capacity increases linearly with the number of BSs providing an effective improvement over the multihop ad hoc network. Therefore, in order to achieve non-negligible capacity gain, the investment in the wired infrastructure should be high enough. It has been shown through outage analysis in [1] that an integration of cellular and multihop communication models results in better relaying and avoids traffic congestion. Deploying relays can clearly help improve the performance of the users near the edges of the cell and has the potential to solve the coverage problems for high data rates in macrocells [7]. TDD is the enabling technology for the multihop design [8]. Hence, by having simultaneous transmission by both BSs and relays, capacity gains can also be achieved in the cellular network. However, a multihop hybrid cellular design also requires extra radio resources for relaying hops and is sensitive to the quality of relaying routes. Therefore, multihop hybrid cellular networks require a welldesigned radio resource allocation strategy in order to secure performance gains. A hybrid architecture, viz., mobile assisted data forwarding (MADF) is proposed in [9], wherein, a multihop relaying system is overlaid on the existing cellular networks. The main objective of this system is to dynamically divert the traffic load from a hot cell (highly loaded cell) to cooler cells (lightly loaded) in its neighborhood. Similarly, a multihop cellular network (MCN) architecture is investigated in detail in [10,11,12], wherein, the end-to-end communication is always between the MS and BS, like in a traditional single-hop cellular network. There has been considerable research work in finding different routing techniques for multihop cellular networks, viz., base assisted ad hoc routing (BAAR), base-driven multihop bridge routing (BMBP), [13] single-interface multihop cellular network routing protocol (SMRP) [14], for different kinds of traffic patterns. These techniques effectively utilize the ad hoc relaying in presence of fixed infrastructure in order to achieve enhanced network capacity. However, it has been shown in [15] that optimum resource allocation in multihop cellular networks, with the objective of throughput maximization with radio resource allocation (TM-RRA) is an NP-hard problem. In fact, the wellknown multiple choice knapsack problem, (MCKP), which is proved to be NP-hard [16], is shown to be a restricted version of TM-RRA [15]. Hence, researchers across the scientific community have worked towards designing suboptimal but efficient heuristic algorithms and architecture designs.

In this paper, a multihop cellular network model is designed such that all communication between the source and destination nodes is routed through the BS. All the mobile nodes communicate with the BS in either single-hop or two hops. However, the focus of this paper is on designing a novel cluster-based architecture for a two-hop cellular network. Section 2 presents the system model and the underlying mechanism of the cluster-based design. Section 3 explains how the system capacity is calculated for such a design. A deterministic clusterbased technique is described in Section 4 and a semi-analytical model is presented for calculating the different interferences and the carrier-to-interference ratio, γ. In addition, the capacity bounds for the semianalytical model and the confidence interval due to lognormal shadowing are also calculated in Section 4. The simulation model and the results of the cluster-based design are presented in Section 5 and Section 6 respectively, along with a performance comparison with other benchmark algorithms. Finally, conclusions are provided in Section 7.

2. SYSTEM MODEL

A multi-cellular system, with a BS at the center of each cell, is considered in the network design. The maximum distance between the BS and the edge of the cell is given by, r. There are 19 hexagonal cells in the coverage area. These 19 cells are arranged such that a center cell is surrounded by six cells in the 1st tier and twelve cells in the 2nd tier. A Protocol Model [17] is considered in the system design in order to reduce the interference. According to this model, a circular exclusion region is defined around every communicating receiver, such that no other transmitter apart from the desired transmitter communicates in this exclusion region. The radius of this circular exclusion region, r_c, is given by the following equation.

$$r_c = (1+\Delta)d_c \tag{1}$$

In the literature [17], r_c is also sometimes termed as the exclusion range. In Equation (1) above, d_c is the distance between the transmitter and receiver of any communicating pair, and $\Delta \geq 0$ is the spatial protection margin, that indicates the ratio of increase of the exclusion range distance to the transmission range distance. Hence, at any time instant in a TDMA system, all receivers of the simultaneously communicating pairs are inherently separated from the unintended transmitters, by at least, the exclusion range distance, i.e., the minimum distance between any receiver and an unintended transmitter is at least $(1+\Delta)$ times the transmission distance of the desired communication pair. It is shown in [18] that under the Protocol Model, the system capacity is maximized when the spatial protection margin of the Protocol Model is around $\Delta=1.0$. Hence, a spatial protection margin of $\Delta=1.0$ is considered throughout this work, while designing the cluster-based architecture for the two-hop cellular network.

The proposed cluster-based design is based on the formation of multiple clusters in every cell. Each cell is initially divided into two layers, viz., inner layer and outer layer.

1) Inner Layer: This is the circular area contiguous to the BS; and the MS in this zone communicate to the BS directly using a single-hop. The distance between the MS and BS in the inner layer is always less than or equal to half the cell edge length, i.e., r/2.

2) Outer Layer: This circular region is located around the inner zone, and the MSs located in this region communicate to the BS in two hops. This area is further divided into several clusters. The MSs within any of the clusters would communicate to the BS via a cluster-head node, called the gateway (GTW). The GTWs are located on the boundary adjoining the inner and outer layer, i.e., at a

distance of r/2 from the BS. Since the radius of the hexagonal cell is r, the maximum distance between the GTW and the MSs in the outer layer would be r/2. Hence, the maximum transmission distance in the cell, i.e., between BS and GTW (inner layer), or between MS and GTW (outer layer) is r/2.

The MSs located in the outer layer are grouped into several clusters. A single cell scenario depicting the schematic of a cluster-based two-hop cellular architecture is shown in Figure 1. There are six circular clusters in each cell. For each of the clusters, a wireless terminal located at the boundary of the inner and outer layer of the cell is selected as a cluster-head node, alternatively known as GTWs. There are six GTWs/cell, each of them located at a distance of r/2 from the BS. Cluster-heads GTW_{1a} and GTW_{1b} are diametrically opposite to each other and are separated by a distance of r, i.e., twice the transmission distance, r/2. The same holds for the cluster-heads GTW_{2a} and GTW_{2b}, and for GTW_{3a} and GTW_{3b}. In practice, the GTWs could be fixed relay stations (RSs), located on the street lamps/roof tops, or, MSs/wireless terminals with their own traffic. In case of fixed RSs/GTWs, they could probably be placed at exactly r/2 from the BS.

However, if the MSs are selected as relays, then the exact location of the relay node would depend on the distribution and the density of the mobile terminals. Hence, the selected GTW node could be located at a distance slightly less or greater than r/2 from the BS, and also, the GTWs would not be equidistant to each other. There would be a small yet noticeable difference in the system performance due to fixed/mobile GTWs, and is explained later in Section 6. In addition, a deterministic cluster-based design is considered in the semi-analytical model, later in this paper, where the six GTWs in the cell are assumed to be both equidistant to each other, located exactly at a distance of half the cell radius from the BS, and most importantly, the MSs in the outer-layer of the cell are assumed to be at a maximum distance of r/2 from the cluster-head GTWs. An important point to be noted is that the number of clusters per cell in the cluster-based design need not be always six. It could be two, four, six, eight, ten or even higher. The only condition is that the number of clusters per cell has to be an even number, due to the basic principle of simultaneous transmission of communication pairs located in the diametrically opposite clusters in the cell. However, in practice there is a limitation that as the number of clusters per cell is increased, the amount of resources that could be given to one cluster decreases. In a very recent work, it is shown [19] that for a cellular network, with six BSs surrounding the central BS, the optimum number of GTWs in each cell that maximizes the system capacity is six.

This justifies the selection of six GTWs in the clusterbased two-hop cellular design. The variation of system capacity for different number of GTWs per cell is shown in Section 6. In addition, the cell region could be divided such that the radius of the inner layer is τr where $0 < \tau < 1$ is the ratio of the inner layer radius to the radius of the cell [20]. In that case, the maximum transmission distance would no longer be restricted to r/2. However, the cluster-based design would be still valid, as would be observed later when the GTWs are selected from the MSs.

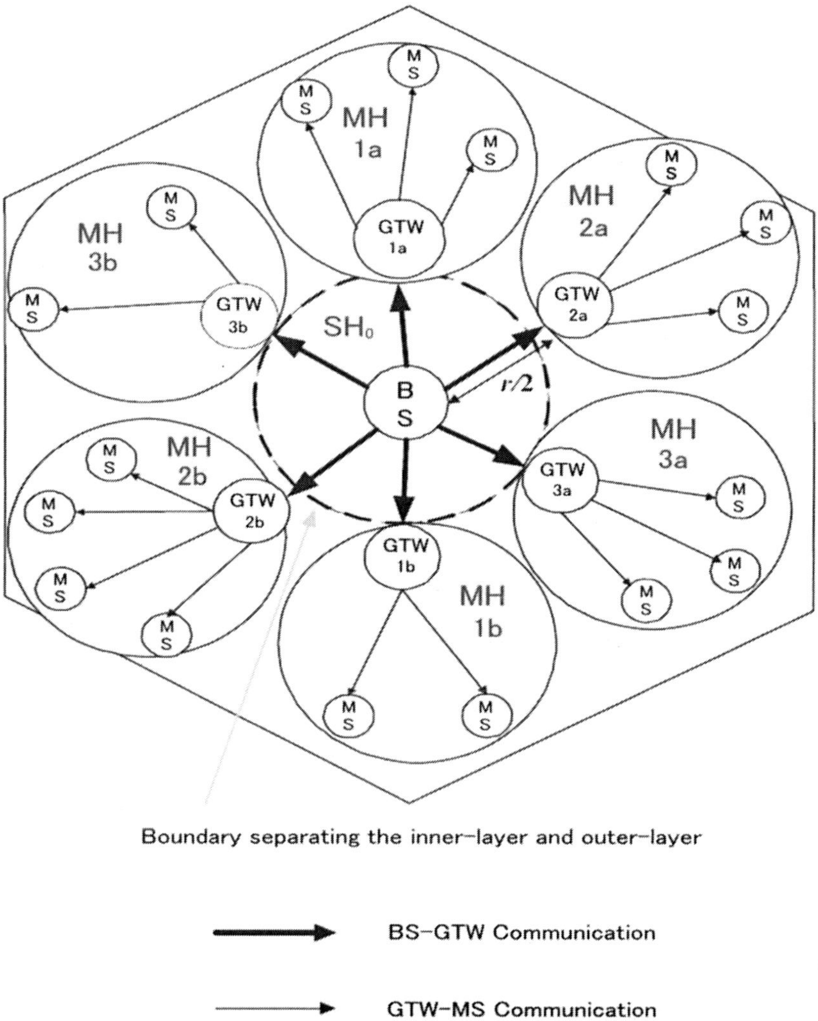

Figure 1. Schematic model of a cluster based two-hop cellular network (downlink).

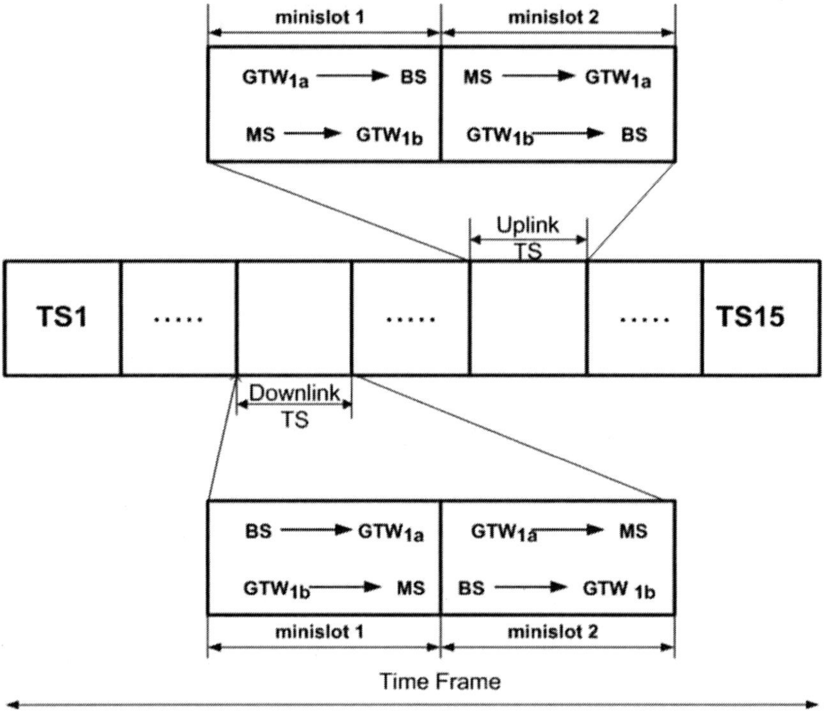

Figure 2. Synchronized resource reuse mechanism for a TDD/TDMA cluster based two-hop cellular network.

In order to understand the complete working mechanism, a conceptual model of the cluster-based two-hop cellular network with six clusters/cell and equidistant GTWs at a fixed distance of r/2 from the BS is considered, and the underlying principle of the synchronized resource reuse technique is described as follows:

1) As shown in Figure 1, SH_0 is the inner-layer (single-hop region) and MH_{1a}, MH_{1b}, ... MH_{3b} are the two-hop clusters in the outer layer. In addition, GTW_{1a}, GTW_{1b} ... GTW_{3b} are the respective cluster-heads for MH_{1a}, MH_{1b} ... MH_{3b}. Each cluster contains a number of MSs. In case of downlink communication between BS and a wireless terminal located in any of the clusters, the BS would communicate to the cluster-head GTW in the 1st hop, and in the 2nd hop, the GTW would communicate to the MSs associated with the corresponding clusters. Similarly, in the uplink, the MSs in any of the clusters communicate with the BS in two hops, wherein, the MS communicates to the GTW in the 1st hop, and the GTW communicates to the BS in the 2nd hop.

2) A TDD/TDMA scheme is considered for the cluster-based two-hop network. For a multihop system with number of hops per link, M≥1, the signal for any hop can be transmitted only for T/M time slot duration, where T is the TS period. Hence, the TS is divided into two minislots for the two-hop links. However, for a wireless node located in the inner layer (SH_0), the communication between the wireless terminal and the BS would take place in single-hop, for the full duration of one TS.

3) The reusability of the resources is increased by allowing two multihop clusters in any cell to occupy the same TS at the same frequency. As shown in Figure 1, the clusters MH1a and MH1b are located at diametrically opposite sides of the BS. The synchronized TDD frame structure for both uplink and downlink is shown in Figure 2. In the downlink, GTW_{1a} can download to the MS in its cluster in a particular time slot. At the same time instant, the BS could download to GTW_{1b} in the opposite cluster of the same multihop cell. It should be noted that both these simultaneously communicating pairs are outside the exclusion region of each other. Similarly, in the next time slot, the GTW_{1b} → MS and BS → GW_{1a} communication takes place simultaneously.

4) In the uplink, the transmitters and receivers of the cluster-based model are reversed, as seen in Figure 2, but the governing principle of the resource reuse technique remains the same. It can be therefore noted that the reuse of the resources can be done independently for both uplink and downlink, using the synchronized resource reuse technique. Hence, this cluster-based design remains valid even for asymmetric traffic, as long as the traffic asymmetry remains the same for all the cells.

5) The given TS resource is also allotted to each of the hexagonal cells in the system. As shown in Figure 3, for both uplink and downlink, the transmitters of all the concurrently communicating pairs in the adjacent cells are beyond the exclusion region of the desired receiver in the intended cell. Hence, as shown in Figure 4, a given TS resource is not only used by two simultaneously communicating pairs in any cell, but also, the same TS resource is reused in every cell. However, it is to be noted that a TS resource given to a communicating pair in a two-hop network is only half the time slot period given to an equivalent single-hop network, as shown in Figure 2. Due to using a resource twice within a cell, however, the cluster-based design effectively results in a frequency reuse factor of one.

6) The GTWs can be considered to be equidistant and located at approximately r/2 from the BS if they are fixed terminals. However, if the GTWs are not fixed, and are selected from the distributed MSs, then, the wireless terminal located at either half the cell radius or closest to half the cell radius (either in the inner-layer or in the outer-layer) is selected as a GTW. Irrespective of whether the selected GTW is in the inner layer or outer layer, the transmission distance of the communicating pair between the BS and GTW would be then different

from r/2. Correspondingly, the transmission distance of the GTW - MS pair would also vary.

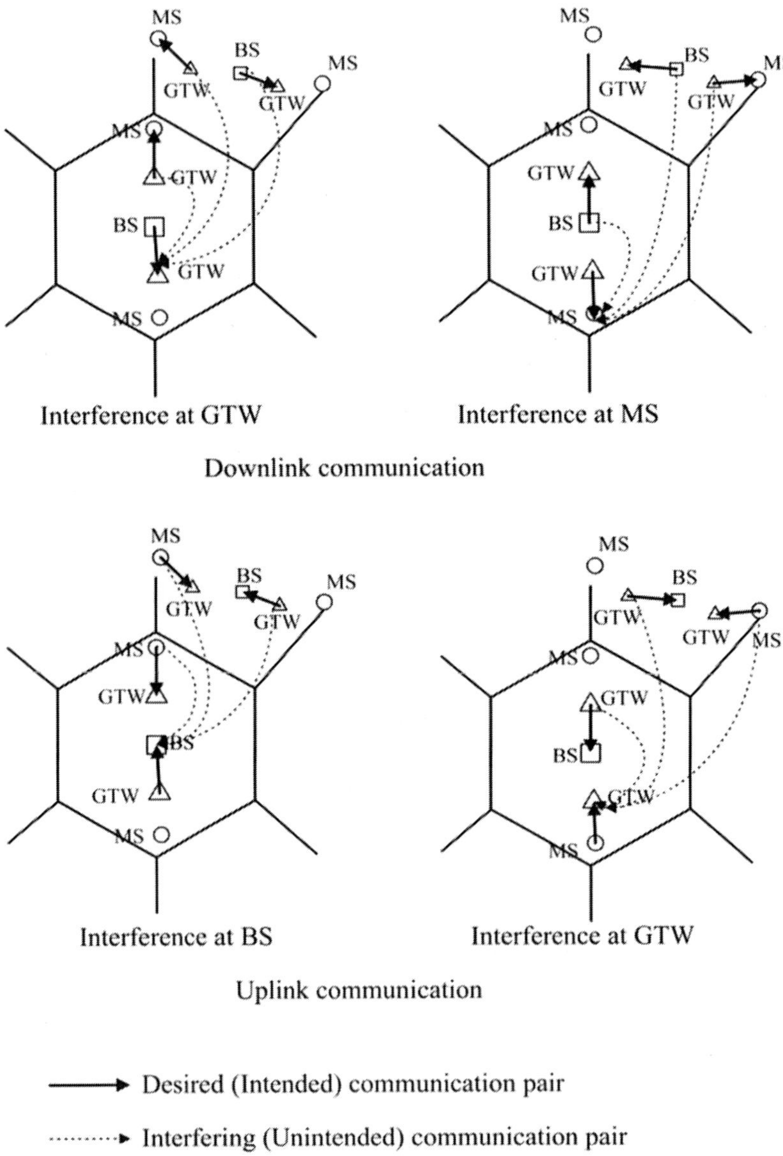

Figure 3. Interference reduction mechanism for cluster based two-hop cellular network using synchronized resource reuse technique.

3. CAPACITY CALCULATION

All the wireless terminals in any cell are assumed to transmit their signals with the same power, P_T. If d_c is the transmission distance between any communicating pair, then the power received, P_R, using a general propagation model is given by:

$$P_R = P_T - (k_1 + 10\alpha \log_{10}(d_c) + \xi_c) \tag{2}$$

where k_1 is a constant that depends on the propagation environment (indoor/urban/suburban), α is the path loss exponent and ξ_c is the shadowing factor across the transceiving pair. In a multi-cell scenario, the given radio resource is utilized by all the cells in the system. The transmitters of all simultaneously communicating pairs in the seven-cell scenario are marked and shown in Figure 4 along with the interference calculation at the receiver gateway (i.e., the receiver of BS → GTW pair) of the center cell (cell 0). The thick arrows in Figure 4, from BS → GTW and GTW → MS in all the cells represent the simultaneous communicating pairs. A reference line (dotted line in the figure) is considered that connects the BS of cell 0, cell 1 and cell 4. The dashed lines from the transmitting BSs and GTWs of cell 2 and cell 3, to the Rx GTW in the center cell indicates the distance of the Rx GTW from other interfering transmitters in cell 2 and cell 3.

In all, the Rx GTW in cell 0 would experience interference from thirteen interferers: two interferers from each of the six adjacent cells and one transmitting GTW (of GTW → MS pair) from cell 0.

For any communicating pair, the inter-cell interference only across six adjacent cells, i.e., the 1st tier of cells is considered. The transmitting interferers from the 2nd tier of cells are very far from the intended receiver, and hence, the interference generated from these transmitters is assumed to be negligible. Therefore, for any communicating pair in this cluster-based model, there is one interferer from own cell and two interferers from each of the adjacent cells. The carrier-to-interference ratio is therefore calculated as follows:

$$\gamma = \frac{10^{-[k_1 + 10\alpha \log_{10}(d_c) + \xi_c]}}{\sum_{i=1}^{N_I} 10^{-[k_1 + 10\alpha \log_{10}(d_i) + \xi_i]}} \tag{3}$$

where d_i is the distance of the desired receiver from the ith interfering entity and N_I is the total number of interfering entities for any receiver in a cluster-based model. ξ_i accounts for shadowing between the desired receiver and the ith interfering transmitter. The capacity in bps/Hz/cell is calculated by finding the system capacity independently over seven cells (center cell and six cells in

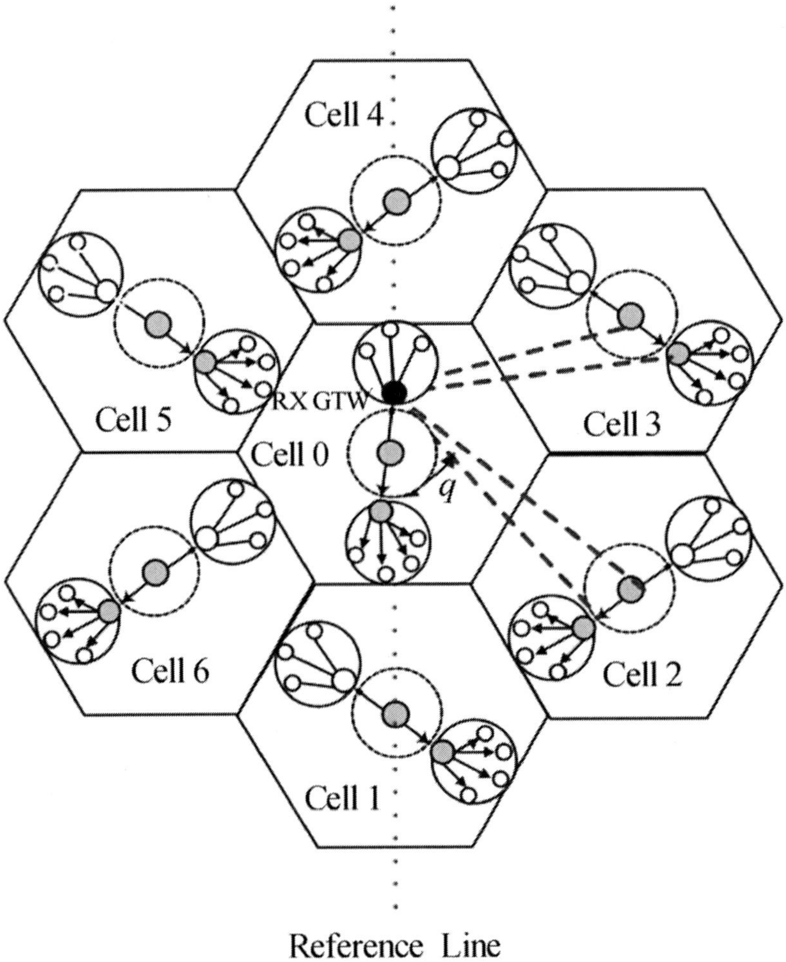

Figure 4. Distance calculation for different interfering entities in the downlink model.

the 1st tier), as shown in Figure 4, and averaging over them. Each cell in the 1st tier is surrounded by six cells out of which three cells belong to the 2nd tier. The traffic in the twelve cells of the 2nd tier only contribute for the intercell interference calculation for the 1st tier of cells. This 2nd tier of cells is necessary to remove the boundary effects while calculating γ for the 1st tier of cells, and hence, the Shannon bound is not calculated for the twelve cells in the 2nd tier. As shown in Figure 2, the data across each communicating pair is transmitted for only half the time slot period in a two-hop system. As a consequence, the

Shannon capacity has to be scaled by a factor of 1/2. Also, in each of the seven cells, there are two simultaneously communicating pairs, and depending on the distance of the interfering transmitters the receivers of these two communicating pairs would have different values of γ. Therefore, the system capacity (of only the two-hop links) is calculated from the Shannon equation as:

$$C = \frac{1}{2N_c} \sum_{i=1}^{N_c} \sum_{j=1}^{N_l} \log_2(\gamma_{ij} + 1) \text{ bps/Hz/ Cell} \qquad (4)$$

where γ_{ij} is the carrier-to-interference ratio of the jth communicating pair in the ith cell. N_l is the number of concurrently communicating pairs in the outer layer that use the same radio resource, in any single cell. For a cluster-based design, two pairs located diametrically opposite to each other communicate simultaneously, i.e., $N_l = 2$. $N_C = 7$ is the number of cells over which the system capacity is calculated. In order to calculate the average per-cell system capacity, the Shannon capacity equation in Equation (4) is summed up over all N_c cells and averaged over them.

4. SEMI-ANALYTICAL MODEL

In order to assess the performance of the synchronized resource reuse technique for the cluster-based design, a semianalytical model is developed for a deterministic cluster-based two-hop network. In this deterministic cluster-based model, the GTWs are fixed and located exactly at a distance of r/2 from the BS. Also, the GTWs are equidistant from each other. Hence, the six GTWs in the cell represent the six vertices of a regular hexagon, with a side length of r/2. In addition, the MSs in the outer layer are assumed to be always located at a distance of r/2 from their respective GTWs; and also, the MSs are assumed to be uniformly distributed between [0°,360°] across the outer layer of the cell. Hence, the distance between the BS and GTW, and also the distance between the MS and its corresponding GTW is fixed. This simplifies the analysis for numerically calculating the Shannon capacity of the cluster-based two-hop network. However, the precise location of the GTW in the cell is determined from the angle made by the BS - GTW pair with the reference line, as shown in Figure 5. q_{11} indicates the angle made by the 1st communication pair, BS → GTW, in the intended cell (cell 0 in this case) with the reference line of cell 1, whereas, q_{11} indicates the angle

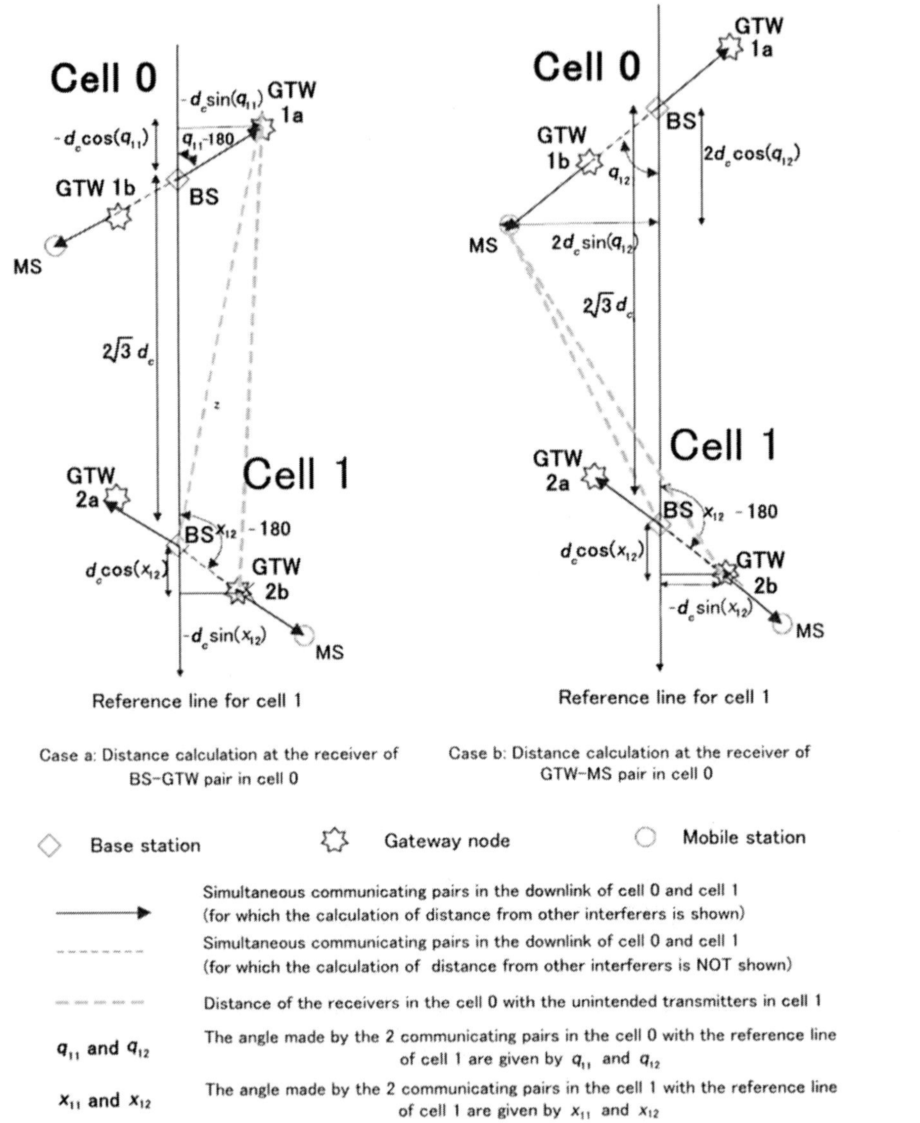

Figure 5. Calculation of distance at the receivers of BS → GTW and GTW → MS pairs (downlink) from the two interfering transmitters of ith cell.

made by the 2nd communication pair, GTW → MS, in the intended cell (cell 0 in this case) with the reference line of cell 1. Similarly, x_{11} and x_{12} indicates the angles made by the 1st interfering communication pair, BS → GTW, and the 2nd interfering communication pair, GTW → MS, (both) in cell 1 with the reference line of cell 1. The precise location of any MS is determined by the angle made by the line joining the MS and its corresponding GTW with the reference line. At the same time, in the deterministic cluster-based model, the transmission distance is always the maximum possible value, r/2.

In Section 6, the capacity results obtained from the deterministic cluster-based model are compared with the simulation model. For both uplink and downlink, this Semi-analytical model first calculates the distance of the intended receiver from all simultaneously communicating intended transmitters. A downlink schematic for a seven-cell cluster-based two-hop model is shown in Figure 4. It should be noted from Figure 4 that only the locations of the BSs are fixed. The distance between two BSs is $\sqrt{3}$ r. Since the transmission distance, d_c=r/2, for the cluster-based model, the distance between two BSs can also be written as $2\sqrt{3}d_c$. In Figure 4, all the transmitters in the seven cells are shaded with gray background. The black circle in the center cell marks one of the desired receivers which would experience interference from other unintended transmitters. The distance between the black circle (desired receiver) and all the gray colored circles (interfering transmitters from the own cell and all the adjacent cells) marks the distance of the different interfering entities. Hence, as shown in Figure 4, the total interference experienced by a receiver depends on the relative distance between this receiver and all its interfering transmitters.

4.1 Carrier-To-Interference Calculation for Downlink.

In the downlink, the communication takes place from BS → GTW and from GTW → MS. Figure 5 (both, case a and case b) shows the simultaneously communicating pairs in cell 0 and cell 1 in the downlink scenario. As seen in Figure 5, there are two simultaneously communicating pairs per cell, i.e., the BS → GTW pair and GTW → MS pair. The receivers of cell 0 would experience interference not only from its own cell, but also from the simultaneously communicating pairs from other cells. The interference experienced by the communicating pairs in cell 0 are calculated as follows:

1) BS → GTW Communication in the Intended Cell:

When the gateway in the intended cell is the desired receiver (say, GTW1a in cell 0 in Figure 5), the distance between this gateway and the interfering transmitters of the adjacent cell are calculated as shown in Figure 5(a). There are two cells, cell 0 and cell 1. Using basic trigonometry, the distance of the communicating receiver in cell 0 from the interfering transmitters in cell 1 is computed. As shown in Figure 5(a), the distance of receiving gateway at cell 0, GTW1a, from the BS of cell 1 is given by:

$$d_{\mathrm{BS}_1}{}^2 = (2\sqrt{3}d_c - d_c\cos(q_{11}))^2 + (d_c\sin(q_{11}))^2 \tag{5}$$

whereas the distance of the unintended transmitting gateway of the cell 1, GTW2b, to the desired gateway receiver in cell 0 is given by:

$$d_{\mathrm{GTW}_1}{}^2 = (2\sqrt{3}d_c + d_c\cos(x_{12}) - d_c\cos(q_{11}))^2$$

$$+ (d_c\sin(x_{12}) - d_c\sin(q_{11}))^2 \tag{6}$$

The angle, q_{11}, is formed between the line joining the communicating pairs, BS→GTW1a in cell 0 with the reference line of cell 1. Similarly, x_{12} is the angle between the line joining the communicating pairs, GTW2b → MS in cell 1, with the reference line of cell 1. The above equations, Equation (5) and Equation (6) could be generalized to calculate the interference coming from the transmitters of all the six adjacent cells into the desired receiver, i.e., the GTW of the intended cell. By changing the reference line for each of the six adjacent cells, the distance of the interfering transmitters from the ith cell can be calculated as follows:

$$d_{\mathrm{BSi}}{}^2 = (2\sqrt{3}d_c - d_c\cos(\theta_{i1}))^2$$

$$+ (d_c\sin(\theta_{i1}))^2 \tag{7}$$

$$= 13d_c{}^2 - 4\sqrt{3}d_c{}^2\cos(\theta_{i1}) \tag{8}$$

whereas the distance of the unintended transmitting GTW to the desired GTW receiver is given by:

$$d_{\mathrm{GTW}_i}{}^2 = (2\sqrt{3}d_c + d_c\cos(\phi_{i2}) - d_c\cos(\theta_{i1}))$$

$$+ (d_c\sin(\phi_{i2}) - d_c\sin(\theta_{i1}))^2 \tag{9}$$

$$= 14d_c{}^2 + 4\sqrt{3}d_c{}^2(\cos(\phi_{i2}) - \cos(\theta_{i1}))$$

$$- 2d_c{}^2\cos(\phi_{i2} - \theta_{i1}) \tag{10}$$

Here,

$$\theta_{i1} = q_{i1} + 60(i-1) \tag{11}$$

is the angle in degrees made by the BS→GTW communicating pair in the intended cell with the reference line of the ith cell, and

$$\phi_{i2} = x_{i2} + 60(i-1) \tag{12}$$

is the angle in degrees made by the GTW → MS in the ith cell with the reference line of the ith cell (Figure 5(a) shows the angle x_{12} made by the GTW2b → MS communicating pair in cell 1, with the reference line of cell 1). It should be noted that all θ and ϕ vary uniformly from [0°, 360°]. In addition, the distance of the intra-cell interfering transmitter is, $d_{owncell}=2d_c$. The carrier-to-interference value at the receiver of any communication pair is therefore given by:

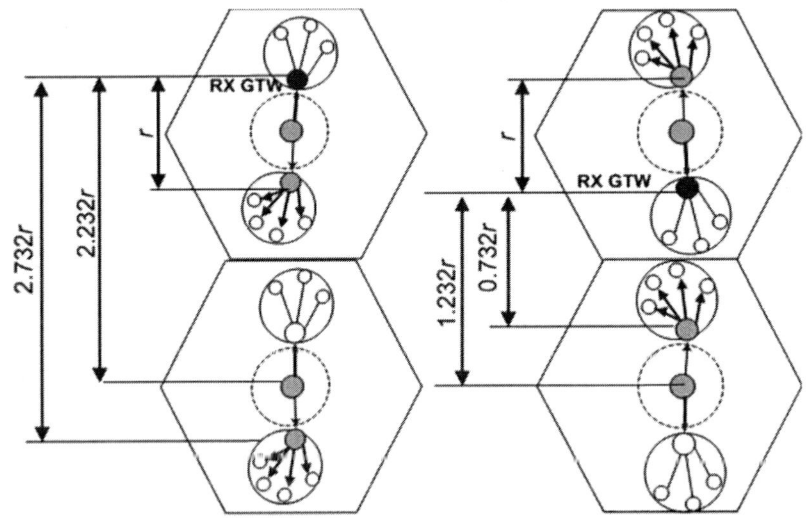

Case a. The inter-cell interferers for the BS-> GTW communicating pair are at the maximum distance. The resulting interference is at its lowest value, and hence, the capacity of the pair is at its highest value.

Case b. The inter-cell interferers for the BS-> GTW communicating pair are at the minimum distance. The resulting interference is it at its highest value and hence, the capacity of the pair is at its lowest value.

Figure 6. Minimum and maximum distance of the inter-cell interfering entities (downlink) from the receiver of the BS → GTW.

$$\gamma = \frac{d_c^{-\alpha}}{(2d_c)^{-\alpha} + \sum_{i=1}^{6} (d_{(GTW)_i})^{-\alpha} + \sum_{i=1}^{6} (d_{(BS)_i})^{-\alpha}} \tag{13}$$

Dividing the numerator and denominator by $d_c^{-\alpha}$ results in:

$$\gamma = \frac{1}{2^{-\alpha} + \sum_{i=1}^{6} (\beta_{(GTW)_i})^{-\alpha} + \sum_{i=1}^{6} (\beta_{(BS)_i})^{-\alpha}} \tag{14}$$

where

$$\beta_{(BS)i}^{\ 2} = 13 - 4\sqrt{3}\cos(\theta_{i1}) \tag{15}$$

$$\beta_{(GTW)i}^{\ 2} = 14 + 4\sqrt{3}\cos((\phi_{i2}) - \cos(\theta_{i1}))$$

$$-2\cos(\phi_{i2} - \theta_{i1}) \tag{16}$$

If the orientation of the GTW is fixed with respect to the BS, then the interference and the capacity of the BS → GTW pair in the semi-analytical model varies only with the location of all the interfering transmitters from own cell and adjacent cell. Hence, the best and worst case for the capacity of BS → GTW pair in the intended cell can be calculated by considering all the interferers to be located at the maximum and minimum distance from the intended receiving GTW of the BS → GTW pair.

Upper bound for capacity:

As can be seen from Figure 6(a), the own-cell interferer (transmitting GTW from the diametrically opposite cluster) is at a fixed distance of r from the intended receiver (GTW of cell 0). Similarly, the interfering transmitters of the adjacent cell (BS and GTW) are at a distance of $d_{BS} = \sqrt{3}r + \frac{r}{2} \approx 2.232r$ and $d_{GTW} = \sqrt{3}r + r \approx 2.732r$ respectively from the intended receiver. It should be observed that these intercell interferers are at a maximum possible distance from the intended receiver; and hence, causes the least interference to the intended communicating pair. The carrier power at the receiver of the BS → GTW pair is given by

$$P_R = P_T - (k_1 + 10\alpha\log 10(r/2)) \tag{17}$$

For a multi-cellular network with r=217m, a propagation model [21] with k_1=37, α=4, and assuming a transmit power of 1W, i.e., 0 dBW, the received power, P_R, would be -118.4 dBW. The BS → GTW pair in the intended cell would experience interference from six adjacent cells. Hence, the upper bound for the capacity of the BS → GTW communicating pair is obtained by substituting the values of these interferences distances into Equation (13) for calculating the resulting interference. The upper bound of the capacity for the BS → GTW communicating pair is then obtained by substituting the resulting interference into Equation (4). The total interference power experienced by the receiving GTW node would be -129.2 dBW. The upper bound for the capacity of BS → GTW pair would be 3.68 bps/Hz.

Lower bound for capacity:

The lower bound for the capacity of GTW → MS pair is obtained by considering the minimum distance of the inter-cell interfering entities. As can be observed from Figure 6(b), the minimum distance of the interfering transmitters (BS and GTW) from the adjacent cell are $d_{BS} = \sqrt{3}r - \frac{r}{2} \approx 1.232r$ and $d_{GTW} = \sqrt{3}r - r \approx 0.732r$ respectively from the intended receiver. The resulting lower bound for the capacity of BS → GTW pair is 0.72 bps/Hz.

2) GTW → MS Communication Pair in the Intended Cell:

For the 2nd active communication pair, GTW → MS, in the downlink of cell 0, the maximum distance of the intended receiver, i.e., the MS of the GTW → MS pair, from the BS is twice the transmission distance. Therefore, as seen in Figure 5, the MS in cell 0 is located at a distance of $d = 2d_c = r$ from the BS, and is distributed uniformly from [0°, 360°]. The distance of the interfering BS and the interfering GTW from the cell 1 is calculated from basic trigonometry as:

$$d_{BS_1}{}^2 = (2\sqrt{3}d_c - 2d_c\cos(q_{12})) + (2d_c\sin(q_{12}))^2 \tag{18}$$

$$d_{GTW_1}{}^2 = (2\sqrt{3}d_c - d_c\cos(x_{12}) - 2d_c\cos(\theta_{12}))^2 + ((d_c\sin(x_{i2}) - 2d_c\sin(\theta_{12}))^2 \tag{19}$$

This equation for calculating the distances of the interfering transmitters from other cells could be written in a generalized form as:

$$d_{BS_i}{}^2 = (2\sqrt{3}d_c - 2d_c\cos(\theta_{i2}))^2 + (2d_c\sin(\theta_{i2}))^2 \tag{20}$$

$$d_{GTW_i}{}^2 = (2\sqrt{3}d_c + d_c\cos(\phi_{i2}) - 2d_c\cos(\theta_{i2}))^2$$

$$+(d_c \sin(\phi_{i2}) - 2d_c \sin(\theta_{i2}))^2 \quad (21)$$

where, $\theta_{i2} = q_{i2}+60(i-1)$ is the angle in degrees made by the GTW → MS pair in the intended cell with the reference line of the ith cell (Figure 5(b) shows the angle

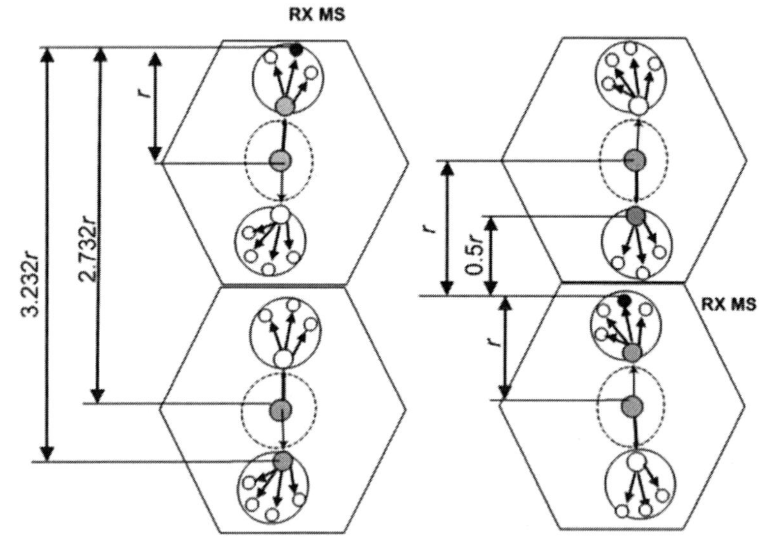

Case a. The inter-cell interferers for the GTW -> MS communicating pair are at the maximum distance. The resulting interference is at its lowest value, and hence, the capacity of the pair is at its highest value.

Case b. The interferers for the GTW -> MS communicating pair are at the minimum distance. The resulting interference is it at its highest value and hence, the capacity of the pair is at its lowest value.

Figure 7. Minimum and maximum distance of the inter-cell interfering entities (downlink) from the receiver of the GTW → MS pair in the intended cell.

q_{12} between the communicating pair, GTW1b → MS with the reference line of cell 1). Equation (20) is simplified and the corresponding equations for $\beta_{(BS)}$ and $\beta_{(GTW)}$ are given as:

$$\beta_{(BS)_i}^{\ 2} = 16 - 8\sqrt{3}\cos(\theta_{i2}) \quad (22)$$

$$\beta_{(GTW)_i}^{\ 2} = 17 + 4\sqrt{3}(2\cos(\theta_{i2}) - \cos(\phi_{i2}))$$

$$-4\cos(\phi_{i2} - \theta_{i2}) \tag{23}$$

It should be noted that the equation for γ remains the same as given in Equation (14).

Upper bound for capacity:

In the semi-analytical model, the MS in the GTW → MS pair is assumed to be at a maximum distance of r/2 from the GTW. The own-cell interferer is the transmitter from the diametrically opposite cluster. Hence, the own cell interferer is at a distance of r from the desired receiver. It can be observed from Figure 7 that the maximum distance of the interferers from an adjacent cell are $d_{BS} = \sqrt{3}r + \frac{3r}{2} \approx 2.732r$ and $d_{GTW} = \sqrt{3}r + \frac{3r}{2} \approx 3.232r$ respectively. Hence, substituting these values into Equation (13) and Equation (4) results in the upper bound for the capacity of the GTW → MS communicating pair, which is 3.91 bps/Hz.

Lower bound for capacity:

Similarly, the lower bound for the capacity of the GTW → MS pair is calculated by knowing the distance of all the interfering entities. The own cell interferer is at a distance of r, whereas the minimum distance of the interferers from any adjacent cell are 0.5r and r, as is shown in Figure 7. Hence, the lower bound of the capacity for the communicating pair is 0.21 bps/Hz.

3) Capacity Bounds for End-to-End Link: In a two-hop system, the capacity of the end-to-end link is limited by the lower value of the capacity of any of the two-hops of the link, i.e., if the capacity of the 1st hop of the link is less than the capacity of the 2nd hop of the link, then the capacity of the end-to-end link is limited by the capacity of the 1st hop of the link [22]. Hence, the capacity of the end-to-end link in the downlink scenario in the cluster-based design is limited by the lesser of the capacity values of BS → GTW pair and GTW → MS pair. Hence, the lower and upper bounds of the capacity for the cluster-based two-hop design is limited by the capacity values of the BS → GTW pair, and is equal to 0.21 bps/Hz/cell and 3.68 bps/Hz/cell respectively.

It should be noted at this stage that the upper bound of the GTW → MS pair would be higher than 3.91 bps/Hz/cell if the distance between the GTW and MS is less than r/2. However, since the upper bound of the downlink capacity for the cluster-based design is limited by the lesser of the capacity values of BS → GTW and GTW → MS pair, the higher values of the capacity obtained by the

GTW → MS pair does not change the capacity result of the cluster-based design.

4) Effect of Lognormal Shadowing on Capacity Bounds: In presence of lognormal shadowing, the bounds for the system capacity would vary. The amount of variation would depend on the standard deviation of the lognormal shadowing. Due to the summation of lognormal variables in the calculation of the interference power (as shown in Equation (3)), it is very difficult to find an exact expression that would reflect the effect of lognormal shadowing on the system capacity. Instead, the effect of lognormal shadowing on the capacity bounds is computed in this paper. Lognormal shadowing with zero mean and standard deviation of 4 dB is considered throughout the analysis [23]. The received power for the carrier signal would vary within a value of 2ζ, i.e., 8 dB, for a confidence measure of 95% [24].

Effect on upper bound of capacity:

The upper bound of the capacity in case of downlink communication is 3.68 bps/Hz/cell, achieved for the BS → GTW communication. The corresponding value of the carrier power and total interference power are -118.4 dBW and -129.2 dBW respectively, resulting in a γ of 10.8dB. An 8 dB variation in the signal strength implies that the carrier power would be between -110.4 dBW and -126.4 dBW for 95% of the cases, i.e., a variation of 6.75% on either side of -118.4 dBW. There are thirteen transmitters that would interfere with the intended communicating pair and the resulting interfering power would be a summation of all these interferers. If the same confidence measure of 95% is to remain for both the carrier signal and the resulting interference signal, then a higher confidence measure should be assumed for each of the interfering entities. A confidence measure of 99.6% for the power received from each of the thirteen independent interfering transmitters would result in a confidence measure of 95% ($0.996^{13} = 0.95$) for the resulting interferers [25]. However, a confidence measure of 99.6% for each of the interfering entities implies that the total interference power would vary by 6ζ, i.e., 24 dB. Hence, the total interference power would be between -105.2 dBW and -153.2 dBW for 95% of the cases, i.e., a variation of 18.55% on either side of -129.2 dBW. Therefore, γ, would experience a variation of 25.3% (6.75+18.5) around 10.8 dB (-118.4+129.2). Hence, the system capacity would vary between 2.75 bps/Hz/cell and 4.61 bps/Hz/cell for 90% of the cases, due to lognormal shadowing. It should be noted that the multiplication of the confidence measures for the carrier power (95%) and for the interference power (95%) results in a confidence measure of 90% for the values of γ and the system capacity.

A significant inference that can be derived from the above calculation is that even for a shadowing with a zero mean and 4 dB standard deviation, the upper bound of the system capacity would vary by 25.3% around its mean, for 90% of the cases. If the environment causes a much higher shadowing, the variation in the upper bound of the system capacity would be still higher. This shows the significance of taking the shadowing into considerations while allocating the resources in a wireless network.

Effect on lower bound of capacity:

The lower bound of the capacity in case of downlink communication is 0.21 bps/Hz/cell, achieved for the GTW → MS communication. The corresponding value of the carrier power and total interference power are -118.4 dBW and -110.3 dBW respectively, resulting in a γ of -8.1 dB. Similar to the upper bound case, a confidence measure of 95% implies a variation of 6ζ = 24 dB in the interference power. Hence, the total interference power would be between -86.3 dBW and -134.3 dBW for 95% of the cases, i.e., a variation of 21.75% on either side of -110.3 dBW. Given that the carrier power would experience a variation of 6.75% on either side of -118.4 dB, the value of γ and the system capacity would experience a variation of 28.5% (21.75% + 6.75%) around 0.21 bps/Hz/cell, for 90% of the cases. Hence, for a lognormal shadowing with zero mean and a standard deviation of 4 dB, the lower bound of the system capacity would vary between 0.15 bps/Hz/cell and 0.27 bps/Hz/cell for 90% of the cases. It can be observed that the absolute effect of lognormal shadowing on the lower bound of the capacity value is not a significant issue, since the lower bound of the system capacity is already a very small value.

4.2. Carrier-to-Interference Calculation for Uplink

In the case of an uplink as well, there exist two simultaneously communicating pairs in the cluster-based model: the MS → GTW and the GTW → BS pairs located at the diametrically opposite clusters. In the case of an uplink, both the transmitters in the cluster-based design: the GTW and the MS, are not fixed, whereas, the receiver of one of the communicating pairs, i.e., the BS of GTW → BS pair is located in a fixed position. This results in a slightly modified expression for β in case of uplink, as compared to the downlink scenario:

1) MS → GTW Communication Pair in the Intended Cell:

For the MS → GTW pair communication, the expression for β is given by:

$$\beta_{(\mathrm{MS})i}{}^{2} = 17 + 4\sqrt{3}(2\cos(\theta_{i2}) - \cos(\phi_{i2})) - 4\cos(\phi_{i2} - \theta_{i2}) \quad (24)$$

$$\beta_{(\mathrm{GTW})i}{}^{2} = 14 + 4\sqrt{3}(2\cos(\phi_{i1}) - \cos(\theta_{i2}))$$

$$-4\cos(\phi_{i1} - \theta_{i2}) \tag{25}$$

where ϕ_{i2} and θ_{i2} are same, as defined for the BS →

GTW communication pair in the semi-analytical model; x_{i2} is the angle made by the GTW_{2b} → MS communicating pair in the ith cell with the reference line of the ith cell; and $\phi_{i1} = x_{i1} + 60(i-1)$, is the GTW → BS communicating pair in the ith cell with the reference line of the ith cell.

2) GTW → BS Communication Pair in the Intended Cell:

Similarly, for the GTW → BS communication in the intended cell, the corresponding β values are:

$$\beta_{(MS)_i} = \sqrt{16 + 8\sqrt{3}\cos(\phi_{i2})} \tag{26}$$

$$\beta_{(GTW)_i} = \sqrt{13 + 4\sqrt{3}\cos(\phi_{i1})} \tag{27}$$

3) Capacity Bounds for End-to-End Link: The lower and the upper capacity bounds for the two communicating pairs in the uplink, GTW → BS pair, and MS → GTW pair can be calculated in a similar manner as is done for downlink. The maximum distance of the inter-cell interferers from the receiver of the GTW → BS pair are 1.732r and 2.732r, resulting in a maximum capacity of 3.32 bps/Hz for the GTW → BS communicating pair. Similarly, the minimum distance of the inter-cell interferers from the receiver of the GTW → BS pair are 0.732r and 1.366r, which results in a minimum capacity value of 0.74 bps/Hz. The minimum and the maximum capacity values for the MS → GTW pair, assuming a constant distance of r/2 between the MS and GTW, are 0.88 bps/Hz and 3.71 bps/Hz respectively. Hence, the upper and lower bounds for the system capacity in the uplink scenario are 3.32 bps/Hz/cell and 0.74 bps/Hz/cell respectively.

4) Effect of Lognormal Shadowing on Capacity Bounds: Similar to the downlink scenario, the lognormal shadowing causes a variation in the bounds of the system capacity in the uplink scenario as well. The upper bound of the system capacity for the uplink scenario (3.32 bps/Hz/cell) experiences a variation of 25.3% for 90% of the cases in presence of lognormal shadowing of 4 dB standard deviation. Similarly, the lower bound of the system capacity (0.74 bps/Hz/cell) would vary by 27.25% for 90% of the cases.

5. SIMULATION MODEL

A simulation model for the cluster-based two-hop cellular network is developed in Matlab. An airport or a campus environment, with a total coverage area of 1 km^2 is considered in the system design. There are 19 cells within a coverage area of 1 km^2. Hence, the distance from the centrally located BS to the edge of the cell, r, is around 130 meters. A propagation model with $k_1 = 37$ and $\alpha = 4$ has been considered in the simulation model. 1000 MSs are uniformly distributed around this network coverage area and each cell is designed to have six clusters. All the MSs that are located in the outer layer of the cell are assigned to any one of the six clusters. This assignment is done depending on the closest distance (lowest path loss, in the presence of lognormal shadowing) of the respective MS to the six GTWs in the cell. This results in a system where there are, on an average, eight MSs per cluster. The GTWs are selected from among the MSs. The MSs selected as GTWs are located at nearly half the cell radius. The exact position of the GTW depends on the distribution of the MSs. A TDMA time frame with 16 TSs has been considered in the simulator design. The simulation model calculates the value of γ and the system capacity for seven cells independently and then takes an average over these seven cells. The network is simulated for two different scenarios. In the 1st case, it is assumed that there is no shadowing. Hence, for this case, $\zeta = 0$ in Equation (2). In the 2nd case, a lognormal shadowing with a zero mean and a standard deviation of 4 dB is considered [23].

For different locations of the GTWs and the MSs (with respect to the reference line), the distance of the desired receiver from the interfering transmitters would vary, which in-turn would vary the γ experienced at the receiver of the communicating pair. The synchronized resource reuse technique ensures that all the interfering transmitters are spatially well-separated in distance. The exact value of γ, and thereby the system capacity value, however, depends on the relative distance between the receiver and other transmitting GTWs and MSs. Hence, the system capacity is plotted as cumulative distribution function (cdf) as can be seen from Section 6. The Shannon capacity obtained from the cluster-based two-hop cellular architecture is compared with the following systems:

1) A single-hop cellular network with no relaying:

There are no relays in this design. In every hexagonal cell, the BS and the MS communicate with each other in single hop, irrespective of whether the MS is located in the inner layer or outer layer.

2) A benchmark relaying algorithm for a two-hop cellular network:

The benchmark algorithms for the two-hop cellular design, introduced in [26], provides three efficient methods for finding the wireless terminals that could act as relays in order to maximize the system capacity. These benchmark algorithms could be either distance-based or path loss-based, as explained below. The path loss based algorithms take the random effects, arising due to shadowing, into account. Hence, in the presence of lognormal shadowing, the path loss based algorithms are superior to distance-based algorithms.

5.1. Distance-Based Benchmark Algorithm

In the two-hop design based on benchmark algorithms, the MSs located in the outer layer of the cell communicate to the BS in two hops, as is the case with the cluster-based model. The GTWs/ relay nodes are selected from the mobile nodes available in the network. Suppose, there are N possible two-hop routes, between the BS and the MS in the outer layer. Then, the selected route, rs, is determined, depending on the transmission distance between the BS and the relay node $d_{c_{n1}}$, and between the relay node and the MS, $d_{c_{n2}}$, for each of the n $\in$ N routes. The three selection schemes of the standard benchmark algorithm for two-hop network [26] are given as follows:

a) shortest total distance (STD) selection scheme:

$$r_s = \min_{\forall n \in N}(d_{c_{n1}} + d_{c_{n2}})$$

b) least longest hop (LLH) selection:

$$r_s = \min_{\forall n \in N} \max(d_{c_{n1}}, d_{c_{n2}})$$ and

c) shortest relaying hop distance (SRD) selection:

$$r_s = \min_{\forall n \in N}(d_{c_{n2}})$$

5.2. Path Loss-based Benchmark Algorithm

In addition to the distance-based benchmark algorithms, [26] also introduced the path loss-based benchmark algorithms. Let $P_{L_{n1}}$ and $P_{L_{n2}}$ denote the path losses in dB associated with the first hop (BS and relay node), and second hop (relay node and MS of the outer layer) respectively, along the nth route where n $\in$ N. Then, the selected route is determined as follows:

a) minimum total path loss (MTP) selection scheme:

$$r_s = \min_{\forall n \in N}(P_{L_{n1}} + P_{L_{n2}})$$

b) least maximum path loss (LMP) selection:

$$r_s = \min_{\forall n \in N} \max(P_{L_{n1}}, P_{L_{n2}})$$ and

c) minimum relaying hop path loss (MRP) selection:

$$r_s = \min_{\forall n \in N}(P_{L_{n2}})$$

In order to have a fair comparison, the source MSs in case of uplink (or the destination MSs in case of downlink) remain the same in all the methods, viz., the cluster-based two-hop design, the three standard benchmark two-hop schemes, and the single-hop non-relaying technique. Also,

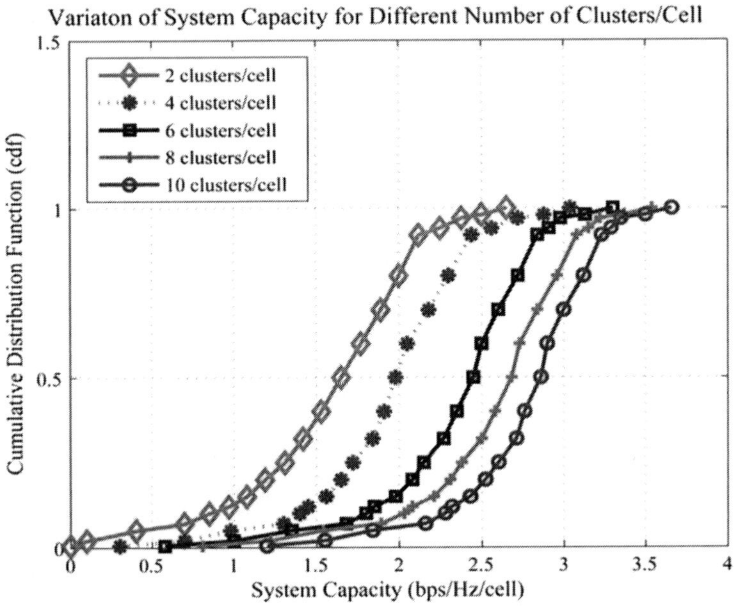

Figure 8. Cumulative distribution function (cdf) of system capacity (average of uplink and downlink) for a two-hop cellular network with different number of clusters/cell.

an interference avoidance model, with an optimum spatial protection margin of Δ = 1.0, is considered for all the different methods. In addition, it should be noted that, in the simulation model, the increase in the overhead due to additional signaling is not considered in any of the two-hop cellular designs. This increase in the overhead in the two-hop design would cause some reduction in the capacity gain with respect to the single-hop cellular network. However, this paper focuses on the different two-hop schemes, and comparing the performance of the two-hop schemes with the single-hop design is not the main focus of this work. Also, it is expected that, the cluster-based architecture with an intelligent resource allocation technique, would require less or same amount of overhead signaling as compared to the benchmark algorithms, for the two-hop cellular network. Hence, the capacity results obtained in this work for the two-hop networks, viz., the cluster-based design and the three benchmark algorithms are directly comparable. In addition, the MSs in the inner-layer of the cluster-based design are not considered in the simulation, for any of the two-hop methods, as well as for the single-hop design. This is because, in both the cluster-based design and the three standard benchmark algorithms, the wireless terminals located in the inner-layer will communicate with the BS directly in single-hop, as in case of single-hop cellular networks. Hence, all these two-hop methods would use the same amount of radio resource for the inner-layer, and hence, the inner-layer design is not considered in this study.

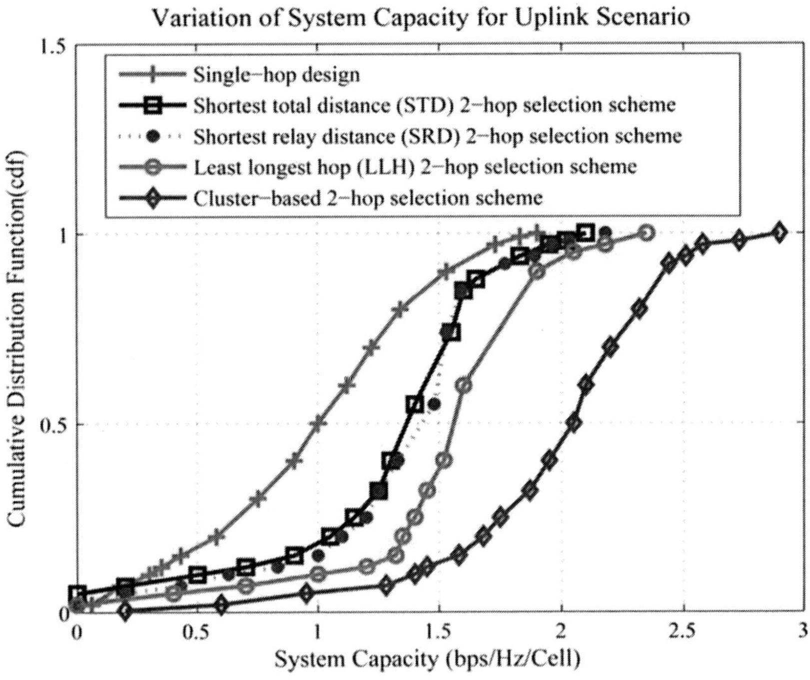

Figure 9. Cumulative distribution function (cdf) of system capacity in the uplink of a multi-cellular network under different transmission schemes (in the absence of lognormal shadowing).

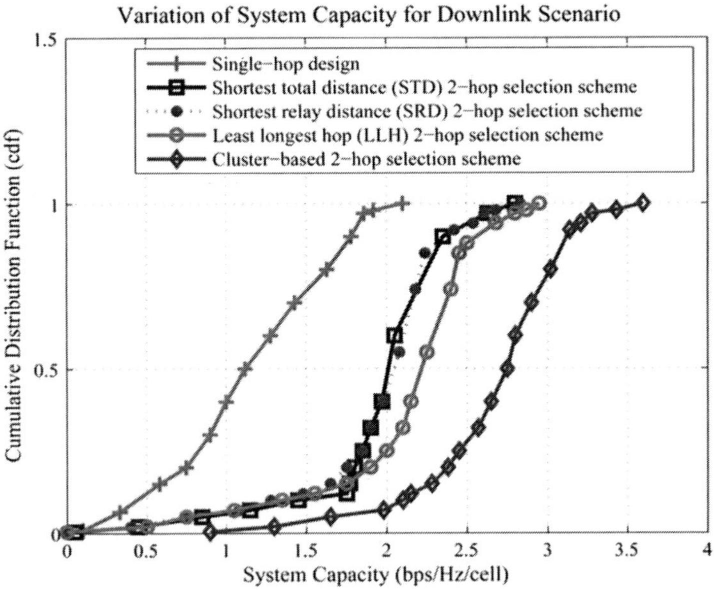

Figure 10. Cumulative distribution function (cdf) of system capacity in the downlink of a multi-cellular network under different transmission schemes (in the absence of lognormal shadowing).

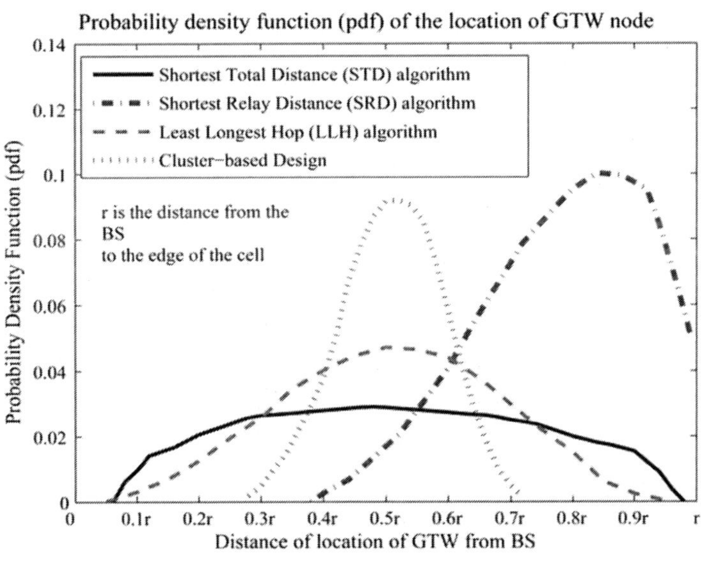

Figure 11. Probability density function (pdf) of the location of GTW for standard benchmark algorithms and the cluster-based design, in the absence of lognormal shadowing.

6. RESULTS

Figure 8 shows the system capacity of the cluster-based two-hop design (average of uplink and downlink results) for different values of clusters per cell. It can be observed that the system capacity shows an increase with an increase in the number of clusters per cell. However, the step size of this increase reduces with an increase in the number of clusters/cell. For example, when the number of clusters per cell is increased from two to four, the expected value of the system capacity is increased from 1.44 bps/Hz/cell to 1.85 bps/Hz/cell, an increase of 0.41 bps/Hz/cell. However, when the number of clusters/cell is increased from four to six, and six to eight, the increase in the expected value of the system capacity is only 0.378 bps/Hz/cell and 0.23 bps/Hz/cell respectively. It should be noted that this increase in the capacity value does not take into account the capacity losses arising due to the increased overhead, as the number of clusters/cell increases. Hence, in all further part of analysis, six clusters/cell are considered, as is also done in [27,28].

Figure 9 and Figure 10 shows the simulation results of the cdf of the system capacity for uplink and downlink scenarios of all different two-hop design methods, in the absence of any lognormal shadowing. The GTWs are selected from among the distributed MSs in the network.

Hence, the GTWs are not located at exactly half the cell radius. It is observed in Figure 10 that in case of downlink in the two-hop cluster-based design, the median of the system capacity (2.71 bps/Hz/cell) is 2.5 times that obtained from the single-hop cellular system with no relaying (1.11 bps/Hz/cell). Similarly, the expected value of the cluster-based design is 2.52 bps/Hz/cell, which is more than twice that obtained from the single-hop cellular network value of 1.12 bps/Hz/cell. More significantly, the cluster-based design shows a superior performance over all three standard benchmark techniques for two-hop cellular network. The capacity behavior of the STD and SRD schemes are nearly similar to each other and their expected values are 1.89 bps/Hz/cell and 1.77 bps/Hz/cell respectively. Hence, the expected value of the system capacity in the cluster-based design is 0.63 bps/Hz/cell and 0.75 bps/Hz/cell better than that obtained from STD and SRD algorithms. The LLH technique provides the best performance out of the three standard benchmark techniques. The expected value of the system capacity for the LLH method is 2.05 bps/Hz/cell, which is higher than the expected value of the capacity obtained from the STD and SRD schemes, but less than the expected value obtained for the cluster-based technique by 0.47 bps/Hz/cell. Similarly, in case of an uplink as well, the expected value of the system capacity for the cluster-based design (1.82 bps/Hz/cell) is greater than the LLH method (1.36 bps/Hz/cell) by 0.46 bps/Hz/cell. In the LLH method, the node that has the minimum value among all longest hops (both between, MS and GTW node, and GTW node and BS) among all possible relay nodes is selected as a relay. Hence, a node located in the vicinity of half the cell radius is selected as a relay, which results in more than one pair utilizing the given resource simultaneously, in any cell. It should be noted that this is similar to the cluster-based design introduced in this paper. However, the significant improvement in the system capacity observed in the cluster-based design is due to the synchronized resource reuse technique proposed in this work that ensures a reuse of the radio resource in every cell. The cluster-based design provides

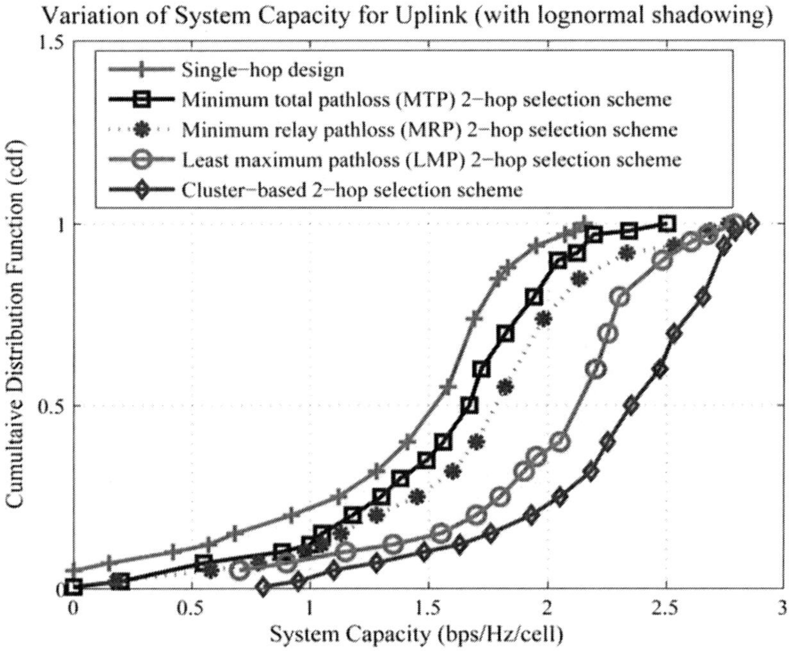

Figure 12. Cumulative distribution function (cdf) of system capacity in uplink, with a lognormal shadowing of zero mean and a standard deviation of 4 dB.

a maximum improvement of 0.8 bps/Hz/cell over the LLH method; and up to 1.4 bps/Hz/cell improvement over the STD algorithm. Figure 11 shows the probability density function (pdf) of the mobile GTWs in the cluster-based design and that of the mobile relay nodes for the benchmark two-hop algorithms. It can be observed from Figure 11 that the pdf of the STD algorithm is almost a straight line, in the range from 0.28r to 0.74r. The SRD algorithm selects the relay that is more towards the cell edge, than at the center of the cell. Hence, the pdf of the SRD algorithm has a non-zero value only after 0.38r. A significant observation that can be made from Figure 11 is that, in case of the LLH method, the mean of the pdf is at 0.5r, same as that of the cluster-based design. Hence, the LLH method outperforms STD and SRD benchmark algorithms. However, the variance of the LLH method is 0.28r, which is twice more than that of the cluster-based design, which has a variance of 0.13r. This implies that in case of LLH method, there is a greater probability of relay nodes being not located in the vicinity of r/2, which results in only one pair in the cell being able to utilize the given resource. It should be noted that if the GTWs are selected from among the MS, then the GTW selection would depend on the distribution of MS. But still, the pattern of the GTWs would remain the same for the different benchmark methods and cluster-based design. Hence, the cluster-based two-hop model, with resource reuse in every cell, gives the best performance in terms of system capacity, as compared to the single-hop non-relaying scenario and the benchmark algorithms for the two-hop cellular network.

The performance of the cluster-based design with MSs as GTWs is then compared with the path loss-based benchmark algorithms, in the presence of lognormal shadowing. It is observed from Figure 12 and Figure 13 that, even in the presence of lognormal shadowing, the performance of the cluster-based two-hop network is superior to all three benchmark techniques, for both uplink and downlink. For example, the expected value of the system capacity for the cluster-based design is 2.18 bps/Hz/cell in case of downlink, and is 0.25 bps/Hz/cell better than LMP (the best performing algorithm among all three benchmark algorithms). Similarly, in case of uplink, the expected value of the cluster-based design is 1.98 bps/Hz/cell, and is 0.09 bps/Hz/cell better than the LMP technique. It should however be noted that, in the presence of lognormal shadowing, the performance of the LMP scheme comes close to the performance of the cluster-based design. This is because, the relays are selected not on the basis of distance measurement, but on the basis of path loss measurement, which vary with lognormal shadowing. The presence of lognormal shadowing results in MSs that are far from half the cell radius, r/2, to be selected as GTWs. As seen from the pdf of the GTW location in Figure 14, in the presence of lognormal shadowing, there is a non-zero probability for a node located beyond 0.8r, to be selected as a GTW. In the absence of lognormal shadowing, the distribution of the GTW is almost symmetric with a mean value of 0.5r. However, in the presence of lognormal shadowing, the pdf of the GTW selection exhibits a long tail, resulting in an expected value of 0.58r. This results in a situation in the cluster-based design, where the exclusion region of a communicating pair in one cell extends to the other cell, and hence, prevents the simultaneous communication of another pair in the adjacent cell. This, in turn, results in a reduction in the gain in the system capacity.

Figure 15 and Figure 16 compare the results of the cluster-based design independently for uplink and downlink, when fixed GTWs are used instead of mobile GTWs. When there is no lognormal shadowing, the expected value of the system capacity, in case of fixed GTWs, is 2.09 bps/Hz/cell (uplink) and 2.76 bps/Hz/cell (downlink) and is greater than the expected value of that obtained

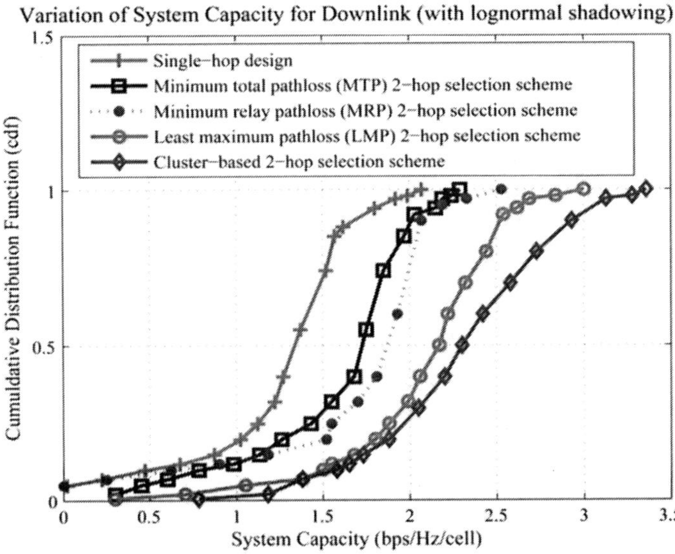

Figure 13. Cumulative distribution function (cdf) of system capacity in downlink, with a lognormal shadowing of zero mean and a standard deviation of 4 dB.

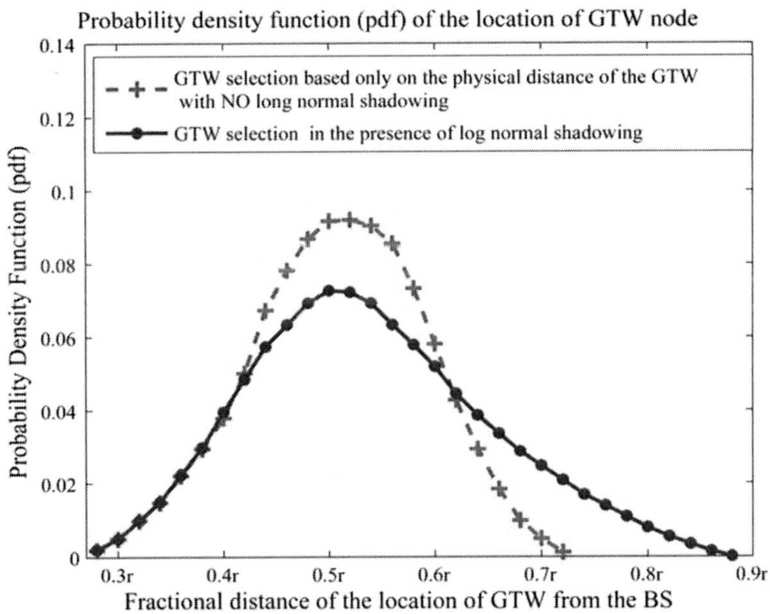

Figure 14. Probability density function (pdf) of the location of GTW in a cell in a multi-cellular network.

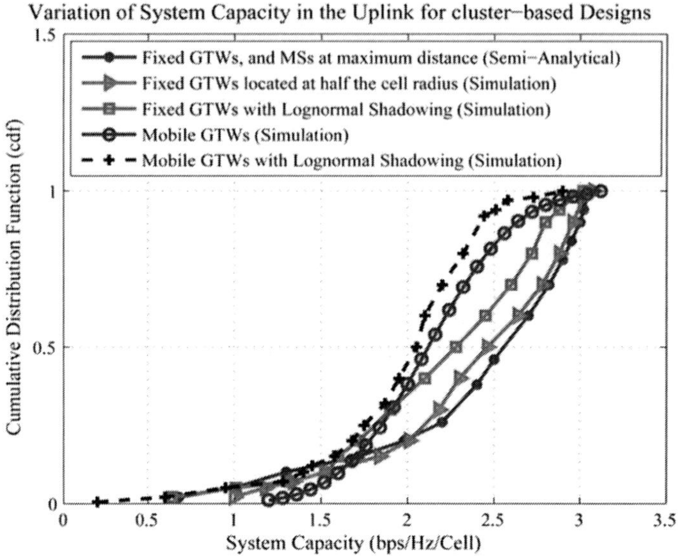

Figure 15. Cumulative distribution function (cdf) of system capacity for different cluster-based scenarios, in case of uplink transmission.

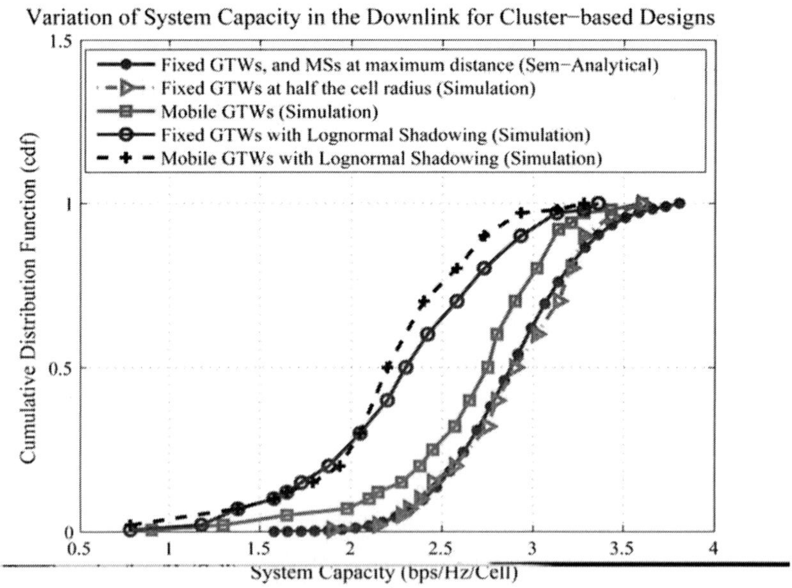

Figure 16. Cumulative distribution function (cdf) of system capacity for different cluster-based scenarios, in case of downlink transmission.

from the mobile GTWs by 0.27 bps/Hz/cell and 0.24 bps/Hz/cell for uplink and downlink respectively. The results observe a similar pattern even in the presence of lognormal shadowing. For example, the expected value of the system capacity in case of uplink is 2.24 bps/Hz/cell for fixed GTWs, and is greater than that of the expected value of the system capacity for mobile GTWs (1.98 bps/Hz/cell) by 0.26 bps/Hz/cell. In addition, it can be observed from Figure 15 (uplink) and Figure 16 (downlink), that the cdf obtained from the semi-analytical model shows a close match with that obtained from the simulation results for cluster-based design with fixed GTWs. The expected values of the system capacity for the semi-analytical model are 2.16 bps/Hz/cell (uplink)

bps/Hz/cell for fixed GTWs, and is greater than that of the expected value of the system capacity for mobile GTWs (1.98 bps/Hz/cell) by 0.26 bps/Hz/cell. In addition, it can be observed from Figure 15 (uplink) and Figure 16 (downlink), that the cdf obtained from the semi-analytical model shows a close match with that obtained from the simulation results for cluster-based design with fixed GTWs. The expected values of the system capacity for the semi-analytical model are 2.16 bps/Hz/cell (uplink) and 2.69 bps/Hz/cell (downlink), and is very close to that obtained from the simulation results for fixed GTWs: 2.09 bps/Hz/cell for uplink and 2.76 bps/Hz/cell for downlink respectively. This is primarily because, in case of semi-analytical model, not only all the GTWs are fixed at a distance of r/2 from the center of the cell, but also all the MSs in the cluster are at the maximum distance of r/2 from the cluster-head GTWs, whereas in the simulation model for fixed GTWs, only the GTWs are fixed at a distance of r/2. Hence, the distance between the cluster-head GTW and the MSs could be any value less than or equal to r/2.

In a significant observation, the upper and lower bounds for the system capacity obtained from the simulation results is quite close to those obtained from the semi-analytical model where all the MSs were assumed to be distributed at a location of r/2 from the GTW. In downlink mode, the lower bound for capacity under the semi-analytical model is 0.21 bps/Hz/cell whereas that obtained from the simulations is 0.22 bps/Hz/cell. The upper bound for the system capacity under the semi-analytical model is 3.68 bps/Hz/cell whereas that obtained from the simulations is 2.91 bps/Hz/cell. Similarly, in the uplink mode, the lower and upper bound for the system capacity under the semi-analytical model is 0.74 bps/Hz/cell and 3.32 bps/Hz/cell respectively; whereas that obtained from the simulations is 0.82 bps/Hz/cell and 2.91 bps/Hz/cell respectively. Hence, it can be concluded that the lower and upper bound results obtained from the simulation model closely match with the results of the semianalytical model, thereby validating the performance of cluster-based two-hop model.

7. SUMMARY AND CONCLUSIONS

In this paper, a novel resource allocation mechanism has been proposed for a two-hop cellular network. The new scheme, known as cluster-based architecture, is designed using a synchronized resource reuse technique. As per this design, each hexagonal cell is divided into two layers, the inner layer and outer layer. All MSs in the inner layer communicate with the BS in single hop, whereas, the MSs in the outer layer communicate with the BS in two hops, using a GTW terminal as relay node. This architecture design, developed under an interference avoidance Protocol Model, results in a frequency reuse ratio of one, whereby, the given resource is used twice in every cell in the system, but for only half the duration of a time slot period. This work, first, shows that the system spectral efficiency of a cellular network can be increased significantly by allowing two-hop communication. Second, it has been found that a hierarchical, co-ordinated approach which essentially means to limit the degrees of freedom for forming the two-hop links does not lower the capacity, but in fact gives higher capacity than compared to state-of-the-art two-hop algorithms. This means that the complexity of the routing problem in such two-hop communication systems can be significantly reduced while the system performance does not have to be compromised.

8. REFERENCES

1. H. Wu, C. Qao, S. De, and O. Tonguz, "Integrated cellular and ad hoc relaying systems," IEEE Journal on Selected Areas in Communication, Vol. 19, No. 10, pp. 2105–2115, October 2001.
2. A. Zemlinanov and G. Veciana, "Capacity of ad hoc wireless networks with infrastructure support," IEEE Journal on Selected Areas in Communications, Vol. 23, No. 3, pp. 657–667, March 2005.
3. H. Venkataraman, H. Haas, S. Yun, Y. Lee, and S. McLaughlin, "Performance analysis of hybrid wireless networks," in Proceedings of IEEE International Symposium on Personal Indoor and Mobile Radio Communications (PIMRC'05), Berlin, Germany, Vol. 3, pp. 1742–1746, 11–14 September 2005.
4. H. W. Chen and M. Guizani, "Next generation wireless systems and networks," in John Wiley & Sons, 2006.
5. O. Dousse, P. Thiran, and M. Hasler, "Connectivity in ad hoc and hybrid networks," in Proceedings of IEEE International Conference on Computer Communications (INFOCOM), Vol. 2, pp. 1079–1088, 23–27 June 2002.
6. B. Liu, Z. Liu, and D. Towsley, "On the capacity of hybrid wireless networks," in Proceedings of IEEE International Conference on

Computer Communications (INFOCOM), San Francisco, USA, Vol. 2, pp. 1543–1552, 30 March–3 April 2003.

7. H. Vishwanathan and S. Mukherjee, "Performance of cellular networks with relays and centralized scheduling," in Proceedings of IEEE Vehicular Technology Conference (VTC Fall'03), Orlando, Florida, USA, Vol. 3, pp. 1923–1928, 6–9 October 2003.
8. H. Haas and S. McLaughlin, Eds., "Next generation mobile access technologies: Implementing TDD," Cambridge University Press, ISBN: 13: 9780521826228, pp. 420, January 2008.
9. X. Wu, S. H. G. Chan, and B. Mukherjee, "MADF: A novel approach to add an adhoc overlay on a fixed cellular infrastructure," in Proceedings of IEEE Wireless Communications and Networking Conference (WCNC'00), Chicago, USA, pp. 549–554, 23–28 September 2000.
10. Y. D. Lin and Y. Hsu, "Multihop cellular: A new architecture for wireless communications," in Proceedings of IEEE International Conference on Computer Communications (INFOCOM) Tel Aviv, Israel, pp. 1273–1282, 26–30 March 2000.
11. R. Ananthapadmanabha, B. S. Manoj, and C. S. R. Murthy, "Multihop cellular networks: The architecture and routing protocol," in Proceedings of IEEE International Symposium on Personal Indoor Mobile Radio Communications (PIMRC'01), San Diego, USA, Vol. 2, pp. 78–82, 30 September–3 October 2001.
12. H. Li, D. Yu, and H. Chen, "New approach to multihop - cellular based multihop network," in Proceedings of IEEE International Symposium on Personal Indoor and Mobile Radio Communications (PIMRC'03), Beijing, China, Vol. 2, pp. 1629–1633, 7–11 September 2003.
13. Y. D. Lin, Y. C. Hsu, K. W. Oyang, T. C. Tsai, and D. S. Yang, "Multihop wireless IEEE 802.11 LANs: A prototype implementation," Journal of Communications and Networks, Vol. 2, No. 4, December 2000.
14. V. Sekar, B. S. Manoj, and C. S. R. Murthy, "Routing for a single interface MCN architecture and pricing schemes for data traffic in multi-hop cellular networks," in Proceedings of IEEE International Conference on Communications (ICC'03), Alaska, USA, Vol. 2, pp. 969–973, 11–15 May 2003.
15. Y. Liu, R. Hoshyar, X. Yang, and R. Tafazolli, "Integrated radio resource allocation for multihop cellular networks with fixed relay stations," IEEE Journal on Selected Areas in Communications, Vol. 24, No. 11, pp. 2137–2146, November 2006.
16. S. Mertello and P. Torth, "Knapsack Problems, Algorithms and Computer Implementation," NewYork: Wiley, 1990.
17. P. Gupta and P. R. Kumar, "The capacity of wireless networks," IEEE Transactions on Information Theory, Vol. 46, No. 2, pp. 388–404, February 2000.

18. H. Venkataraman, S. Sinanovic, and H. Haas, "Variation of spatial protection margin in multihop wireless networks," in Proceedings of IEEE International Symposium on Personal Indoor Mobile Radio Communications (PIMRC'07), Athens, Greece, 3–6 September 2007.
19. H. Venkataraman, S. Nainwal, and P. Shrivastava, "Optimum number of gateways in cluster-based two-hop cellular networks," AEU Journal of Electronics and Communications, Submitted and under review, http://nextgenwireless.daiict.ac.in/Sub_Papers/AEU_May_2008_Hrishikesh_Venkataraman.pdf, 2008.
20. S. D. Z. Dawy and I. Oikonomidis, "Coverage and capacity enhancement of CDMA cellular systems via multihop transmission," in Proceedings of IEEE Global Communication Conference (GLOBECOM), San Fransisco, USA, pp. 1147–1151, 1–5 December 2003.
21. 3rd Generation Partnership Project (3GPP), Technical Specification Group Radio Access Network, "Selection procedures for the choice of radio transmission technologies of the UMTS," 3GPP TR 30.03U, May 1998. Retrieved November 30, 2006, from:http://www.3gpp.org/ftp/Specs/htmlinfo/3003U.htm.
22. J. Cho and Z. Haas, "On the throughput enhancement of the downstream channel in cellular radio networks through multihop relaying," IEEE Journal on Selected Areas in Communications, Vol. 2, No. 7, pp. 1206–1219, September 2004.
23. IST-2003-507581 WINNER, "D7.6 winner system concept description," Retrieved July 15, 2007, from https://www.istwinner.org/DeliverableDocuments, October 2005.
24. R. A. Fisher, "Statistical methods and scientific inference," in Oliver and Boyd, Edinburgh, 1956.
25. I. Hacking, "Logic of statistical inference," in Cambridge University Press, Cambridge, 1965.
26. V. Sreng, H. Yanikomeroglu, and D. D. Falconer, "Relay selection strategies in cellular networks with peer-to-peer relaying," in Proceedings of IEEE Vehicular Technology Conference (VTC Fall'03), Orlando, USA, Vol. 3, pp. 1949–1953, 4–9 October 2003.
27. P. li, M. Rong, T. Liu, and D. Yu, "Interference modeling and analysis in two-hop cellular network with fixed relays in FDD mode," in Proceedings of IEEE Wireless Communications, Networking and Mobile Computing (WiMob), Montreal, Canada, pp. 177–182, 22–24 August 2005.
28. T. Liu, M. Rong, H. Shi, D. Yu, Y. Xue, and E. Schulz, "Reuse partitioning in fixed two-hop cellular relaying network," in Proceedings of IEEE Wireless Communications Networking Conference (WCNC), Las Vegas, USA, pp. 177–182, 3–6 April 2006.

Chapter 2

Energy-Efficient Crowdsensing of Human Mobility and Signal Levels in Cellular Networks

Paweł Foremski, Michał Gorawski, Krzysztof Grochla * and Konrad Polys

Institute of Theoretical and Applied Informatics, Polish Academy of Sciences, ul. Baltycka 5, Gliwice 44-100, Poland

ABSTRACT

The paper presents a practical application of the crowdsensing idea to measure human mobility and signal coverage in cellular networks. Currently, virtually everyone is carrying a mobile phone, which may be used as a sensor to gather research data by measuring, e.g., human mobility and radio signal levels. However, many users are unwilling to participate in crowdsensing experiments. This work begins with the analysis of the barriers for engaging people in crowdsensing. A survey showed that people who agree to participate in crowdsensing expect a minimum impact on their battery lifetime and phone usage habits. To address these requirements, this paper proposes an application for measuring the location and signal strength data based on energy-efficient GPS tracking, which allows one to perform the measurements of human mobility and radio signal levels with minimum energy utilization and without any engagement of the user. The method described combines measurements from the accelerometer with effective management of the GPS to monitor the user mobility with the decrease in battery lifetime by approximately 20%. To show the applicability of the proposed platform, the sample results of signal level distribution and coverage maps gathered for an LTE network and representing human mobility are shown.

Keywords: crowdsensing; energy efficiency; mobility models; network modeling; mobility measurements

1. INTRODUCTION

Current mobile phones are becoming multi-purpose computers for which voice transmission is just one of the functions provided to the user. Smartphones are powerful measuring devices, equipped with multiple sensors and a large data storage capacity. They also allow one to wirelessly transmit the measurement results to any location on the Internet.

Smartphones are carried by the users to virtually any location they visit. Therefore, a smartphone seems to be a natural device for mobile measurements in order to realize the idea of crowdsensing. The mobile crowdsensing is a new sensing paradigm based on the power of various mobile devices. It is used to acquire local knowledge through sensor-enhanced mobile devices, where individuals use them to collectively share data and to extract information to measure and map phenomena of common interest [1]. Multiple applications have been proposed to use a mobile phone as a platform for the measurements and gathering of information about, for instance, human mobility, locations of interest [2], air quality [3] or wireless network connectivity [4].

Crowdsensing enables carrying on extensive measurements covering a large area with very limited costs, but requires user participation. Three obstacles to crowd scaling have been identified by Xiao at al. [5] using the analysis of papers that describe crowdsensing applications. According to [5], the barriers are: heterogeneity of the used hardware (the developed software needs to be available for the phone of the user willing to participate), the burden imposed on the users and bandwidth demands. The heterogeneity of the used hardware can be easily overcome by the implementation of the crowdsensing application on multiple types of devices, and the bandwidth demands can be minimized using compression or using WiFi transmission instead of cellular communication. Thus, the burden imposed on users becomes the most important limiting factor. In this work, we try to further analyze it by running a questionnaire among a group of 100 people and looking into more details of those limits.

We identified that people are willing to participate in a crowdsensing experiment only if the burden in terms of change in their behavior is minimal. We concentrated on using the smartphone as a platform for the sensing application. Our analysis showed that the burden can be interpreted as the requirement of interacting with the crowdsensing application (e.g., denoting the current activity) and the requirement of charging the phone more often. We show how to overcome those limitations by implementing an application for monitoring of the wireless signal level and human mobility. In crowdsensing, it is possible to minimize the burden imposed on the user by merging and performing on-line analysis of the data provided by multiple sensors available on the smartphone. Different sensors have different energy footprints: for example, the accelerometer requires very low power, while the GPS requires more power. By performing continuous analysis of the values provided by the low power sensors, such as the accelerometer, and the signal strength of the received cellular network, we were able to detect moments in which it is required to enable the high power sensors, such as the GPS. The proposed method can be easily extended to enable or disable other sensors, like the microphone to monitor the noise level. It also allows for identifying changes in human mobility to detect states, like walking, running, driving, etc., and correlating those states with other measurements. Below, we list the major contributions of our paper:

- We present a novel platform for crowdsensing location and signal strength data (Section 4) and use it to collect preliminary data on human mobility and signal coverage in an LTE network (Section 6).
- We improve the energy-efficient GPS tracking, for example by using signal strength of GPS satellites to avoid unnecessary battery drain, and show how to implement it for crowdsourcing (Section 5).
- In an online survey of 100 people, we found that 37% of our respondents would agree to participate in crowdsourcing, but they expect a minimum impact on their battery lifetime and phone usage habits (Section 3).
- By running a two-week experiment on various Android devices, we found that, on average, our tracking application reduced the battery lifetime by 20% (Section 5.3 pt 1).
- By analyzing real data on 272,000 searches for a GPS fix (acquiring the radio signal from satellites to calculate the position) on mobile platforms, we found that 92% of searches stopped in under 30 s, and 84% of platform users experienced a daily GPS duty cycle below 5%, which roughly corresponds to one hour of movement a day (Section 5.3 pt 2).

The rest of the paper is organized as follows: in Section 2, we review the related literature; in Section 3, we analyze the barriers stopping people from participation in crowdsensing experiments; in Section 4, we describe a proposed energy-efficient crowdsensing measurement application; in the following section, we provide details regarding the GPS management to minimize the energy usage and describe the measured energy usage; we continue in Section 6 by showing sample results of six experiments carried out using the proposed platform to measure human mobility and to map LTE network signal coverage; Section 7 concludes the paper.

2. RELATED WORK

The crowdsensing idea has been well defined and presented in [1] and in the recently-published book [6]. The survey in [7] shows that using mobile phones for sensing has a very large potential due to a big number of devices that can be used. During the last few years, a number of crowdsensing applications and platforms have been described in the literature, like OpenSignal [8], LifeMap, APISENSE—a platform providing a solution to build and deploy crowd-sensing applications for collecting experimental datasets [9] or MOSDEN—a scalable mobile collaborative platform for opportunistic sensing applications [10].

The problem of motivating people to participate in a crowdsensing experiment is presented in [11], with game theoretic analysis of human behavior and the proposition of an auction-based approach for incentives. Koutsopoulos [12] extends this approach, providing the calculation of optimal payment allocation that minimizes the cost of compensation provided for participants. However, the compensations are not always possible, and a large number of research applications of the crowdsourcing idea rely on people voluntarily helping in the research project [13] or being given non-financial incentives. The participants may be motivated by serious games, like Foursquare [14], or functions provided by the application only when the data are provided by the user (such as improved navigation accuracy [15]).

The amount of effort provided by a crowdsensing participant is another important factor influencing the willingness of participation in the research. The barriers have been discussed in [5] showing that people are more likely to help when the amount of effort required from them is minimal and when no extra cost, for example in the form of an additional data transfer, is imposed. The barriers limiting the application of crowdsensing in a business to business relation are discussed in [16]. The problem of scarce resources limiting the possibility of using the mobile phone for measurements is described in [17], with the battery capacity shown as the main factor. A similar problem is discussed in [18], where the authors propose an incentive mechanism for a mobile crowdsensing scheduling; however, the mechanism is only research carried out through simulations, and it is not implemented in a real environment.

The problem of energy awareness in crowdsensing has been analyzed recently in a few papers. In [19], a piggyback mechanism is proposed that collects the data when the sensors are used by an application. This approach significantly lowers the energy usage, but decreases the number of performed measurements and does not allow for tracking the device mobility fully. The energy-efficient location API described in [20] allows one to predict the location based on the assumption that most of the users repeat the same path every day, but it decreases the accuracy and may introduce significant errors in case the user does not follow his or her habits. Osin et al. [21] proposed a method to enable GPS for location measurement based on the 3D accelerometer, achieving energy savings of up to 27%. A similar method is used in [22], but with emphasis on the improvement of the location accuracy. In this paper, we further extend this approach to significantly decrease the energy usage, as is described in Section 4. Dong Zhao et al. [23] propose a novel model of opportunistic coverage, along with a measurement methodology to estimate the coverage quality and an adaptive sampling mechanism to improve energy efficiency. The collected user's mobility traces are used to select which user should execute the sensing action. The method proposed in [23] minimizes the energy utilization by coordination between nodes on larger time scales, without detection of movement, while we concentrate on local optimization for each node in short time scales by the detection of changes in node location. Kjaergaard et al. in [24] present an energy-efficient system for tracking trajectory, in which the goal is to track straight segments of a path instead of individual coordinates. The paper describes a system built on the EnTracked position tracking algorithm [25], extending it to trajectory simplification and three new modes of operation: heading-aware (which employs the compass), distance-aware (which employs GPS data) and movement-aware (which employs the accelerometer). The authors show significant improvement in energy savings comparing to the EnTracked algorithm, while keeping the location error within the requested error bound. However, comparing to our work, the paper considers Nokia smartphones, assumes constant device orientation for the heading-aware mode and does not support the accelerometer while biking and driving. In our paper, we show a crowdsensing application that leverages GPS signal strengths and activity recognition for a more robust use of sensors commonly found on modern smartphones

In a recent series of works by H. Xiong et al. [26,27,28,29,30], the authors discuss the energy usage caused by data transfer in mobile crowdsensing systems. The papers describe methods that optimize participant selection in a crowdsensing experiment, so that the amount of redundant work is minimized while fully covering a geographical area of interest. The authors exploit an interesting idea of piggybacking the data over already established connections. By predicting the likeliness of voice calls in given times of interest, they distribute the tasks only to selected participants. The authors also propose the methods of minimizing the number of tasks allocated to users. Comparing to our work, we propose to provide each user with their full location history as a participation incentive; hence, we do not apply any selection mechanism, and the proposed crowdsensing platform provides slightly different functionality; we minimize the energy usage due to data transfer by using a single bulk transmission with the data of the whole day. However, the piggybacking technique may be combined with our proposal to further reduce the energy usage or to improve platform responsiveness by shorter upload cycles.

The energy minimization by selecting which user should enable sensing functions is also discussed in [31]. The authors propose the use of comprehensive sensing techniques in crowdsensing scenarios to select users according to a proposed novel algorithm to reduce the user contribution. The system is tested on real datasets with good results; however, there are no data of the testing system in a real-time running application. Furthermore, the nature of the BX Tracker application forces the user to run it constantly (if no data traces are to be complete), so as in above cases, such a mechanism is not needed. Kyon-mo Yang and Sung-Bae Cho in [32] develop a low power context-aware system using modular Bayesian networks. The proposed system uses a Bayesian network, altered using a Markov boundary. Such an approach reduces the energy usage and enables eliminating the GPS usage. Experiments, performed using five Samsung S5 smartphones used by students, gave promising results: the accuracy of GPS + 4 sensors was highest; however, using only four sensors without the GPS was only 5% less accurate, while consuming 50% less power. The method proposed in this paper gives better energy optimization, and the system described in [32] detects only context-aware information and does not collect mobility data, thus realizing a slightly different function.

The mobility measurement using crowdsensing is presented by Bin Guo et al. [33], who propose an interesting idea of cross-space public information reposting, tagging and sharing in their FlierMeet project. Although the project itself is not similar to our solution, the use of GPS while tagging is an interesting one. Potentially, the FlierMeet project can be a source of mobility traces. However, the authors do not mention anything about the energy-saving techniques they use.

3. ANALYSIS OF BARRIERS FOR PARTICIPATION IN CROWDSENSING

To correctly identify the potential barriers limiting the willingness of users in taking a part in crowdsourcing experiments, we have carried out an online survey. It has been responded to by 100 users over a period of 10 days. We present the results below.

(1) Battery life: We asked the respondents about their battery lifetime during a typical daily usage, after the phone is fully charged. Figure 1 presents the results for different platforms. The battery lifetime differs, with the most popular answer of “2 days” for Android and Windows Phone, “1 day” for iPhone and “3–4 days” for other devices. The smallest battery life has been reported by iOS users: one day. The answers allow one to estimate the energy budget that a crowdsourcing application may use without influencing the phone charging habits of the users.

(2) Crowdsourcing opportunities: 37 respondents agreed to anonymously share their data for scientific purposes, which shows a significant potential for mobile crowdsourcing. Figure 2a presents the likeliness for sharing various types of data (a multiple choice question). As expected, people tend to object to collecting private data, like their location, which highlights the importance of incentives and proper data handling in such cases.

(3) Expected impact on battery life: Among the 37 respondents who agreed to participate in crowdsourcing, 76% (28 people) agreed to reduce their battery life due to the tracking application running in the background. However, people expect the battery impact to be as low as possible, preferably to reduce their battery lifetime by less than one hour. See Figure 2b for details.

(4) Expected impact on usage patterns: Among the 37 respondents, 51% (19 people) agreed to do additional tasks on their phones during the day, for example manually starting and stopping applications, entering data, etc., but only 35% (13 people) agreed to do this once a day; and only one respondent agreed to do this more frequently. Generally, people expect no additional duties due to crowdsourcing or as little additional work as possible.

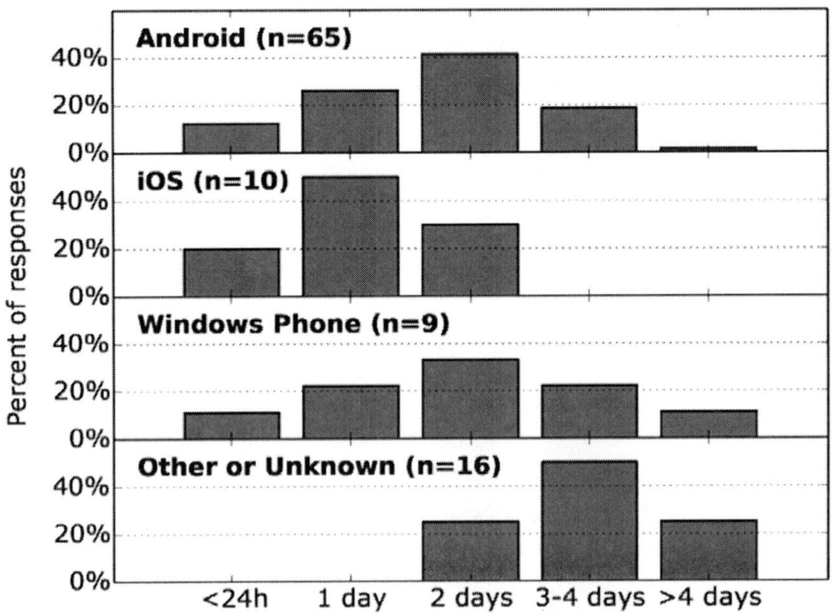

Figure 1. Single choice question: "In a typical daily usage, how long is your mobile phone able to work?"

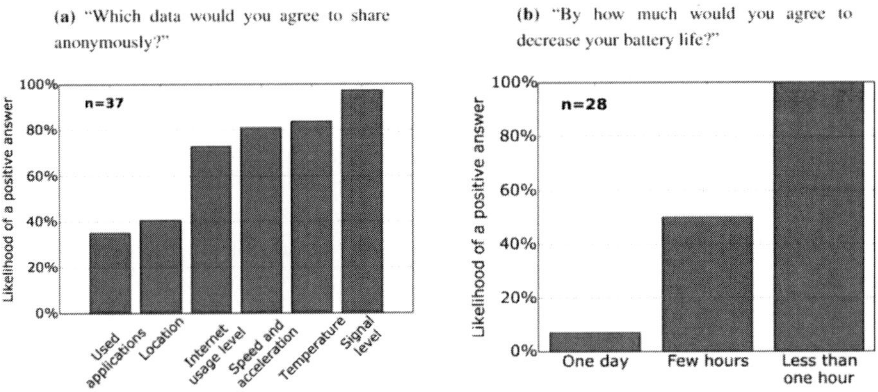

Figure 2. Multiple choice questions among: (**a**) respondents who agreed to participate in crowdsourcing; and (**b**) respondents who agreed to a shorter battery life due to data collection.

The results of the survey provide guidelines for the design of a mobile crowdsourcing platform, which should perpetually execute the measurements, with small battery usage and no additional tasks to perform during the day.

4. BX TRACKER: AN ENERGY-EFFICIENT CROWDSENSING PLATFORM

To answer the need for real and accurate data that could be used to monitor the state of a cellular network or other measurements that can be executed using mobile phones, we have developed a novel BX Tracker platform. It was developed to minimize the barriers for participation identified in the previous section. BX Tracker is an energy-efficient crowdsensing application for Android smartphones and tablets. We designed it to accurately and perpetually track the user location and cell network characteristics without significant battery drain, which is an important factor for crowdsensing. Our platform also provides a web service that allows users to access their own data.

The main difficulty in the development of the application was the limitation of the energy used by the application. A crowdsensing measurements needs to be correlated with the location information. Currently-available tools, proposed in [4,9], execute measurements in constant intervals and acquire location information using, e.g., GPS for each measurements. This method is energy consuming, as the GPS needs to be enabled for each measurement. The novel approach implemented in the BX Tracker platform overcomes this limitation by constantly monitoring the accelerometer (which uses a few orders of magnitude less energy than GPS) and enabling the location measurements only when the user is moving. It allows limiting the amount of time at which the GPS is enabled, thus decreasing the overall energy usage.

Since deploying a test version in 2013, the system has tracked over 40,000 km of human movements from over 50 unique user identifiers. The collected data were successfully applied to modeling cellular networks: in [34], we proposed a novel human mobility model based on real data. Since September 2014, our mobile application has been publicly available on Google Play, under the name "BX Tracker" [35]. The system is now fully operational and used worldwide. We plan to convey experiments on cellular networks with a large user base.

Below, we describe the architecture and implementation of our platform and comment on the incentives for crowdsourcing location data.

4.1. Architecture

The overview of our platform is depicted in Figure 3. The system is divided into two parts: the client side, where the BX Tracker application runs on possibly many Android devices, and the server side, where the collected data are processed and stored for later usage. The client application constantly acquires the data, applies basic pre-processing and filtering and uploads it to the server every 3 h over a secure Internet connection. The server immediately puts the uploaded data into a work queue that is later processed in a separate thread (it uses the producer-consumer scheme). Typically, the data will be available to the user a minute after the upload. In this step, the server converts raw data to KML (Keyhole Markup Languagage) files, computes the total distance traveled each day and extracts the names of visited cities. Finally, the data are stored in a database that is accessible to users through a web interface.

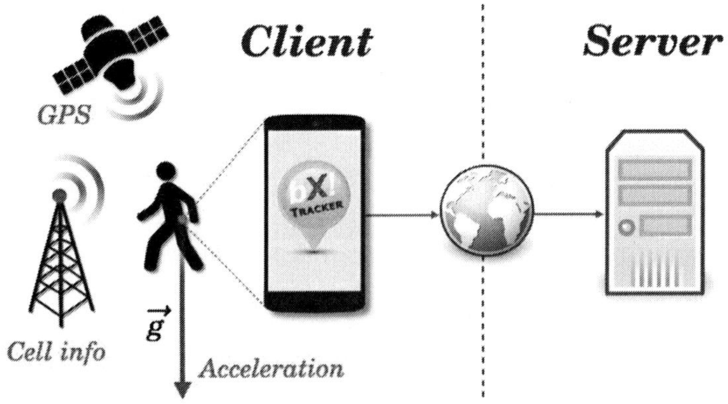

Figure 3. System overview. Pre-processed sensor data collected on mobile devices is periodically uploaded to the server.

The reason for extracting additional information from the location data is two-fold: to attract the users with useful facts about their life and to help system operators quickly find interesting data, for example all users that traveled more than 1000 km last month. To collect the names of visited cities, we adopted machine learning and the Google Maps API. First, we apply the DBSCAN (Density-based spatial clustering of applications with noise) algorithm [36] to find clusters of geographical coordinates for a particular day, with parameters of ε=5×10−3 and the minimum number of points equal to five. For each cluster centroid, plus the first and the last location of a given day, we query the Google Maps geocoding service [37] which converts geographical coordinates to street addresses. Finally, we select the most specific result and store the city name. To speed up the process, we also adopt caching for popular coordinates.

We ensure user privacy by assigning each user a completely random anonymous identifier (the "ID") on the first run of the BX Tracker application. We adopted the UUID (Universally Unique Identifier) as the format for IDs, which makes them long enough for simple user authentication in our web service. However, the physical geographical coordinates are not anonymized. Note that proposing a dependable anonymization scheme even for coarse-level location data is problematic [38], especially if we want to retain the details of mobility patterns for each user. Thus, we store intact location data in the form that makes person identification as difficult as possible, that is we assign completely random identifiers to platform users. In such a setup, a sufficiently high number of users makes finding a particular person troublesome. Finally, having the original location data is essential for providing a useful web service.

4.2. Android Application

The client application runs on Android 2.3.3 (Gingerbread, released 2011) through 5.0 (LolliPop, released 2014), which constitutes about 99% of all Android installations [39] as of 2015. Android had almost an 85% share of the worldwide smartphone market in Q2 2014 [40], so we believe that BX Tracker is available for the vast majority of mobile devices sold nowadays.

The application was implemented in Java, using the Android SDK, as a "foreground service", that is a Service process that constantly runs in the background and has a minimum chance of being killed by the operating system due to a limited amount of resources. Android OS saves battery power when it is able to turn the CPU off during inactivity periods; hence, we designed the service as an event handler instead of constantly checking for changes. Thus, at startup, the service registers for necessary events and then sleeps to conserve the battery power. We also provide the user with a graphical user interface (an Activity), which runs in a separate process and can be safely killed when needed.

The inter-process communication between the service and the user interface is realized using the Android Intent API, whereas the intra-process communication inside the service is realized using a custom callback system. The Android OS acts as a proxy between the application and the hardware and as a source of real-time clock interrupts. The BX Tracker architecture is briefly presented in Figure 4. The service part consists of many subsystems that will be described in detail in Section 5.

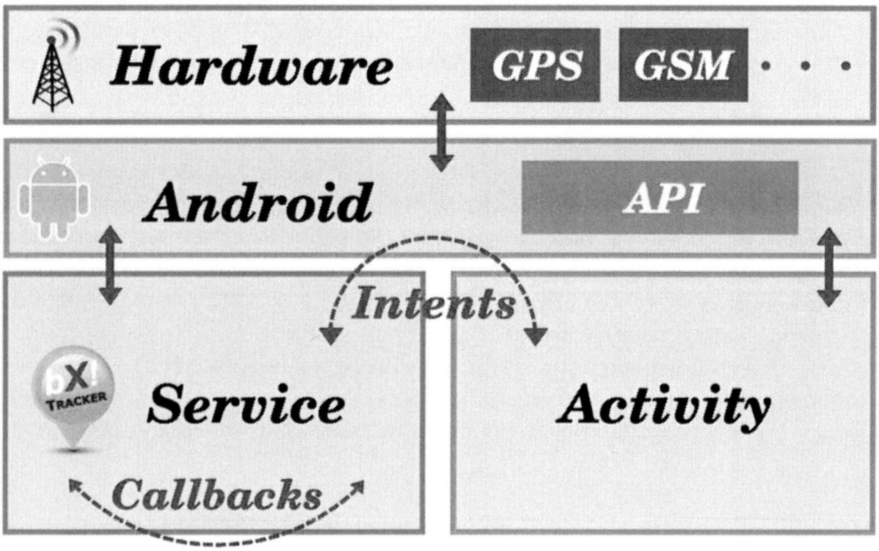

Figure 4. Architecture of the BX Tracker application.

From the technical point of view, the most important feature in the BX Tracker application is user mobility monitoring. We rely on the location tracking subsystem that is able to work in a perpetual, accurate and unobtrusive manner, as will be explained in Section 5.

4.3. Collecting Data

Our platform can easily be modified to collect any data that needs geospatial features, like the popularity of public roads, frequent locations of voice calls or even city noise pollution [41], but we chose to measure human mobility and signal levels. See Table 1 for a listing of collected mobility data and Table 2 for signal data. Note that the data points of location and signal data do not always have the same values in the now column, because they are triggered by different events. Collected data are stored in CSV files named after the current date (e.g., 20140926-signal.txt), plus one text file with the details about the user device and the Android version (e.g., Sony D5503 4.4.4). On upload, all files are compressed as a ZIP archive and sent to the server over an HTTPS connection.

BX Tracker tracks location using only the GPS, because it is simple and universal. The collected data are stored in the columns lat, long, alt and acc. In addition to the GPS, the Android OS also provides the network-based location, and more recently, Google introduced its Fused Location Provider (FLP) [42]; however, these techniques depend on Wi-Fi for fine-grained results, which is not always available. Moreover, although FLP aims at fusing the best of the GPS with the network-based location, its high accuracy mode induces a 7.25% battery drain per hour, which effectively makes it infeasible for perpetual location tracking. Thus, we developed a new technique for GPS tracking, which is more universal than the network-based location and more energy-efficient than FLP (for our use case). However, we acknowledge that our technique described in Section 5 depends on a subsystem of FLP, Google Activity Recognition, which is used for the activity column. Note that the GPS works reliably only outside of buildings, but this is acceptable for our experiments: when the GPS signal is lost, the last known location will usually approximate the coordinates of the building in which the user is currently.

Table 1. Mobility data collected by BX Tracker.

Column Name	Description
`now`	Date and time of this data point (ISO 8601, UTC)
`lat`	Location: latitude
`long`	Location: longitude
`alt`	Location: altitude (m)
`acc`	Estimated location accuracy (m)
`speed`	Top speed in the last 2 min relative to `now` (km/h)
`activity`	Google Activity Recognition: `in-vehicle`, `on-bicycle`, `on-foot`, `still`, `unknown`
`age`	Time between measuring the location and `now` (s)

Table 2. Signal level data collected by BX Tracker.

Column Name	Description
now, lat, long, acc, age, speed, activity	The same as in Table 1
type	Cellular network type, e.g.: LTE, EDGE (Enhanced Data rates for GSM Evolution), HSPA (High Speed Packet Access), HSPA+
signal	Signal level (dBm)
netop	Mobile Country Code (MCC) and Mobile Network Code (MNC)
cid	Cell ID (CID) of the connected base transceiver station (BTS)
lac	Location Area Code (LAC)
signal-age	Time between measuring the signal level and now (s)

Mobile power saving techniques have a deep impact on the collected data. The Android OS aggressively tries to put the device into "deep sleep", which can delay events, stop data collection for several hours or even completely remove the tracking process. As a work-around, we used a foreground service that runs in a dedicated process and uses the Android alarm manager for scheduling important tasks. Another important factor is the effect of the device screen being off. A popular power saving method is to disable all network interfaces in such situations. This could disrupt background monitoring if location tracking relied on Internet connectivity. For some rare hardware designs, the power source connected to the screen also drives the sensor chips, which makes the accelerometer unreliable on these devices. However, it is easy to detect such a situation and, thus, to mark collected data as faulty. Finally, our experiments showed that on many Android smartphones, the mobile signal level is not updated (or even reported) when the screen is off. Hence, by default, BX Tracker collects the signal data (Table 2) only when the screen is on; the location data (Table 1) are collected when the screen is either on or off. In case one needs fine-grained signal data, BX Tracker can periodically switch on the screen for a short period of time. On some devices, such a technique does not incur additional battery drain.

Unobtrusive and energy-efficient location tracking is the key requirement for mobile, location-aware crowdsensing systems that need perpetual user monitoring. We will show this in the next subsection. We highlight that our work can be generalized and used as a novel method in the field of crowdsourcing. The BX Tracker platform is an example of a successful application of our idea to measuring cellular networks on Android devices.

4.4. Attracting Users

In order to attract the users, we provide a free GPS tracking application on Google Play and a free web service available at [43].

First of all, BX Tracker can be used as an ordinary GPS tracking tool for many tasks, like recording a walking tour. Compared to existing applications on the market, BX Tracker is designed to be energy-efficient by design; hence, the user does not need to control the tracking process manually. The application is free of charge and ads and runs on almost all Android platform versions.The web service provides the users with useful information on their travels. For example, Figure 5 presents an example of the month view: it lists days in a given month one by one, with information on the total distance traveled on a particular day and on the visited cities. The BX Tracker web service allows for storing the entire location history for years. Moreover, users can download KML files with their trips for each day and view the recorded routes online in a map browser, as presented in Figure 5. Finally, we provide the user with basic travel statistics: total monthly distance, weekday average and weekend average.

Figure 5. Exemplary screenshot of the web service. The image was cropped for clarity.

Many other incentives are possible for crowdsourcing systems that track location data. For example: geotagging photos taken with DSLR cameras, tracking time spent in various places or suggesting better options for everyday routes. The signal strength data can be used to pinpoint locations of base stations and places with a good signal for high-quality voice calls.

5. ENERGY-EFFICIENT GPS TRACKING

In this section, we show how to combine the GPS with the accelerometer to implement energy-efficient location tracking on mobile devices. We give a general idea using BX Tracker, which was designed for Android platforms, as an example.

5.1. Motivation and Idea

The original motivation for perpetual location tracking is human mobility modeling and location prediction, an important topic for cellular networks. Such research on human movements requires long-range, accurate and continuous location data of many individuals, which is a scarce resource. Some data are available in the CRAWDAD (Community Resource for Archiving Wireless Data At Dartmouth) database (for example [44,45,46]) and through the OpenStreeMap project [47], but none of the published datasets were suitable for our model presented in [34]. Moreover, we considered collecting the required data using off-the-shelf GPS trackers as inadequate. Apart from the costs, such an approach imposes additional duties on the volunteers and scales poorly. Thus, we decided to collect the data using mobile crowdsensing, which seems the natural choice nowadays [48].

Almost all new smartphones support location using various techniques. Apart from the traditional GPS location, smartphones offer network-based location, which employs wireless base stations as reference points. However, as already pointed out, such a technique is unreliable: in some regions, its accuracy can be as low as a few kilometers. On the other hand, popular GPS tracking applications are designed for short-term usage and, thus, require a manual start-stop procedure. As shown in Section 3, crowdsourcing volunteers expect minimal battery impact and no additional tasks to do. Hence, we propose a location tracking algorithm optimized for crowdsourcing data for modeling cellular networks.

The location can be tracked by having the GPS constantly on, but this could reduce the battery lifetime from 284 h down to 12 h [21]. On the other hand, people on average spend only an hour a day traveling [49]. Thus, our basic idea is to switch off the GPS chip whenever possible. In order to do that, we monitor the phone accelerometer for various activities, observe the signal strengths of

GPS satellites, dynamically adjust the fix interval and track the user speed history.

In [50], Galeana-Zapién et al. proposed to solve the energy saving problem with a middleware architecture and a runtime environment that directly interfaces with application programming interfaces and device sensors, to manage the duty cycle process based on energy and context aspect. In comparison to that work, our solution optimizes energy usage and generates mobility and signal strength data in a single Android application. Moreover, we use the accelerometer to manage the GPS and granularity of measurements, and we monitor the signal strengths of GPS satellites to switch GPS off inside buildings. These techniques assure an energy-efficient usage of location sensors without third party optimizers.

5.2. Implementation

A more detailed architecture of BX Tracker is presented in Figure 6. The system is divided into subsystems responsible for tracking various phenomena (visible on the left-hand side), and the Android Service part that controls these subsystems and receives callbacks (right-hand side; see Section 4.2). Note that the GSM subsystem, which collects signal quality data, is optional with respect to the overall system architecture. It can be replaced with another data collection subsystem, as explained in Section 4.3, or omitted. In Section 6, we show practical applications for the additional data. Below, we describe the crucial processes implemented in our location tracking method.

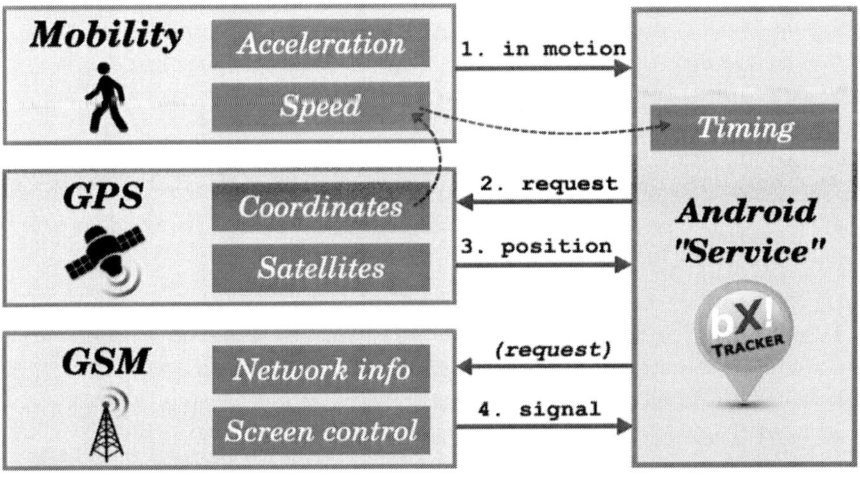

Figure 6. Implementation of the BX Tracker application.

(1) Mobility tracking: Let us start with an observation that people have long periods of time during the day in which they do not change their location, for example: sleeping, staying at home, working at the office [48]. It is unnecessary to periodically check for the GPS position when no significant movement is recorded by the smartphone sensors. Another observation is that once people start moving, we can monitor the maximum speed attained in the last few minutes to ensure that they are still mobile. Thus, it is unnecessary to check the sensors until people stop moving for some time. We used these two simple observations as the motivation to build the Mobility subsystem (see Figure 6): it aims at detecting whether the user is currently more likely to be in motion or to be stationary. When the user is in motion, we request a new GPS fix. However, we also obey a minimum time pause between successive requests, as will be explained later.

The physical activity of a smartphone user can easily be detected using an accelerometer. This is a well-known research topic [51,52,53,54,55,56], so we adopted an off-the-shelf solution available in the Google Play Services software library, v. 6.1: Google Activity Recognition [42]. Once enabled, it will periodically send notifications about the detected activity [57] “in vehicle”, “on bicycle”, “on foot”, “running”, “walking”, “tilting” and “still”. The expected period between the updates can be configured, and each notification contains a confidence level. In BX Tracker, we set the interval to 10 s, and we skip the updates with the confidence level below 54%, which was verified experimentally to give the best results. Finally, when we receive a certain activity update of “tilting” or “still”, we consider the user to be stationary. We map the other classes to simple representations: see activity in Table 1.

The mobility tracking subsystem also passively listens to GPS location updates. Whenever the user moves more than 35 m in less than 2 min (about 1 km/h), the instantaneous speed and the distance between the two consecutive updates are recorded in a circular buffer with a timestamp value. The buffer is later used to retrieve the maximum speed and the maximum distance in the last 2 min, relative to the query time. If the speed is above 4.5 km/h or the distance is larger than 200 m, the user is considered to be in motion. Note that if no new GPS positions are received, the speed and distance data will be invalidated with time.

(2) GPS control: The task of the GPS subsystem is to handle requests for a GPS fix in an battery-efficient manner. We observed that it is often unnecessary to leave the GPS active for periods longer than 30 s. Such an amount of time is long enough to discern between a meaningful position search that is going to succeed (e.g., good GPS signal, outside of a building) and a process that is finally going to fail (e.g., GPS started inside an office room). Data collected from users confirmed this: in Section 5.3 pt 2, we analyze the time our algorithm needed to stop the GPS. The reason could be either obtaining a valid fix or observing a poor signal. For the vast majority of examined cases, the GPS was

stopped in less than 20 s, with almost all fix attempts finishing in less than 1 min.

Based on the observations above, we implemented a GPS status monitor. Whenever we request a GPS position fix from the Android OS, we also register for updates on the GPS satellite signal strengths, which should be available each second. For each signal strength report, we calculate a "GPS score": S, a metric that quantitatively estimates the quality of the GPS signal at the current location:

$$\begin{aligned} S &= \max X + \tfrac{1}{2}|X| \\ X &= \{x_i : x_i \geq 10\} \end{aligned} \qquad (1)$$

where *xi* are the signal-to-noise ratios of the GPS satellites [58]. We experimentally obtained a threshold value for S: if the maximum S value for the last 15 s was below 30.5, the fix is likely to fail, and thus, the GPS chip can safely be shut down to conserve the battery power. On the contrary, if this value is above the threshold after 30 s since the GPS start, we let the process continue, and we recheck it every 15 s. Finally, the GPS is unconditionally stopped after 5 min, which was verified experimentally to work well. As a work-around for Android-specific issues with the satellite signal reporting, we restart the GPS fix and the GPS reporting processes every minute, with no impact on the presented procedure.

We also implemented simple filtering and processing of GPS results. On a new GPS fix, if the last successful fix was more than 5 min ago, we ignore the first two results and let the GPS run. This was verified experimentally to reduce the noise in GPS paths on several smartphones. For the same reason, we also ignore the results with an accuracy worse than 100 m, with at most 10 tries to obtain a higher-quality result. If the result passes the presented filtering procedure, we stop the GPS and store the new location. We also calculate the instantaneous speed and the traveled distance, as mentioned in the previous point.

(3) Proper timing: Let us make another observation that different types of mobility require different time pauses between successive requests for a GPS fix. For example, if our requirement is 100-m accuracy and the user walks, the pause can be around 30 s, because it is unlikely that the user will be able to walk more than 100 m in such an amount of time. On the other hand, if the user drives a car in a city center, the pause time should be around 10 s or less, assuming an average speed of 36 km/h. Finally, if the user drives a car on a highway, we can increase the pause time, as highways are designed to avoid

sharp road curves. The GPS track on a highway can be approximated with a straight line connecting two distant positions.

The Service part of the system presented in Figure 6 is responsible for imposing a pause regime on the GPS subsystem, using the information on user activity and speed. When the Mobility subsystem reports that the user is mobile, a GPS fix request will be generated only if a sufficient amount of time, d, has passed since the last successful fix. See Table 3 for the pause times used in BX Tracker, which were derived experimentally.

Table 3. Dependence of GPS pause time on user activity and speed.

Activity	*d* (s)	
in-vehicle	15	$v > 60\frac{km}{h}$
	10	
on-bicycle	20	
on-foot	30	
Otherwise	60	$v < 6\frac{km}{h}$
	10	$v > 60\frac{km}{h}$
	$360/v$	

(4) Collecting additional data: If possible, the event of a new GPS location should trigger collecting additional data; in the case of BX Tracker, the data about the mobile signal strength and network information. In such a case, an error in the geographical position is minimized, and the geospatial features of the measured phenomena are accurate. However, as previously noted, the power saving functionality of smartphones can disable some of the phone features when the device screen is off, which can be a common situation when the user is in motion. Hence, the subsystem responsible for recording additional data can either turn on the screen for a short period of time or fall back on opportunistic data collection, in which the data are stored only if the screen was on at the time of measurement. In BX Tracker, the user can manually choose between active and passive mobile data collection methods. In active mode, the Service subsystem will request turning on the screen and measure the signal data each time the user moves to a new location, plus an additional measurement will be made each minute disregarding user mobility. Collecting the mobile network signal is one of the additional BX Tracker features that is meaningful for both practical and scientific use. This mechanism for sampling the network signal is used in the BX Network application [59]. This application automatically samples both the Wi-Fi network (if accessible) and

the mobile network and switches them according to network efficiency. The data collected from BX Tracker was used in [60,61], where the authors research the characteristics of signal strength changes in buildings and the possibilities of simulating this phenomena.

5.3. Battery Impact

In order to verify the effectiveness of our battery saving method, we evaluated BX Tracker on real mobile devices. Our goal was to learn how the perpetual location tracking affects the battery life and how it aligns with user expectations. We ran an experiment with a group of volunteers, and we analyzed the real data stored in our BX Tracker platform.

(1) Two-week experiment: First, we measured battery usage on 12 real devices carried everyday by a group of 10 people. We asked them to install a battery monitoring application and to use their devices in a typical way for one week, without BX Tracker. Then, we asked our volunteers to install and run BX Tracker for one more week, while still monitoring the battery usage. We also asked the group not to change their phone usage habits during the whole experiment. Of course, we acknowledge that we could not fully control all of the factors that impact the battery usage and that our sample group was small. Hence, the obtained results only approximate the real battery impact of BX Tracker, averaged for different mobile devices.

In order to monitor the battery usage, we provided the experiment group with a custom application. Whenever a phone charger was connected to the device, the application logged information about the percentage of the battery that was used, relative to the most recent event of the phone charger being disconnected. In other words, we monitored the amount of energy that was used after each time the battery was re-charged. We also logged the date and time of measurement, the time since the last re-charging and the information if the BX Tracker service process was running. We filtered the data by ignoring the measurements for which: (i) the user re-connected the charger quicker than after 2.5 h; (ii) the user re-charged the battery in the morning after <10 h, that is after sleep; and (iii) the battery level decreased by less than 5%. After two weeks, we collected the information and analyzed the change in battery usage for each device separately. For four devices, the logged data did not indicate that the BX Tracker process was running or we received very few measurements (1–2), so we skipped them. For the rest of the eight devices, please see Table 4 for the obtained results.

Table 4. Impact of BX Tracker on battery consumption, as measured on various mobile devices. BD: average battery drain per hour; BL: estimated battery lifetime on a full charge; E: estimated energy usage.

Device	Battery	No Tracking		Tracking		Difference		
	(mAh)	BD (%/h)	BL (h)	BD (%/h)	BL (h)	BD (%/h)	BL (h)	E (mWh)
Galaxy S2 Lite	1500	2.4	42	3.9	26	+1.5	−38%	+83
Galaxy S3	2100	4.2	24	5.0	20	+0.8	−16%	+62
Galaxy S4	2600	3.5	29	5.3	19	+1.8	−34%	+170
Xperia Z1 C	2300	1.9	53	2.4	42	+0.5	−21%	+43
Xperia Z2	3200	1.4	71	1.5	67	+0.1	−7%	+12
Nexus 5	2300	3.2	31	3.7	27	+0.5	−14%	+43
ME302KL No. 1	6760	0.6	167	0.7	140	+0.1	−16%	+28
ME302KL No. 2	6760	0.7	152	0.8	130	+0.1	−15%	+30
					Average:	+0.7	−20%	+59

The first six devices are the popular smartphones made by Samsung, Sony and LG, whereas the last two devices are Asus tablet devices, with a much larger battery capacity. We consulted manufacturer data to obtain the declared battery capacity and assumed that a 100% battery level corresponds to the declared values. The BD metric presented in the table measures the average battery drain per hour, and the BL metric estimates the battery lifetime on a full charge, given the BD value: $BL=100/BD$. The E metric roughly estimates the real energy usage, assuming the average battery voltage of 3.7 V, which is common: $E=BD/100\cdot3.7C$, where C is battery capacity.

As shown in the table, BX Tracker on average decreased the battery lifetime by 20%. For the most popular battery lifetime in our survey, two days, this corresponds to decreasing the lifetime from 48 h down to 38 h. This still could be enough to give the user an acceptable phone experience, but in our survey, the respondents expected a decrease in battery lifetime by less than one hour or less than a few hours (see Figure 2b). However, our results show that even with aggressive energy optimizations, it is difficult to completely meet user expectations. From the technical point of view, the results show that it is possible to perpetually track the GPS position on an Android smartphone without draining the battery in a dozen of hours. We also estimated the real energy usage to enable comparison between smartphones and tablets, which differ in the battery capacity and, thus, have a different battery drain per hour. BX Tracker on average used 59 mWh, with a possible outlier (the Galaxy S4), which was probably caused by some other applications being activated on the phone in the second week of our experiment.

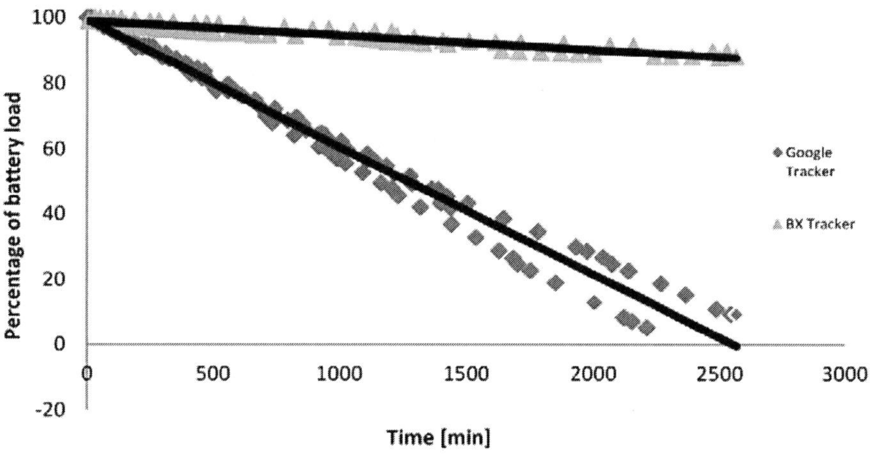

Figure 7. Comparison of battery use for BX Tracker and Google Tracker.

The next experiment compared the energy utilization of the proposed BX Tracker application to a popular location tracking application: Google Tracker. The tests were performed on two identical Asus ME302KL tablets, both with the same software, not connected to the Internet and not running any additional applications in the background (the difference in their discharge characteristics without tracking application was lower than 1%. Google Tracker was installed on one device and BX Tracker on the other device. The devices were turned on only during travel and keeping them always on with active trackers. The cumulative results of the test repeated twice are shown on Figure 7. After 2500 min of tests, the BX Power used less than 20% of the battery capacity, while the device running Google Tracker has been completely discharged.

(2) Collected data: We also analyzed the real data stored in the BX Tracker database, collected for one year by 44 different user identifiers. The data corresponds to over 272,000 events of searching for a GPS fix. We analyzed two factors for the tracked devices: the daily duty cycle of the GPS chip, presented in Figure 8, and the time between starting and stopping the GPS fix procedure, presented in Figure 9.

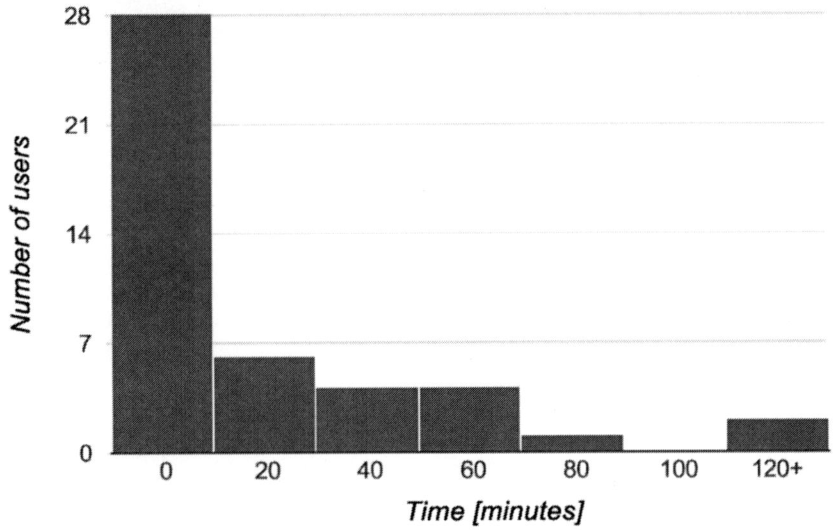

Figure 8. Total time of GPS activity per day. Measurements on 44 unique user identifiers.

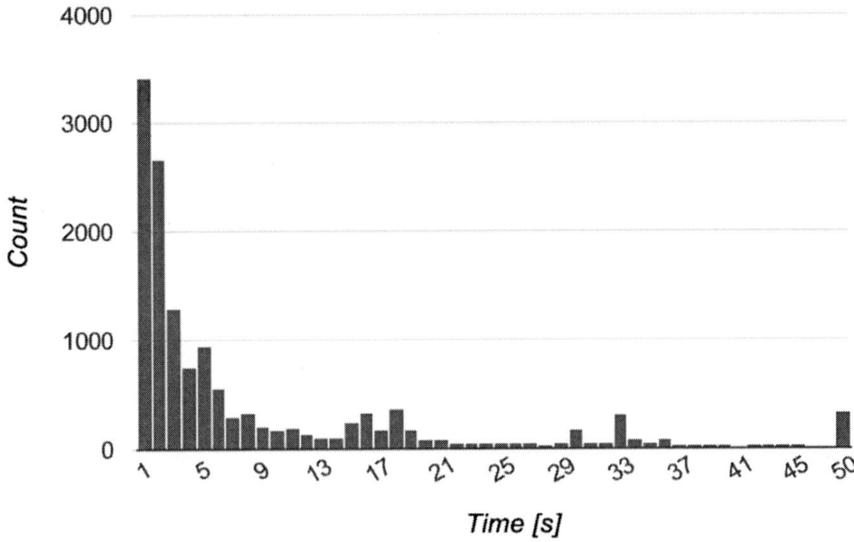

Figure 9. Time of GPS activity before providing a position fix or being disabled due to poor signal. Random 5% sample from 272,000 measurements on 44 unique user identifiers.

For the data presented in Figure 8, the majority of users experienced the total time of the GPS chip being on for less than 20 min in 24 h, which corresponds to a 1.4% duty cycle. Moreover, 84% of users experienced a duty cycle of 4.2% (60 min) or less, and 95% of users experienced a duty cycle of 8.3% (2 h) or less. The average total daily activity of the GPS was 29 min (2.0% duty cycle).

For the data presented in Figure 9, in 89% of the cases, the GPS was stopped in less than 20 s. For 92% of the samples, the process ended in less than 30 s, and for 97% of the samples, the process ended in less than 45 s. Only 1.7% of the GPS queries were finalized in more than 1 min.

These experimental data show the vast potential for optimizing GPS usage, comparing to a situation in which it is constantly enabled and searches for a position fix as long as possible. The data confirm our algorithm design presented in the previous subsection. As a more general conclusion, people spend the vast majority of their time stationary, as shown in [49], and briefly demonstrated by our Experiment 1 in the next section.

6. SAMPLE RESULTS

The experiments below demonstrate the capabilities of BX Tracker and evaluate its usefulness in the crowdsensing research. We present example data obtained by using our mobile application to show its versatility and potential in gathering mobility and network signal data. The data used in the experiments below were pre-processed to select adequate daily logs and merged into a single CSV file. Further filtering, analysis and basic visualization were performed in Microsoft Excel. We have measured the reference signal received power (RSRP) value reported by the LTE terminal (mobile phone or tablet) to represent the received signal level. The LTE signal strength heat map was generated using Google Maps, Python scripts, Gnuplot and Photoshop software. The density heat map of human mobility was created using Google Maps JavaScript API v3.

(1) Experiment 1: The first experiment analyzed a single user's mobility during a five-day workweek and a two-day weekend collected on a smartphone. The data show the distinctive characteristics of human mobility, which were applied in [34] and described in [62]: repeated cyclic actions during the workweek and longer travels during the weekend. In seven days, the user traveled 27 km during the workweek and 191 km during the weekend. This gives characteristics similar to the Levy walk observed in [62]. Table 5 gives detailed information about the user activity during the whole measurement period.

Table 5. Mobility of a single user during one week (Experiment 1).

Activity	Duration	Percentage
	Workweek	
still	114:21:53	94.19%
in-vehicle	1:00:00	0.82%
on-foot	6:03:09	4.98%
	Weekend	
still	40:20:46	84.78%
in-vehicle	3:50:45	8.08%
on-foot	3:23:49	7.14%
	Total weekly values	
still	154:42:39	91.54%
in-vehicle	4:50:45	2.87%
on-foot	9:26:58	5.59%

An example of the user trace during a weekend trip is presented in Figure 10.

The data gathered using BX Tracker can be used for modeling various aspects of cellular networks; for example, discovering human mobility patterns, researching new mobility models [63], forecasting user behaviors in mobile networks [64], planning wireless network resource allocations [65], and more.

(2) Experiment 2: The experiment demonstrates local variations of the signal strength in an LTE network. We used three Asus ME302KL tablets equipped with mobile Internet SIM cards (registered in LTE Network 1) and one Samsung Galaxy S4 smartphone (registered in LTE Network 2). We were collecting the data for 18 days in the same office building. The data showed variations in the LTE signal strength, both for tablets and for the mobile phone. Figure 11 presents the signal level for all tablet devices in the same office building. The two significant peaks correspond to two nearby office rooms.

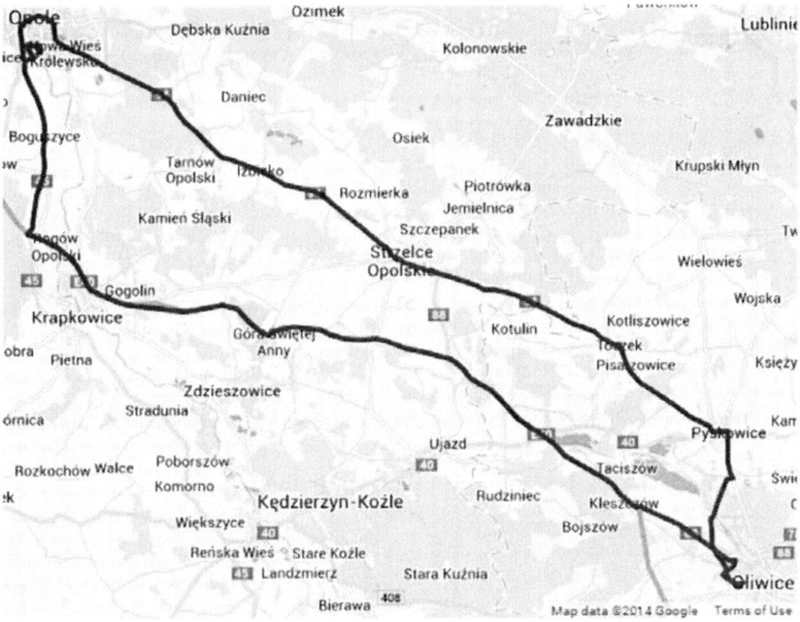

Figure 10. Daily GPS trace of a single user (Experiment 1). Map data ©2014 by Google.

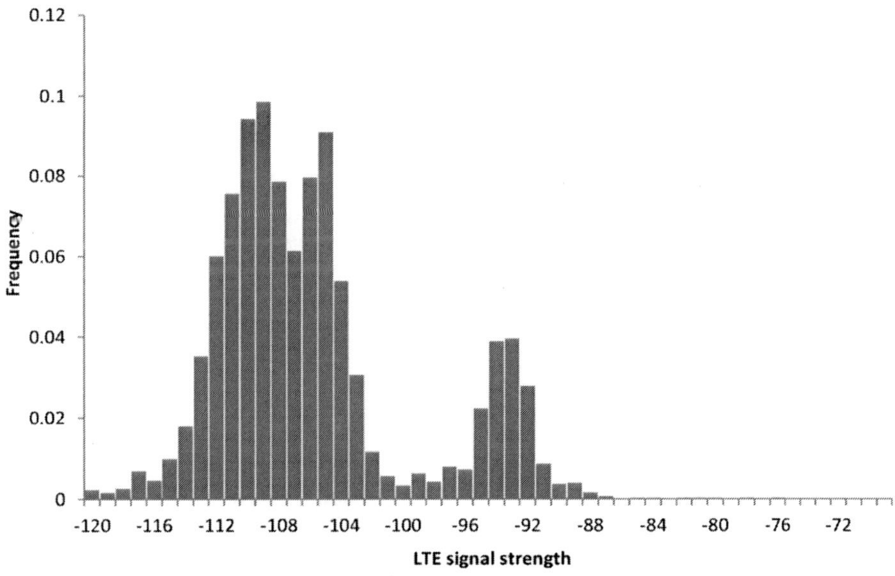

Figure 11. Signal strengths for all evaluated tablet devices (Experiment 2).

Figure 12 presents the data collected on the smartphone (blue) and on a single tablet device (red), but placed in the same office spots. We observed substantial differences in stability of the signal, with the tablet signal resembling the normal distribution and the smartphone signal being much more stable.

(3) Experiment 3: The experiment demonstrates variations in the LTE signal for a single office room. For 10 days, the same tablet device was placed for 5–6 h in different spots inside the same office room. Figure 13 shows the data for two different room spots, whereas Figure 14 shows the total signal variations for the whole experiment.

(4) Experiment 4: The experiment demonstrates variations in the LTE signal for a smartphone carried by an average office worker. The movement traces were collected for two days: the first day series involved traveling around many office rooms, but located on the same floor; the second day series involved many floors and the usage of an elevator. The data comparing both series are presented in Figure 15.

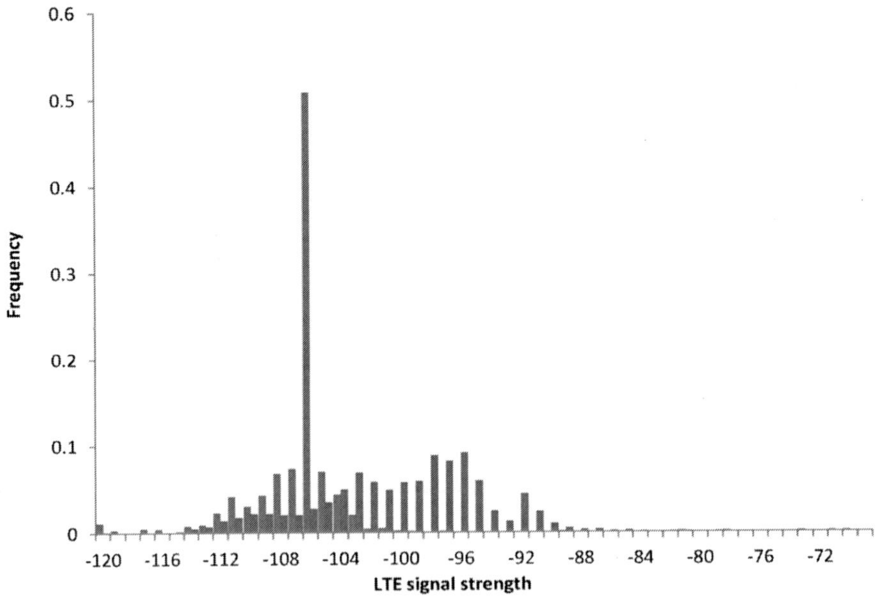

Figure 12. Comparison of signal strengths for a smartphone (blue) vs. a tablet (red) used in the same office room (Experiment 2).

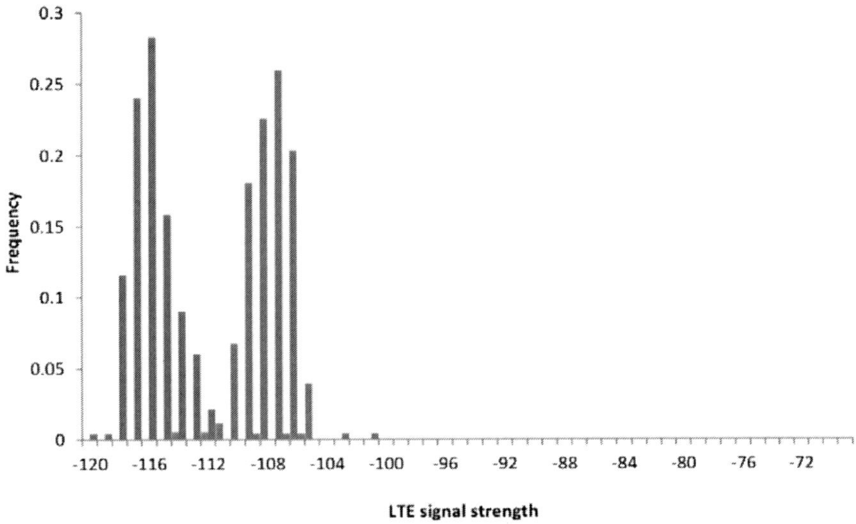

Figure 13. Signal strengths for two different locations in the same office room (Experiment 3).

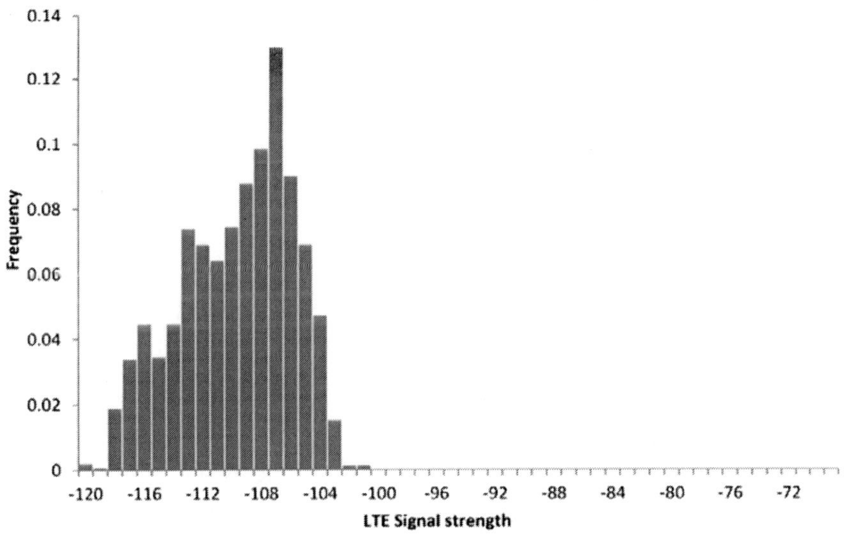

Figure 14. Signal strengths for 10 different locations in the same office room (Experiment 3).

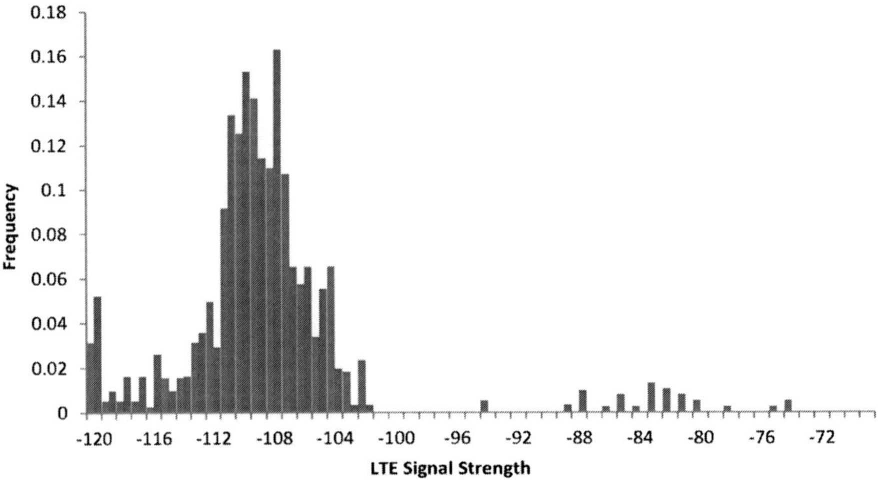

Figure 15. Signal strengths for movements on a single floor (blue) vs. many floors (red). (Experiment 5)

Figure 16. Density heat map of user locations (Experiment 5). Map data ©2014 by Google.

Table 6. Dependence of connection technology on the movement pattern (Experiment 4).

Value	LTE	EDGE	HSPA	HSPA+
Series I: single floor				
Total time	03:37:44	00:04:02	00:40:57	00:15:09
Average time between changes	00:12:06	00:01:00	00:01:39	00:01:05
% of connections	78.36%	1.45%	15.04%	5.45%
Series II: many floors, elevator				
Total time	02:56:42	00:13:24	01:49:07	00:07:03
Average time between changes	00:14:43	00:03:21	00:04:36	00:00:53
% of connections	57.69%	4.38%	35.63%	2.30%
Total				
Total time	06:34:26	00:17:26	02:30:04	00:22:12
Average time between changes	00:13:25	00:02:11	00:03:07	00:00:59
% of connections	67.53%	2.98%	25.69%	3.80%

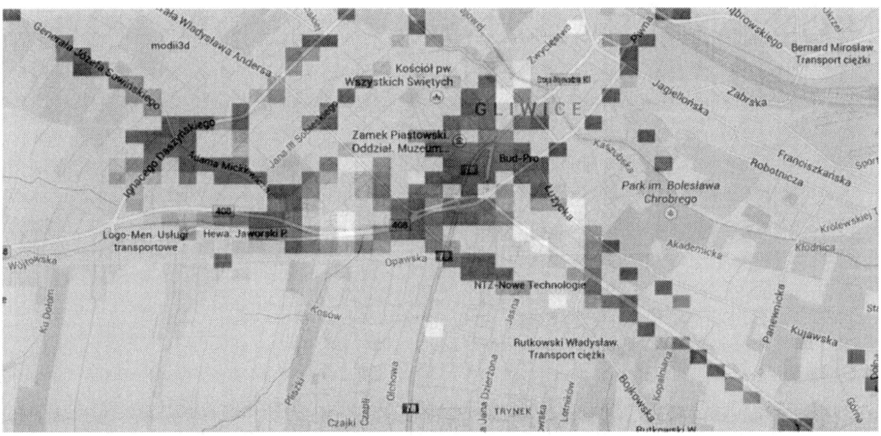

Figure 17. LTE signal heat map (map data ©2014 by Google).

We calculated the time between connection technology changes (e.g., LTE to HSPA+); see Table 6. The main difference between the series is that in Series II, when the movement was more intensive, we can observe a lower percentage of LTE connections: it dropped from 78% down to 58%. However, the actual time between technology changes increased from 12 to almost 15 min.

(5) Experiment 5: The experiment visualizes popular locations of our test users for seven days in Gliwice, Poland. The heat map presented in Figure 16 has a few high density points, which correspond to work places, homes and other popular places, like shopping malls and restaurants.

(6) Experiment 6: The experiment visualizes real LTE signal strengths measured around the center of Gliwice. The resulting heat map is presented in Figure 17. The area is similar to the density heat map shown in Figure 16: we can observe that the highly popular places are not necessarily the areas with the best signal strengths.

7. CONCLUSIONS

In this paper, a novel crowdsensing platform oriented towards minimization of the barriers limiting human engagement has been proposed. Our online survey confirmed that the battery drain and the requirement of interaction with the application are important aspects discouraging people from participation in a crowdsensing experiment carried through a smartphone application. We showed that these factors can be limited through the implementation of an energy-efficient GPS tracking and automatic detection of human activity by monitoring the accelerometer. The proposed BX Tracker application can gather information on the user location while reducing the device battery lifetime by only 20%. Moreover, 84% of our platform users experienced a GPS duty cycle of <5%, which corresponds to less than one hour per day. In comparison to traditional GPS tracking applications, the presented solution offers a 5% lower energy consumption, as is shown in Figure 7. This is achieved by enabling the GPS sensor only when the device is moving, by monitoring GPS satellite signal strengths to avoid using the GPS inside buildings and by adjusting the time pauses between successive GPS measurements.

We also presented various applications of the developed crowdsourcing platform in measuring human mobility and cellular signal levels. The results shown in Section 6 prove that BX Tracker can be used to efficiently collect data about the signal level distribution, maps of signal coverage and data about human mobility applicable to modeling cellular networks.

ACKNOWLEDGMENTS

This work was funded by the Polish National Centre for Research and Development, under Research Grant No. LIDER/10/194/L-3/11/: Project "Optimization and load balancing in next generation wireless networks", http://zosb.iitis.pl.

AUTHOR CONTRIBUTIONS

P. Foremski: idea and implementation of the BX Tracker system (energy-efficient GPS tracking and crowdsensing platform); analysis of survey results and the battery usage data. M. Gorawski: gathering data for experiments, data analysis and interpretation, conducted and described the experiments, worked on conclusions. K. Grochla: defined the scope of the paper, worked on the Introduction and literature review, coordinated the measurements and helped with the data analysis. K. Polys: performed part of the measurements described in Section 5 and analyzed its results.

CONFLICTS OF INTEREST

The authors declare no conflict of interest.

REFERENCES

1. Ganti, R.K.; Ye, F.; Lei, H. Mobile crowdsensing: current state and future challenges. IEEE Commun. Mag. **2011**, 49, 32–39.
2. Chon, Y.; Lane, N.D.; Li, F.; Cha, H.; Zhao, F. Automatically characterizing places with opportunistic crowdsensing using smartphones. In Proceedings of the 2012 ACM Conference on Ubiquitous Computing, Pittsburgh, PA, USA, 5–8 September 2012; pp. 481–490.
3. Dutta, P.; Aoki, P.M.; Kumar, N.; Mainwaring, A.; Myers, C.; Willett, W.; Woodruff, A. Common sense: participatory urban sensing using a network of handheld air quality monitors. In Proceedings of the 7th ACM Conference on Embedded Networked Sensor Systems, Berkeley, CA, USA, 4–6 November 2009; pp. 349–350.
4. Farshad, A.; Marina, M.K.; Garcia, F. Urban WiFi characterization via mobile crowdsensing. In Proceedings of the IEEE Network Operations and Management Symposium (NOMS), Krakow, Poland, 5–9 May 2014; pp. 1–9.
5. Xiao, Y.; Simoens, P.; Pillai, P.; Ha, K.; Satyanarayanan, M. Lowering the barriers to large-scale mobile crowdsensing. In Proceedings of the 14th Workshop on Mobile Computing Systems and Applications, Jekyll Island, GA, USA, 26–27 February 2013.
6. Brabham, D.C. Crowdsourcing; Mit Press: Cambridge, MA, USA, 2013.

7. Lane, N.D.; Miluzzo, E.; Lu, H.; Peebles, D.; Choudhury, T.; Campbell, A.T. A survey of mobile phone sensing. IEEE Commun. Mag. **2010**, 48, 140–150.

8. Open Signal Inc. Available online: http://www.opensignal.com (accessed on 15 October 2014).

9. Haderer, N.; Rouvoy, R.; Ribeiro, C.; Seinturier, L. APISENSE: Crowd-Sensing Made Easy. ERCIM News **2013**, 93, 28–29.

10. Jayaraman, P.P.; Perera, C.; Georgakopoulos, D.; Zaslavsky, A. Efficient opportunistic sensing using mobile collaborative platform mosden. In Proceedings of the 9th IEEE International Conference on Collaborative Computing: Networking, Applications and Worksharing (Collaboratecom), Austin, TX, USA, 20–23 October 2013; pp. 77–86.

11. Yang, D.; Xue, G.; Fang, X.; Tang, J. Crowdsourcing to smartphones: incentive mechanism design for mobile phone sensing. In Proceedings of the 18th ACM Annual International Conference on Mobile Computing and Networking, Istanbul, Turkey, 22–26 August 2012; pp. 173–184.

12. Koutsopoulos, I. Optimal incentive-driven design of participatory sensing systems. In Proceedings of the IEEE INFOCOM, Turin, Italy, 14–19 April 2013; pp. 1402–1410.

13. Haderer, N.; Rouvoy, R.; Seinturier, L. A preliminary investigation of user incentives to leverage crowdsensing activities. In Proceedings of the IEEE International Conference on Pervasive Computing and Communications Workshops (PERCOM Workshops), San Diego, CA, USA, 18–22 March 2013; pp. 199–204.

14. Foursquare Inc. Available online: http://www.foursquare.com (accessed on 15 October 2014).

15. Ramani, S.V.; Tank, Y.N. Indoor Navigation on Google Maps and Indoor Localization Using RSS Fingerprinting. Int. J. Eng. Trends Technol. **2014**, V11, 171–173.

16. Simula, H.; Vuori, M. Benefits and barriers of crowdsourcing in B2B firms: Generating ideas with internal and external crowds. Intern. J. Innov. Manag. **2012**, 16.

17. Faggiani, A.; Gregori, E.; Lenzini, L.; Luconi, V.; Vecchio, A.; di Ingegneria dell'Informazione, D. Smartphone-based crowdsourcing for network monitoring: Opportunities, challenges, and a case study. IEEE Commun. Mag. **2014**, 52, 106–113.

18. Han, K.; Zhang, C.; Luo, J. Truthful Scheduling Mechanisms for Powering Mobile Crowdsensing. CoRR **2013**, PP.

19. Lane, N.D.; Chon, Y.; Zhou, L.; Zhang, Y.; Li, F.; Kim, D.; Ding, G.; Zhao, F.; Cha, H. Piggyback CrowdSensing (PCS): Energy efficient crowdsourcing of mobile sensor data by exploiting smartphone app opportunities. In Proceedings of the 11th ACM Conference on Embedded Networked Sensor Systems, Roma, Italy, 11–15 November 2013.

20. Duribreux, J.; Rouvoy, R.; Monperrus, M. An Energy-efficient Location Provider for Daily Trips; RR-8586; HAL: Lyon, France; 28; August; 2014.

21. Oshin, T.O.; Poslad, S.; Ma, A. Improving the Energy-Efficiency of GPS based location sensing smartphone applications. In Proceedings of the 11th IEEE International Conference on Trust, Security and Privacy in Computing and Communications (TrustCom), Liverpool, UK, 25–27 January 2012; pp. 1698–1705.

22. Ofstad, A.; Nicholas, E.; Szcodronski, R.; Choudhury, R.R. Aampl: Accelerometer augmented mobile phone localization. In Proceedings of the First ACM International Workshop on Mobile Entity Localization and Tracking in GPS-less Environments, San Francisco, CA, USA, 14–19 September 2008; pp. 13–18.

23. Zhao, D.; Ma, H.; Liu, L. Energy-efficient opportunistic coverage for people-centric urban sensing. Wirel. Netw. **2014**, 20, 1461–1476.

24. Kjærgaard, M.B.; Bhattacharya, S.; Blunck, H.; Nurmi, P. Energy-efficient trajectory tracking for mobile devices. In Proceedings of the 9th International Conference on Mobile Systems, Applications, and Services, Bethesda, MD, USA, 28 June–1 July 2011; pp. 307–320.

25. Kjærgaard, M.B.; Langdal, J.; Godsk, T.; Toftkjær, T. EnTracked: Energy-efficient Robust Position Tracking for Mobile Devices. In Proceedings of the 7th ACM International Conference on Mobile Systems, Applications, and Services, Wroclaw, Poland, 22–25 June 2009; ACM: New York, NY, USA, 2009; pp. 221–234.

26. Xiong, H.; Zhang, D.; Wang, L.; Gibson, J.P.; Zhu, J. EEMC: Enabling Energy-Efficient Mobile Crowdsensing with Anonymous Participants. ACM Trans. Intell. Syst. Technol. **2015**, 6, 1–26.

27. Xiong, H.; Zhang, D.; Wang, L.; Chaouchi, H. EMC3: Energy-efficient Data Transfer in Mobile Crowdsensing under Full Coverage Constraint. IEEE Trans. Mob. Comput. **2015**, 14, 1355–1368.

28. Wang, L.; Zhang, D.; Pathak, A.; Chen, C.; Xiong, H.; Yang, D.; Wang, Y. CCS-TA: Quality-Guaranteed Online Task Allocation in Compressive Crowdsensing. In Proceedings of the 2015 ACM International Joint Conference on Pervasive and Ubiquitous Computing (UbiComp'15), Osaka, Japan, 7–11 September 2015.

29. Xiong, H.; Zhang, D.; Chen, G.; Wang, L.; Gauthier, V. CrowdTasker: Maximizing coverage quality in Piggyback Crowdsensing under budget constraint. In Proceedings of the 2015 IEEE International Conference on Pervasive Computing and Communications (PerCom), St. Louis, MO, USA, 23–27 March 2015; pp. 55–62.

30. Zhang, D.; Xiong, H.; Wang, L.; Chen, G. CrowdRecruiter: Selecting Participants for Piggyback Crowdsensing Under Probabilistic Coverage Constraint. In Proceedings of the 2014 ACM International Joint Conference on Pervasive and Ubiquitous Computing, Seattle, WA, USA, 13–17 September 2014; ACM: New York, NY, USA, 2014; pp. 703–714.

31. Xu, L.; Hao, X.; Lane, N.; Liu, X.; Moscibroda, T. More with Less: Lowering User Burden in Mobile Crowdsourcing through Compressive Sensing. In Proceedings of the 2015 ACM International Joint Conference on Pervasive and Ubiquitous Computing (UbiComp'15), Osaka, Japan, 7–11 September 2015.

32. Yang, K.M.; Cho, S.B. A Non-GPS Low-Power Context-Aware System using Modular Bayesian Networks. In Proceedings of the MOBILITY 2014, The Fourth International Conference on Mobile Services, Resources, and Users, Paris, France, 20–24 July 2014.

33. Guo, B.; Chen, H.; Yu, Z.; Xie, X.; Huangfu, S.; Wang, Z. FlierMeet: Cross-space Public Information Reposting with Mobile Crowd Sensing. In Proceedings of the 2014 ACM International Joint Conference on Pervasive and Ubiquitous Computing: Adjunct Publication, Seattle, WA, USA, 13–17 September 2014; ACM: New York, NY, USA, 2014; pp. 59–62.

34. Gorawski, M.; Grochla, K. The real-life mobility model: RLMM. In Proceedings of 2013 IEEE Second International Conference on Future Generation Communication Technology (FGCT), London, UK, 13–14 November 2013; pp. 201–206.

35. BX Tracker. Available online: https://play.google.com/store/apps/details?id=com.bxlabs.tracker (accessed on 1 September 2015).

36. Ester, M.; Kriegel, H.P.; Sander, J.; Xu, X. A density-based algorithm for discovering clusters in large spatial databases with noise. In Proceedings of the Second International Conference on Knowledge Discovery and Data Mining (KDD-96), Portland, OR, USA, 2–4 August 1996; Volume 96, pp. 226–231.

37. Google Maps Geocoding. Available online: https://developers.google.com/maps/documentation/geocoding/ (accessed on 19 September 2014).

38. Zang, H.; Bolot, J. Anonymization of location data does not work: A large-scale measurement study. In Proceedings of the 17th ACM Annual International Conference on Mobile Computing and Networking, Las Vegas, NV, USA, 19–23 September 2011; pp. 145–156.

39. Google. Android Developers: Dashboards. Available online: https://developer.android.com/about/dashboards/index.html (accessed on 19 September 2014).

40. Corporation, I.D. IDC: Smartphone OS Market Share 2014, 2013, 2012, and 2011. Available online: http://www.idc.com/prodserv/smartphone-os-market-share.jsp (accessed on 19 September 2014).

41. Rana, R.K.; Chou, C.T.; Kanhere, S.S.; Bulusu, N.; Hu, W. Ear-phone: An end-to-end participatory urban noise mapping system. In Proceedings of the 9th ACM/IEEE International Conference on Information Processing in Sensor Networks, Stockholm, Sweden, 12–15 April 2010; pp. 105–116.

42. Kadous, W.; Ganesh, J. Beyond the Blue Dot: New Features in Android Location. Available online: https://developers.google.com/events/io/sessions/325337477 (accessed on 19 September 2014).

43. BX Labs. Available online: http://www.bx-labs.com/tracker (accessed on 1 September 2015).

44. Rhee, I.; Shin, M.; Hong, S.; Lee, K.; Kim, S.; Chong, S. CRAWDAD data set ncsu/mobilitymodels. Available online: http://crawdad.org/ncsu /mobilitymodels/ (accessed on 19 September 2014).

45. Ficek, M. CRAWDAD Data Set Ctu/personal. Available online: http://crawdad.org/ctu/personal/ (accessed on 19 September 2014).

46. Chon, Y.; Talipov, E.; Shin, H.; Cha, H. CRAWDAD Data Set Yonsei/lifemap. Available online: http://crawdad.org/yonsei/lifemap/ (accessed on 19 September 2014).

47. OpenStreetMap, Public GPS Traces. Available online: https://www. openstreetmap.org/traces (accessed on 15 October 2014).

48. Isaacman, S.; Becker, R.; Cáceres, R.; Martonosi, M.; Rowland, J.; Varshavsky, A.; Willinger, W. Human Mobility Modeling at Metropolitan Scales. In Proceedings of the 10th International Conference on Mobile Systems, Applications, and Services, Ambleside,

UK, 25–29 June 2012; ACM: New York, NY, USA, 2012; pp. 239–252. [Google Scholar]

49. National Travel Survey: 2012. Available online: https://www.gov.uk/government/uploads/system/uploads/attachment_data/file/243957/nts2012-01.pdf (accessed on 14 October 2014).

50. Galeana-Zapién, H.; Torres-Huitzil, C.; Rubio-Loyola, J. Mobile Phone Middleware Architecture for Energy and Context Awareness in Location-Based Services. Sensors **2014**, 14, 23673–23696.

51. Kwapisz, J.R.; Weiss, G.M.; Moore, S.A. Activity recognition using cell phone accelerometers. ACM SigKDD Explor. Newsl. **2011**, 12, 74–82.

52. Ravi, N.; Dandekar, N.; Mysore, P.; Littman, M.L. Activity recognition from accelerometer data. In Proceedings of the Seventeenth Annual Conference on Innovative Applications of Artificial Intelligence (IAAI-05), Pittsburgh, PA, USA, 9–13 July 2005; Volume 5, pp. 1541–1546.

53. Brezmes, T.; Gorricho, J.L.; Cotrina, J. Activity recognition from accelerometer data on a mobile phone. In Distributed Computing, Artificial Intelligence, Bioinformatics, Soft Computing, and Ambient Assisted Living; Springer: Berlin, Germany, 2009; pp. 796–799.

54. Shoaib, M.; Bosch, S.; Incel, O.D.; Scholten, H.; Havinga, P.J.M. Fusion of Smartphone Motion Sensors for Physical Activity Recognition. Sensors **2014**, 14, 10146–10176.

55. Morillo, L.M.S.; Gonzalez-Abril, L.; Ramirez, J.A.O.; de la Concepcion, M.A.A. Low Energy Physical Activity Recognition System on Smartphones. Sensors **2015**, 15, 5163–5196.

56. Khan, A.M.; Siddiqi, M.H.; Lee, S.W. Exploratory Data Analysis of Acceleration Signals to Select Light-Weight and Accurate Features for Real-Time Activity Recognition on Smartphones. Sensors **2013**, 13, 13099–13122.

57. Google Activity Recognition. Available online: https://developer.android.com/reference/com/google/android/gms/location/DetectedActivity.html (accessed on 19 September 2014).

58. GPS satellites. Available online: http://developer.android.com reference/android/location/GpsSatellite.html#getSnr() (accessed on 19 September 2014).

59. Foremski, P.; Grochla, K. LTE or WiFi? Client-side Internet link selection for smartphones. In Computer Networks; Springer International Publishing: Gewerbestrasse, Switzerland, 2015; pp. 43–53.

60. Gorawski, M.; Grochla, K.; Polys, K. Variability of LTE Signal in Indoor Environment. In Proceedings of the Internet in the Information Society, 10th Scientific Conference, Dabrowa Gornicza, Poland, 24–25 September 2015.

61. Gorawski, M.; Grochla, K.; Polys, K. Comparison of Measurements in Simulation Model of Indoor Signal Level Variation in LTE Networks. In Proceedings of the Eighth EAI International Conference on Simulation Tools and Techniques, Athens, Greece, 24–26 August 2015.

62. Rhee, I.; Shin, M.; Hong, S.; Lee, K.; Chong, S. On the Levy-walk Nature of Human Mobility: Do Humans Walk like Monkeys? IEEE/ACM Trans. Netw. **2011**, 9, 630–643.

63. Gorawski, M.; Grochla, K. Review of Mobility Models for Performance Evaluation of Wireless Networks. In Man-Machine Interactions 3; Springer International Publishing: Gewerbestrasse, Switzerland, 2014; pp. 567–577.

64. Barcelo-Arroyo, F.; Gorawski, M.; Grochla, K.; Martín-Escalona, I.; Połys, K.; Ribeiro, A.G.; Sofia, R.; Zola, E. New Trends in Mobility Modelling and Handover Prediction. In Wireless Networking for Moving Objects; Springer International Publishing: Gewerbestrasse, Switzerland, 2014; pp. 88–114.

65. Giambene, G.; Yahiya, T.A.; Grochla, K.; Połys, K. Resource Management and Cell Planning in LTE Systems. In Wireless Networking for Moving Objects; Springer International Publishing: Gewerbestrasse, Switzerland, 2014; pp. 177–197.

Chapter 3

A 181 GOPS AKAZE Accelerator Employing Discrete-Time Cellular Neural Networks for Real-Time Feature Extraction

Guangli Jiang, Leibo Liu *, Wenping Zhu, Shouyi Yin and Shaojun Wei

Institute of Microelectronics, Tsinghua University, Beijing 100084, China

ABSTRACT

This paper proposes a real-time feature extraction VLSI architecture for high-resolution images based on the accelerated KAZE algorithm. Firstly, a new system architecture is proposed. It increases the system throughput, provides flexibility in image resolution, and offers trade-offs between speed and scaling robustness. The architecture consists of a two-dimensional pipeline array that fully utilizes computational similarities in octaves. Secondly, a substructure (block-serial discrete-time cellular neural network) that can realize a nonlinear filter is proposed. This structure decreases the memory demand through the removal of data dependency. Thirdly, a hardware-friendly descriptor is introduced in order to overcome the hardware design bottleneck through the polar sample pattern; a simplified method to realize rotation invariance is also presented. Finally, the proposed architecture is designed in TSMC 65 nm CMOS technology. The experimental results show a performance of 127 fps in full HD resolution at 200 MHz frequency. The peak performance reaches 181 GOPS and the throughput is double the speed of other state-of-the-art architectures.

Keywords: AKAZE; binary feature descriptor; feature extraction; hardware architecture; VLSI implementation

1. INTRODUCTION

Recently, visual sensor networks (VSNs) have attracted considerable attention in the research field. They are able to collect, process and communicate visual information in their surrounding environments [1]. They are already available in a wide range of visual applications, such as environmental monitoring [2] and virtual reality [3]. Real-time visual feature extraction is a key requirement for VSNs. The visual feature is a fundamental algorithm that is widely used in many high-level visual applications, such as augmented reality and object recognition [4,5,6,7,8,9]. Furthermore, the local processing of image data can significantly reduce the communication burden. Complex vision analysis can also provide VSNs with high levels of intelligence. These algorithms (such as feature extraction) enable VSNs to collaborate by exchanging detected features and collectively estimating the behavior of the captured object [10].

It is difficult to directly embed and locally process in real time for applications on VSNs because the feature extraction algorithm is computationally intensive [11]. Numerous efforts have been made to accelerate these vision algorithms by VLSI implementation. Huang et al. implemented SIFT [12] on a parallel architecture [13] that separately used interactive components for detector and descriptor tasks. The accelerator reached 30 fps for VGA images at 100 MHz frequency. In this system, the descriptor generation module accounted for 89% of the total time and became the bottleneck. Chiu et al. proposed a layer parallel SIFT feature and implemented it on a hardware architecture [14]. The design used an integral image technique to accelerate the scale pyramid build. The proposed system achieved 30 fps for images with different resolutions. As the image size increased, the number of descriptors was reduced due to the limited memory bandwidth. More recent studies have paid closer attention to the acceleration of the descriptor part. Jiang et al. introduced a real-time SIFT accelerator with a task-level parallel and pipeline technique [15]. In this design, a window dividing method was proposed to avoid sample patch rotation in the descriptor. This technique reduced computational complexity and gained a 15× speed increase for descriptor generation. The system processed a 512 × 512 image in 6.55 ms. Other hardware designs are based on binary descriptors that show significant efficiency improvements. In [11], Wang et al. proposed a real-time FPGA-based embedded system architecture that employed a SIFT detector and BRIEF [16] descriptor. This architecture achieved 60 fps for 720 p video. In general, most of the current designs are based on traditional features, such as SIFT or SURF. These features employ float descriptors that may cause a communication burden for the VSNs [17]. For designs based on binary descriptors, high throughput was realized at the expense of lower robustness in some aspects of the transformation due to the BRIEF descriptor [11,18].

In order to achieve real-time performance, the design for the current research is based on the accelerated KAZE (AKAZE) feature [19]. The AKAZE feature employs nonlinear scale space and a binary descriptor, which provides a considerable trade-off between speed and accuracy. Although AKAZE embeds a recent numerical scheme in order to accelerate the scale pyramid build, the computational complexity still poses a challenge for the hardware design. Also, considerable memory burden still exists due to the data dependency in the pyramid build and descriptor generation.

In light of the above considerations, this paper proposes a flexible embedded system architecture for real-time feature extraction. The architecture embeds a different kind of block-serial scheme in order to reduce the hardware cost and a pixel-level parallel scheme to increase throughput. It achieves a high throughput while maintaining a comparable performance to the original AKAZE algorithm. The contributions of this work are as follows:

- The authors believe this to be the first feature extraction design based on the AKAZE algorithm. The AKAZE feature was mapped to an octave-serial architecture (OSA) primarily consisting of a two-dimensional pipeline array. It decreases the hardware resource requirement and also provides sufficient flexibility for the various application fields, characterized as different image resolutions, precision and power consumption.
- A substructure consisting of a block-wise discrete-time cellular neural network (B-DTCNN) is presented. It decreases memory demand through the reduction in data dependency.
- A hardware-friendly descriptor, termed the robust polar binary descriptor (RPB), is presented. The polar arrangement of the sample pattern, combined with a simplified technique to realize rotation invariance, greatly decreases the memory burden and computational complexity.

Section 2 analyzes the AKAZE algorithm in order to determine possible obstacles. It also introduces the hardware-friendly descriptor, RPB. Section 3 presents the hardware design details. Section 4 presents the experimental and simulation results. Section 5 provides the conclusions.

2. ALGORITHM OPTIMIZATION

This section provides a brief introduction to the AKAZE feature. Further analysis demonstrates the advantages and disadvantages of AKAZE to achieve real-time performance for hardware solutions. Finally, a hardware-friendly binary descriptor is introduced to reduce memory cost and computational complexity.

2.1. AKAZE Overview

The AKAZE feature contains three major stages: nonlinear pyramid build, key point location and binary descriptor generation. In the first stage, AKAZE employs the Perona-Malik (P.M.) equation [20] to build a nonlinear scale pyramid. In order to construct different sublevels in the pyramid, this method diffuses the original image to a series of increasing scale levels using Equation (1), where I is the image luminance and k is the contrast factor. In order to accelerate the diffusion process, AKAZE adopts a fast explicit diffusion (FED) scheme [20,21] that approximates the solutions by iterations. Each iteration can diffuse the image with a small-scale step. Based on variable scale-steps (rather than constant), FED greatly reduces the number of iterations:

$$\frac{\partial I}{\partial t} = div\left(\frac{\nabla I}{1+|\nabla I|^2/k^2}\right) \tag{1}$$

The key points are located in the second stage. Once the pyramid is constructed, the determinant of the normalized Hessian matrix is computed (Equation (2)). Next, the local maxima at each sublevel are picked out as candidate key points. In order to search the scale extremes, a potential point is compared with other candidate key points within a σ × σ window from sublevel i − 1 to i + 1:

$$H_{norm} = \sigma^2 \begin{vmatrix} I_{xx} & I_{xy} \\ I_{xy} & I_{yy} \end{vmatrix} \tag{2}$$

The third stage of AKAZE introduces a modified-local difference binary (MLDB) descriptor. Typically, MLDB descriptor generation is divided into three steps: main orientation estimation, sample patch rotation and binary descriptor generation. Firstly, MLDB estimates the main orientation using a SURF-like method based on the histogram. Then, the sample pattern is rotated according to the direction of the key point. Finally, the binary descriptors are generated through comparison among the grids in three channels (one luminance and two rotated first-order derivatives).

2.2. AKAZE Analysis

Brief runtime analysis of AKAZE is conducted on a PC platform (Intel i5-3210M @2.5 GHz) in order to optimize the configuration. A “boat” image from the

Oxford dataset [22] is used for the AKAZE algorithm testing. Figure 1 shows the time usage of each stage. The first stage (nonlinear pyramid build) is the most time-consuming component, taking approximately 40% of the total time. The third stage requires 0.034 ms to calculate a descriptor. The PC platform can only achieve 3 fps and struggles to meet real-time performance for high-resolution images.

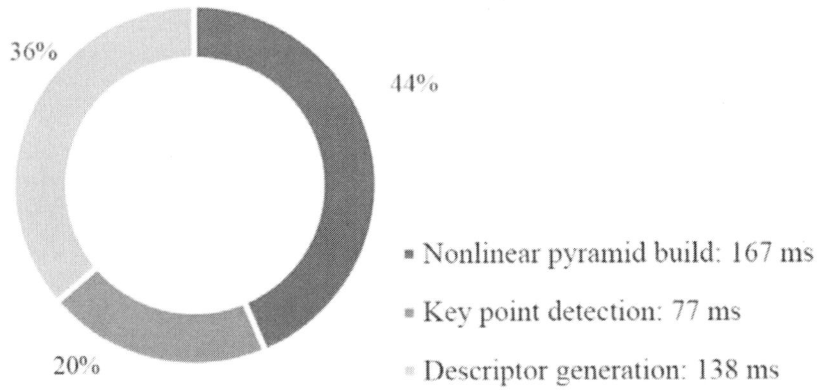

Figure 1. Runtime analysis of AKAZE on PC platform (Intel i5-3210M @2.5 GHz).

Further analysis of AKAZE shows one main advantage and two significant disadvantages for the hardware design. As an advantage, AKAZE provides a novel simplified method to find spatial and scale extremes. It divides the process into two different steps and introduces candidate key points as intermediate data. The memory demand can be reduced because there is no need to align data for scale extreme searching in the adjacent sublevels. However, there are still two significant challenges for achieving real-time performance. The first challenge is a nonlinear pyramid build. Several nonlinear filters must be processed in series in order to build four sublevels within one octave (Figure 2a). A nonlinear filter contains various iterations (Figure 2b), each having 17 arithmetic operations (Figure 2c). Due to the generation process of the cascade, the increase of sublevels and octaves may lead to unacceptable hardware resource usage under the current architecture design [11,13,23]. Also, the frame-level intermediate data in a nonlinear filter generate large memory demand.

The second challenge is the descriptor generation. The AKAZE uses a traditional 3-stage process flow. Similar hardware designs [13,14,24] indicate that this design is a bottleneck for system throughput. Also, it poses challenges for hardware resources (especially memory demand). Firstly, the pixels around a key point are read out in the first and third stages. This not only encumbers the algorithm pipelining in the hardware, but also causes significant random memory access.

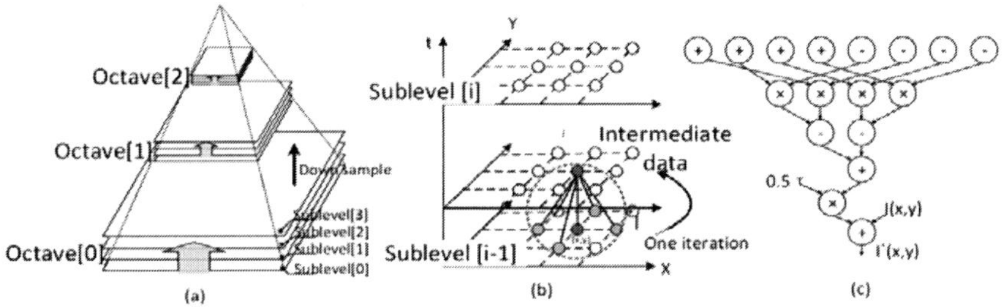

Figure 2. Nonlinear scale pyramid build process: (**a**) scale pyramid build process; (**b**) process to generate next sublevel (various iterations); (**c**) operations to update one pixel within an iteration.

Recent studies [11,15,23] show that random access is the main barrier to achieving real-time performance. This challenge is difficult to resolve only using hardware design. Also, the data random access leads to excessive buffer caching for the intermediate data around a key point. Additionally, for rotation invariance, the coordinates (x, y) need to be rotated according to the main direction, θ, in the second stage (Equation (3)) as well as the first-order derivative in the third stage, which results in numerous triangle functions:

$$\begin{bmatrix} x' \\ y' \end{bmatrix} = \begin{bmatrix} cos\theta & -sin\theta \\ sin\theta & cos\theta \end{bmatrix} \begin{bmatrix} x \\ y \end{bmatrix} \tag{3}$$

2.3. AKAZE Optimization

In this section, a hardware-friendly binary descriptor is proposed. It aims to overcome the memory burden challenge in the hardware. The challenge results from the number and order of random access, which is difficult to overcome only using hardware design. The PRB descriptor is introduced in order to find a method to speed up generation. It consists of a polar sample pattern combined with local integral images. It successfully converts most random access into regular access. Also, a simplified method to estimate the main orientation is proposed based on the polar sample pattern. It further reduces random memory access as well as computational complexity. In addition, the PRB descriptor easily pipelines in hardware because it only accesses the integral image once. Details of the RPB descriptor are as follows.

Descriptor generation is always the bottleneck in hardware design. This general process can be divided into three stages: main orientation estimation, patch rotation and local descriptor generation. Analysis shows that the process contains significant random memory access in the first and third stage, which poses a challenge to achieve high throughput (see Section 2.2). Also, the random scatter of key points brings extra complexity to the control. Huang [13] proposed an interactive architecture with a co-processor for descriptor generation that successfully overcame the control complexity of the random distribution. However, 89.7% of the time was used by the descriptor generation. Several hardware solutions [11,13,15] were proposed to accelerate the descriptor generation. Although substantial improvements were achieved, the processing speed was still unable to match key point detection. As such, the current research introduces a hardware-friendly descriptor.

The RPB descriptor has three major stages. Firstly, according to the sample pattern, it reads out all the grid values for later use. It contains two channels: the grey value and the gradient norm. The main orientation is estimated using the grid values of the outermost layer in the first channel. Secondly, invariance to rotation is achieved by reordering the sample values according to the main direction. Thirdly, pseudo-random pairs are selected for comparison. The two key features that focus on simplification of rotation invariance are as follows:

Sample Pattern: The RPB descriptor employs a novel sample pattern of polar arrangement for rotation invariance through reordering and less random memory access. Figure 3a shows the three layers of the sample pattern. The arrangement of each layer is similar to DAISY [25] and has various radiuses and distances from the center point. For each layer, there are sixteen grids on the circle that matches the number of discrete angles for reordering.

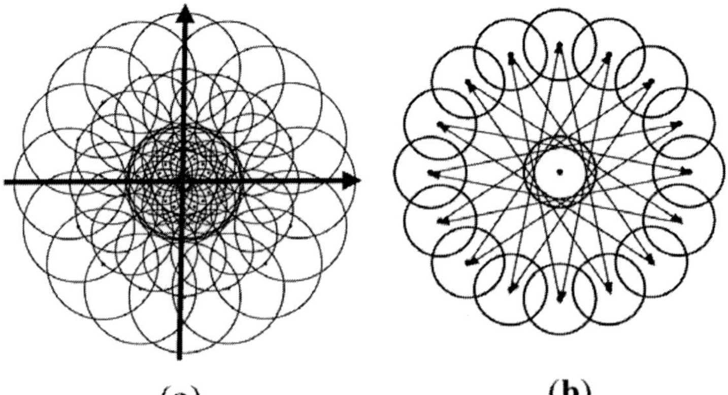

Figure 3. (**a**) Sample pattern containing three layers, where each layer has 17 sample grids with polar arrangement; (**b**) Pairs used for main orientation estimation based on the outermost layer.

Main Orientation: The RPB descriptor adopts a simplified gradient method similar to BRISK [26] for rotation invariance. Equation (4) shows that the angle vector, A, is estimated through the sum of gradients of specific pairs in the set P. The two-dimensional vector, p, contains the grid coordinates and I(p) is the grey value of the first channel. The sixteen selected pairs are symmetrically distributed to avoid a division operation (Figure 3b):

$$A = \sum_{(p_i, p_j) \in P} (p_i - p_j)\left[I(p_i) - I(p_j)\right] \quad (4)$$

Based on the hardware-oriented optimization, the RPB descriptor reduces both the memory burden and computational complexity. In terms of memory usage, random access decreases by 68.7% (650 to 240 per key point). Based on the integral image, the polar sample pattern changes most of the random access into regular access. In addition, the proposed main orientation estimation method does not add extra access to the integral image memory, thereby removing approximately 40% of the memory burden. Also, the optimized method to estimate the main orientation significantly decreases computations, especially in the triangle computation (Table 1). For robustness, the RPB descriptor achieves similar matching accuracy compared with the original algorithm; however, accuracy is slightly less in the rotation and brightness tests (See Section 4.1).

Table 1. Comparison of operations to estimate main orientation.

Operations	Proposed	Original (SURF-Like)	BRISK (OpenCV)
Extra Random Access	0	218	240
Trigonometric Function	1	452	1
Addition/Subtraction	80	1746	4350
Multiplication	32	452	1741
Lookup Tables	32	109	3480

3. PROPOSED HARDWARE ARCHITECTURE

This section introduces details of the hardware architecture based on the optimized AKAZE feature. The proposed system consists of three main functional parts (Figure 4). In order to achieve high throughput with a reasonable hardware cost, three techniques (OSA, B-DTCNN and PRB) are utilized in the design.

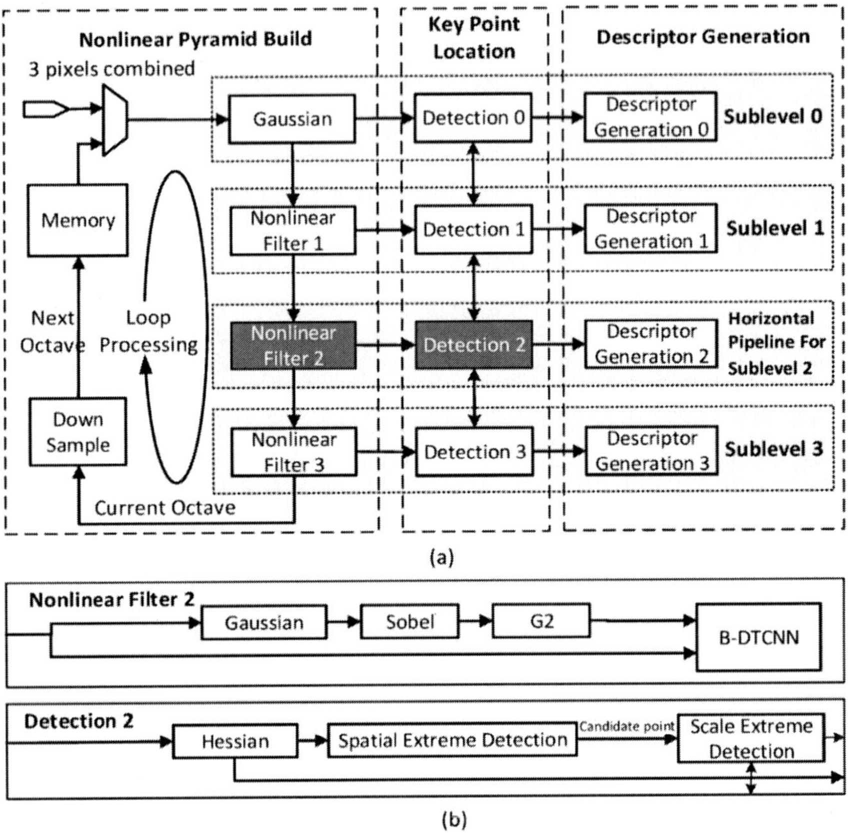

Figure 4. (**a**) Proposed OSA architecture with three main stages due to the sublevels' process in pyramid; the different octaves are processed in series with various pixels processed in parallel in each module for efficiency; (**b**) Typical block diagrams of a nonlinear filter module and detection module used to illustrate timing design.

3.1. Octave-Serial System Architecture

This section introduces the OSA system architecture (Figure 4). A two-dimensional pipeline array is applied to make full use of computational similarities among the different octaves. The OSA architecture is able to reduce hardware resources as well as provide additional flexibility to image size and speed-precision tradeoffs in real-time.

The number of octaves directly affects matching accuracy. However, multiple octaves result in extra hardware costs. In [11,13,23], similar functional modules, such as key point location and the descriptor generation module, are repeated in parallel to process data in the second or third octave. The current analysis shows that multiple octaves does not significantly increase throughput. For example, the second, third and fourth octave only add 25%, 6.25% and 1.56% extra pixels, respectively. Therefore, processing octaves in series significantly reduces hardware costs with only a minimal sacrifice in speed.

Figure 4a depicts the whole OSA architecture, which can be considered as a two-dimensional pipeline array. The vertical pipeline constructs a scale pyramid through cascaded nonlinear filters. It continuously reads in the stream of the original or down-sampled image and sends four sublevels within one octave to the next stages. There are four horizontal pipelines and each processes one sublevel: (1) locate key points based on the Hessian response of adjacent levels, which sequentially searches space and scale extremes; (2) generate descriptors depending on the local integral image. Figure 5 shows the processing flow. Unlike the general block-wise scheme, this block-partitioning scheme penetrates into the inner scale pyramid. The image data of the different octaves are divided into several trunks (Figure 5a) and combined in a ribbon-like form for processing (Figure 5b). The accelerator moves to process one chunk in the next octave (in the memory) after finishing two chunks in the current octave. In this way, the proposed system processes the data of different octaves in series on the same hardware. Figure 5c is an example of a two-octave pyramid in which there are four sublevels within one octave and each sublevel is processed by one horizontal pipeline. Also, overlap between the adjacent chunks can be added. Furthermore, the off-chip memory can cache one more down-sampled trunk if the key points near the block border (approximately 15–20 pixels) are necessary in practice; otherwise, these points are ignored in the descriptor generation stage. The timing sequence is as follows. Most of the modules, such as linear filter (Gaussian, Sobel and Hessian) and detection, are capable of processing three pixels per cycle. However, the B-DTCNN and descriptor modules are bottlenecks in the structure. In the current design, the B-DTCNN processing ability is 1.64 pixels per cycle, which determines the system throughput (see Section 3.2). For each descriptor module, 51 cycles are required to generate a descriptor. Through the buffer and discarding schemes, the descriptor module does not stall the pipeline (see Section 3.3). Due to the loop-processing scheme, the whole system can run at 127 (200 M × 1.64/(1920 × 1080 × 1.25)) fps in full HD, with two octaves in pyramid.

The OSA architecture has two main advantages. Firstly, an OSA-based system provides considerable flexibility in image size, precision and power consumption. Due to the block-wise method, the system proposed can easily process images in different resolutions such as 720 p, 1080 p and 4 K. The OSA-based system is more robust to image scale change due to the handling of multiple octaves. The power consumption can be reduced by approximately 25% when shutting off the last horizontal pipeline. Secondly, having a similar hardware cost, the system provides an extra 54% throughput compared with previous works [11,13,24]. Due to the change from octave-parallel order to octave-serial order, the similar hardware cost provides a higher degree of pixel-level parallelism within one octave. When the pixel-parallelism degree doubles, the second and third octave only yield 25% and 6.25% extra pixels processing, respectively (due to the octave-serial order). Hence, the processing speed sharply increases. Moreover, due to its design flexibility, other feature extraction algorithms, such as SIFT, can be mapped to the proposed architecture. The architecture also provides efficiency and flexibility to algorithms that use linear filters to construct pyramid. Although the OSA architecture greatly accelerates key point detection, the throughput is still limited by the descriptor generation module and the nonlinear filter module. These challenges are addressed in the following sections.

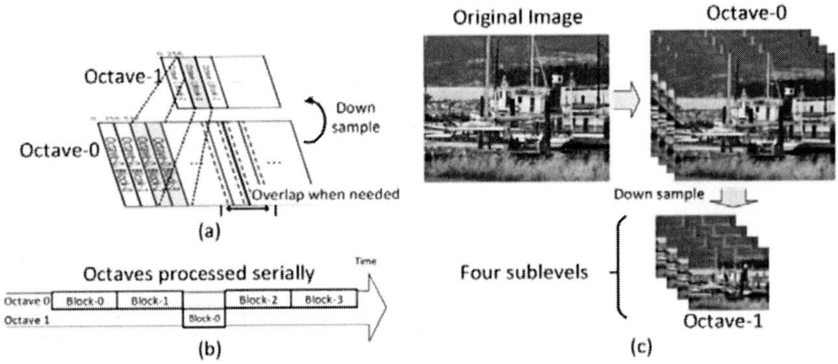

Figure 5. (**a**) Block-partitioning scheme in a nonlinear scale pyramid; (**b**) Specific order to process octaves in series, where different blocks in multi-octaves are divided and combined in a ribbon-like form to fit the octave-serial system architecture; (**c**) An example of a two-octave pyramid.

3.2. Block-Serial DTCNN

This subsection presents a substructure named B-DTCNN that significantly decreases memory demands brought about by the nonlinear pyramid construction. An analysis of the relationship between the Gaussian convolution and nonlinear diffusion provides a block-wise serial strategy. The strategy adopts elastic overlap between blocks in order to gain a better trade-off between accuracy and hardware costs. Finally, the corresponding hardware structure is presented, which contains a Ping-Pong structure to accelerate throughput.

The DTCNN architecture [27] is used to construct the pyramid because of its local connection and parallelism. Due to frame-level data dependency in iterative filters, the intermediate data cache results in a significant memory demand (see Section 2.2). Therefore, it could only use block level rather than whole image processing. Furthermore, the nonlinear pyramid requires a series of images at different run-time stages. In [28], a sequential DTCNN architecture was proposed based on FPGA. The computing task of pixels in an iteration was mapped to one cell where the images at different run-times could be easily passed to the next stage. However, the buffers between the different stages resulted in long latency and memory costs. Inspired by the above design, this paper proposes a B-DTCNN, in which pixels within a block are processed in parallel.

The removal of frame-level dependency is based on the following analysis. When the flow function, g, becomes a constant function, an anisotropic diffusion Equation (5) evolves to linear diffusion Equation (6) and the nonlinear scale pyramid becomes a Gaussian scale pyramid:

$$\frac{\partial I}{\partial t} = div(g(|\nabla I|) \cdot \nabla I) \tag{5}$$

$$\frac{\partial I}{\partial t} = div(\nabla I) \tag{6}$$

The linear diffusion equation can be solved two ways. One way is the Gaussian template convolution with the original images, where the evolution time, t, is mapped to the scale parameter, σ, in Equation (7). The other way is a numerical scheme identical to anisotropic diffusion (such as an explicit scheme). The first solver shows that the longer the distance between the pixel and center, the less it contributes to the result; the corresponding pixels have little impact when the distance is more than a given threshold. Moreover, the threshold distance is proportional to the evolution time, e, in each stage:

$$t_i = \frac{1}{2}\sigma_i^2, \ i = 0,1,2... \tag{7}$$

The block layout (Figure 6) has three key parameters: overlapping pixels, O, which directly affect diffusing accuracy; block width, W; and block height, H, all of which determine the hardware cost and system throughput. In general, the mean square error (MSE) is used to evaluate the difference between the original pyramid, I, and the optimized I′(O, W, H).

$$\textit{Diffusing Distortion}: MSE(I, I'(\mathrm{O}, \mathrm{W}, \mathrm{H})) \tag{8}$$

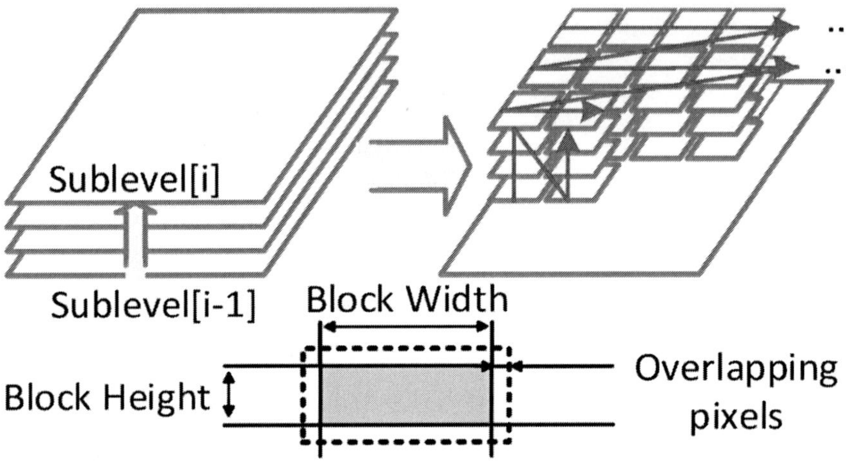

Figure 6. A block-serial scheme in B-DTCNN where the frame-level data dependency is deconstructed to block-level one. The blocks of different stages choose elastic overlapping pixels due to time steps for different trade-offs between accuracy and hardware costs because the block configure determines the size of the PE arrays.

Figure 7 shows data flow and hardware implementation. From the view of PE array, two symmetric register arrays are used to construct a Ping-Pong structure (Figure 7a). In each iteration, the ALU array executes the arithmetic operation and exchanges the internal temporary data with one register array when the other register array switches data with the external modules. Inside a PE, there are four registers to cache grey values and nonlinear coefficients of two channels separately and to communicate data with PEs in North, South, East and West (NEWS) directions (Figure 7d). Each iteration requires four cycles. After several iterations, the block of data is sent to the next stage for further processing. The speed of the nonlinear filter block is determined by the number of iterations and the size of the processing array, which further determines the throughput of the vertical pipeline. For the timing consideration, Figure 7b shows an example of the B-DTCNN module in the first nonlinear filter (5 × 9 array). Due to the diffusion parameter, it only has one overlapping pixel. Therefore, it requires 12 cycles (3 iteration) and outputs 21 (3 × 7) pixels. The speed is 1.75 (21/12) pixels/cycle. For each row, “padding” pixels are used on the border and idle cycles are used to synchronize the other two nonlinear filters. Hence, the unified processing speed is 1.64 pixels/cycle.

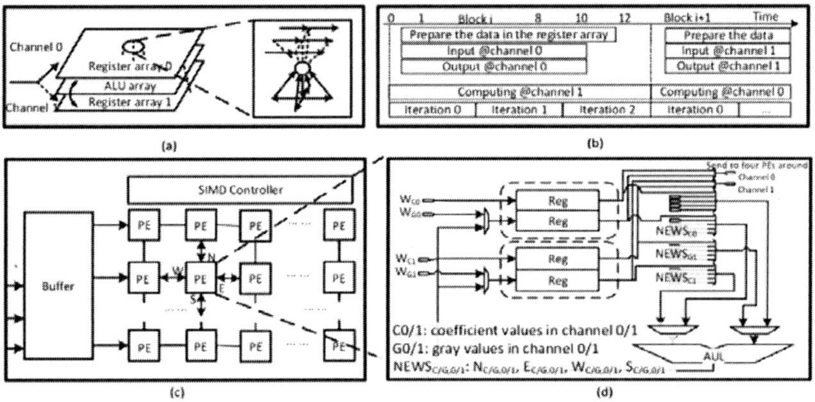

Figure 7. (**a**) Data flow of the Ping-Pong operations. The arithmetic operations and data preparation are done in parallel; (**b**) Typical timing sequence to process one block; (**c**) Diagrams of B-DTCNN array in which PEs are connected in North, South, East and West (NEWS) directions to communicate data during iterations; (**d**) Inner structure of each PE that caches coefficient values and gray values of two channels in local registers.

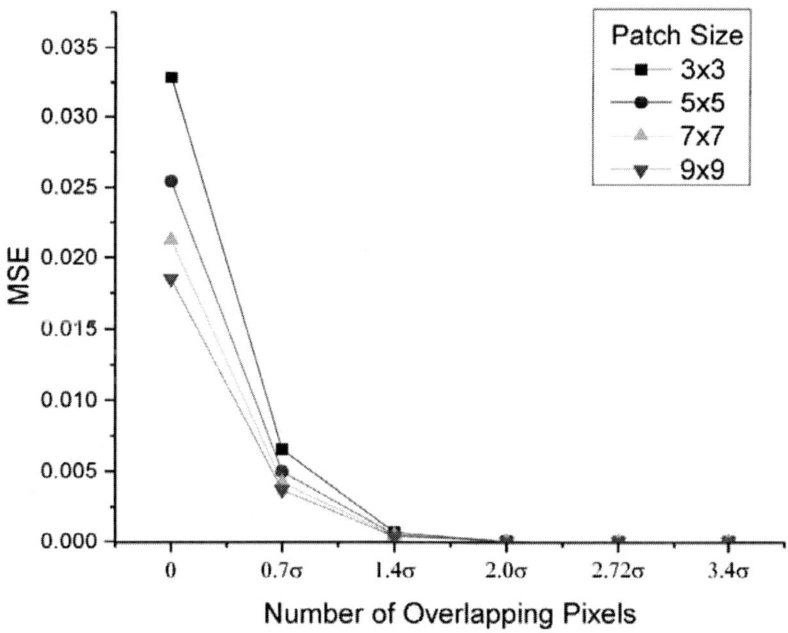

Figure 8. Error between the diffused images using two kinds of iterations (full-frame iteration and independent block-level iteration). The error quickly drops with the increase of overlapping pixels.

The pyramid distortion dramatically decreases when the overlapping pixels increase (Figure 8). In general, the overlapping pixels are set to more than 2σ to minimize the error (in the order of 10^{-5}), which conforms to the solver of the Gaussian convolution (where the radius of the Gaussian template is always 2σ ~ 3σ). In the current hardware design, the memory demand decreases by approximately 80% compared with the design in [28]; furthermore, there is only 13.7% redundant pixel computation in the image boundary.

3.3. Robust Polar Binary Descriptor Module

This subsection introduces the descriptor generation module. The structure is based on the proposed hardware-friendly RPB binary descriptor. The hardware/software co-design successfully reduces the random memory access number and adjusts the access order. Hence, the RPB descriptor can be mapped to a three-stage pipeline. Finally, this module is able to generate the descriptor at relatively high speeds.

Figure 9c shows that the descriptor generation module primarily consists of a three-stage pipeline in the hardware. Firstly, two channels of one sublevel (Figure 9a) are used to construct the integral image and both results are combined. This operations reduce the memory random access. The memory block consists of four partitions to further speed up the random access, especially for the integral image [23]. Also, the circular data dependency around a key point results in significant memory demand. Therefore, the integral images are compressed. One way is through module-N arithmetic based on [29]. The other way is through general rounding and saturation operations with minimal accuracy loss. Secondly, all the grid values are read out and then buffered in cache. The main orientation is estimated based on the first sixteen grid values. Thirdly, the rotation invariance is achieved by reordering the sixteen grid values in each layer. Then, comparisons are performed in order to generate the final binary descriptor (Figure 9c). The following analysis shows that the proposed structure has strong robustness against random memory scatter access. The whole process can be treated as a three-stage pipeline. The most time-consuming stage is to read the sample grid from memory. The operation requires 51 cycles. The average speed to feed a row in memory for an integral image is 156 cycles. The key points are abandoned when the number of points exceeds 10 in three continuous rows (each containing 256 pixels) within one sublevel. According to the test on the Oxford datasets, the extreme case does not occur. Within one sublevel, the average number of key points in each row is less than one. Therefore, the descriptor generation module speed matches the detection speed in most cases.

Due to the co-design of hardware and software, the descriptor generation module achieves real-time performance. The optimized method to estimate the main orientation enables the process to be pipelined. The memory burden is reduced by the integral image and the novel sample pattern. The random access is further accelerated by the embedded memory partition scheme for the integral images. For the whole architecture with four modules, the peak generation speed is 12.5 cycles per key point, which matches the detection speed in most cases.

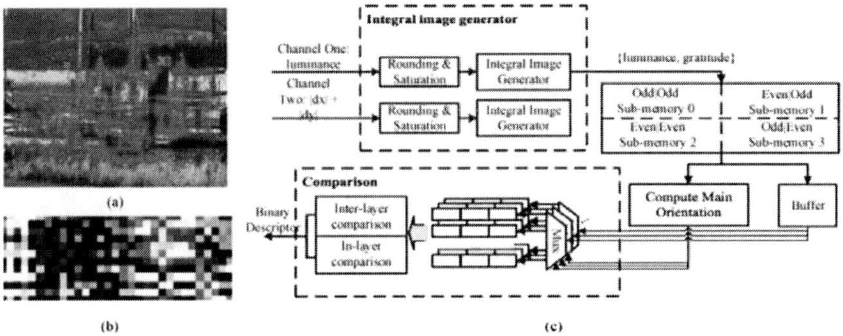

Figure 9. (**a**) An example of the input sublevel with key points detected; (**b**) An example of the binary descriptors in which each grid represents eight binary bits; (**c**) Block diagram of descriptor generation module containing three main stages: integral image build, grid value read-out with main orientation estimation, and grid values comparison.

4. SIMULATION AND VERIFICATION

In this section, the performance of the optimized AKAZE is tested using detector repeatability and matching accuracy metrics. The hardware implementation details are also presented (focusing on system throughput and hardware cost).

4.1. Feature Accuracy

Firstly, a brief introduction to the test configuration is provided. The pyramid parameter is selected due to the flexibility with octaves that result from the proposed architecture. There are two octaves, each with four sublevels. In the benchmark, fixed-point precision is adopted in the optimized AKAZE with various bit widths in the different stages (while maintaining the original floating-point precision in the original one). In order to assess the accuracy of the different transformations, image sets from the Oxford datasets are adopted for algorithm evaluation. The benchmark criteria is introduced below. The repeatability [30] is selected to measure the ability of the detector to extract repeatable key points. The overlap error is less than 0.4. A third-part framework [31], used in [32], is applied to evaluate the descriptor. In the benchmark, matching accuracy is mainly used to show the robust quality of the key points, which combines the detector and descriptor.

Figure 10 shows the feature-matching results that more intuitively demonstrate the simulation results. Quantitative results are provided for comparison (Figure 11 and Table 2). The results show that the optimized algorithm obtains similar performance compared with the original AKAZE algorithm. Table 2 indicates that, on average, the matching accuracy of the proposed feature only marginally drops (approximately 1%–2%). Figure 11 also shows these results. The matching accuracy in the rotation change drops a little at the boundary of each discrete angle for the proposed RPB descriptor (Figure 11i). Also, there are significant changes in brightness when the accuracy declines (Figure 11j) because of the constant contrast factor, k, in Equation (1) rather than an adaptive one as in the original AKAZE. The scaling change improvement can be attributed to the adjustment of the scale parameter in the second octave. The chosen scale parameter is similar to SIFT.

Figure 10. Feature matching results.

Table 2. Comparison of average matching accuracy.

		Brightness	Gaussian	Viewpoint	Rotation	Scaling
Boat	AKAZE	87.47%	93.29%	78.53%	95.51%	83.03%
	Proposed	83.70%	93.78%	78.95%	93.99%	82.30%
Trees	AKAZE	88.97%	89.45%	72.66%	90.74%	80.26%
	Proposed	88.07%	92.77%	76.75%	92.57%	82.55%
Bikes	AKAZE	90.03%	95.34%	77.17%	91.01%	82.89%
	Proposed	88.03%	94.69%	76.81%	88.32%	81.76%
Bark	AKAZE	93.24%	95.92%	76.80%	92.14%	86.18%
	Proposed	90.88%	95.20%	76.53%	93.66%	84.37%
Graf	AKAZE	88.37%	95.96%	77.92%	93.23%	85.00%
	Proposed	84.06%	95.64%	77.65%	91.01%	82.69%
Leuven	AKAZE	90.80%	96.36%	78.73%	89.27%	84.83%
	Proposed	89.21%	94.86%	78.86%	85.50%	82.58%
Ubc	AKAZE	89.43%	93.25%	79.24%	94.43%	82.33%
	Proposed	88.25%	94.28%	81.48%	91.52%	83.68%
Wall	AKAZE	87.98%	89.06%	77.64%	93.05%	82.81%
	Proposed	85.09%	91.68%	76.48%	95.09%	83.58%

4.2. Hardware Performance

The proposed architecture is mapped to hardware using TSMC 65 nm CMOS technology. Due to lack of AKAZE-based hardware designs, the current design is mainly compared with state-of-the-art works based on SIFT or SURF, which have similar workflows (see Table 3). Figure 12 shows a comparison of hardware costs. Figure 13 shows the layout of the die, which is generated by the integrated circuit compiler (ICC). Table 4 provides a summary of the hardware resources.

The multiplication and addition are taken into account for the measurement of peak performance [33]. For the linear filter module, there are 3n − 1 fixed-point operations per pixel, where n is the template size. For the B-DTCNN module, there are 17 operations/pixel for every iteration. Each descriptor generation module is capable of calculating three key points per row. Hence, the speed to process a key point is 0.047 (4 × 3/256) key points per pixel. For each key point, there are 31 operations to estimate the main orientation and 306 operations to construct the sample grid. The operations required to generate descriptors are 20 (2 + 0.047 × (306 + 31)) operations/pixel. Table 5 provides a summary of the number of operations in each stage. In total, the whole structure reaches 181 (127 × 1920 × 1080 × 1.25 × 550) GOPS. In order to compare the processing speed, the throughput is normalized to the unit frequency and the equivalent gate in which the memory is converted to gate according to TSMC 65 nm technology. Figure 12d and Table 3 show that the proposed system is nearly twice as fast as comparable state-of-the-art works. Figure 12c shows that the high throughput can be attributed to the proposed RPB descriptor, which is no longer the bottleneck in the hardware.

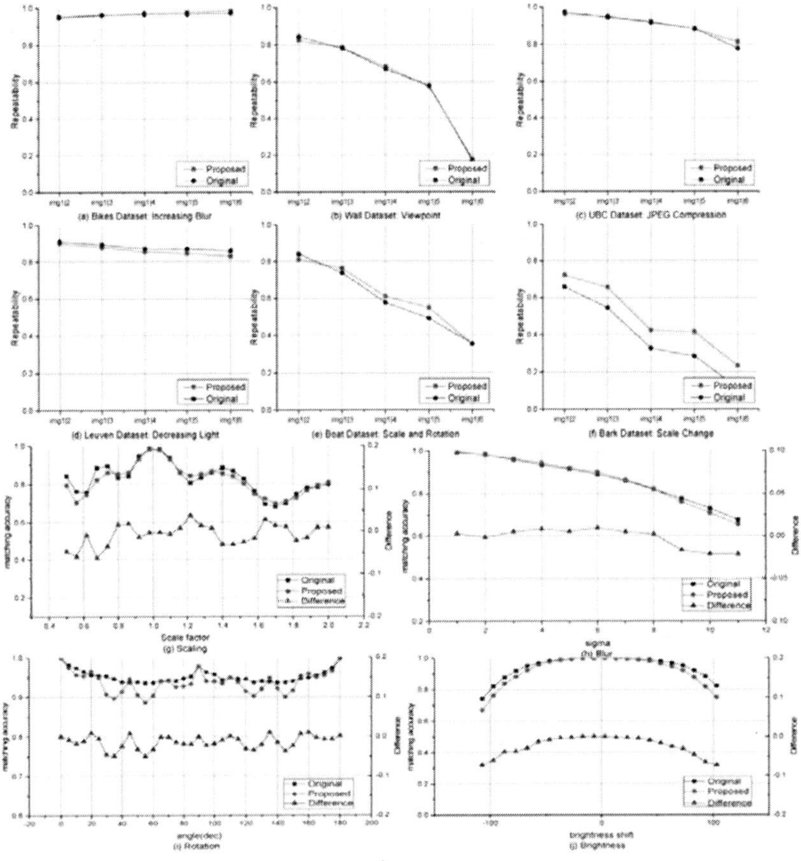

Figure 11. Comparison of detector repeatability (**a–f**) and matching accuracy (**g–j**).

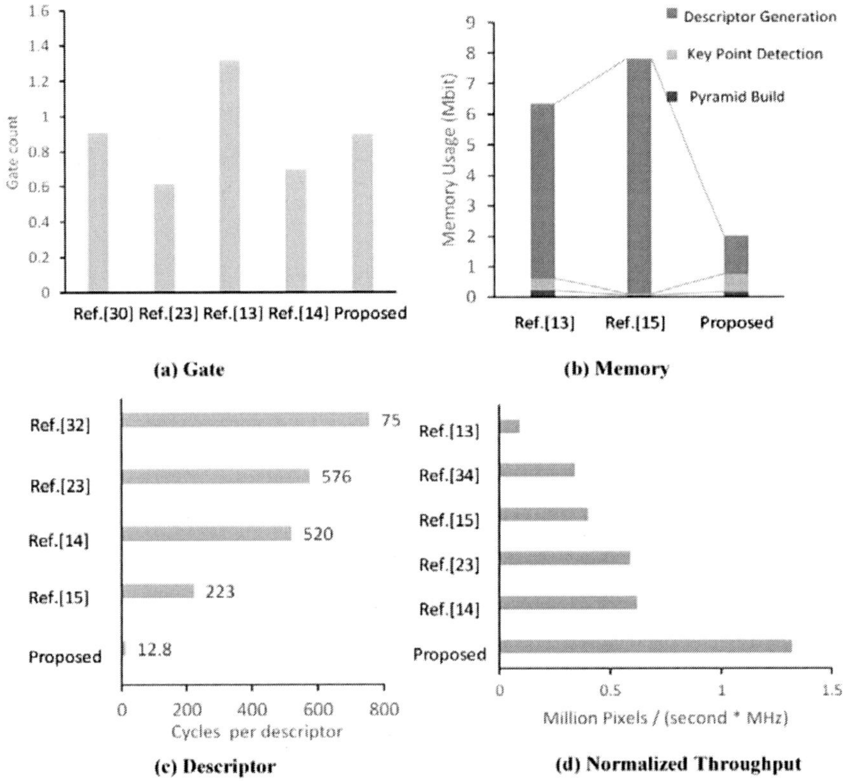

Figure 12. Comparison of hardware cost (**a**,**b**) and process speed (**c**,**d**).

Table 3. Comparison of performance.

	Reference [23]	Reference [34]	Reference [13]	Reference [14]	Reference [15]	Proposed	Proposed
Approach	SURF	SURF	SIFT	SIFT	SIFT	AKAZE	AKAZE
Pyramid [1]	(3, 4)	(1, 5)	(3, 6)	(2, 4)	(2, 4)	(2, 4)	(3, 4)
Platform	ASIC	ASIC	ASIC	ASIC	FPGA	ASIC	AISC
Frequency	200 MHz	27 MHz	100 MHz	100 MHz	100 MHz	200 MHz	200 MHz
Memory	3.2 Mb	-	5.73 Mb	0.55 Mb	7.8 Mb	2.12 Mb	2.12 Mb
Gate	0.6 M	-	1.32 M	0.7 M	-	0.95 M	0.95 M
Resolution	1920 × 1080	640 × 480	640 × 480	1920 × 1080	512 × 512	1920 × 1080	1920 × 1080
Speed	57 fps	30 fps	30 fps	30 fps	153 fps	127 fps	121 fps
Throughput	0.118 G	0.0092 G	0.0092 G	0.062 G	0.040 G	0.263 G	0.251 G
Throughput per gate [2] (Pixels/s@1 MGate)	31.2 M	-	1.31 M	49.8 M	-	85.9 M	81.8 M
Throughput per frequency (Pixels/s@1 MHz)	0.59 M	0.34 M	0.092 M	0.62 M	0.40 M	1.3 M	1.2 M

[1] Pyramid parameter is (octaves, sublevels); [2] The memory is converted to gate according to TSMC 65 nm CMOS technology.

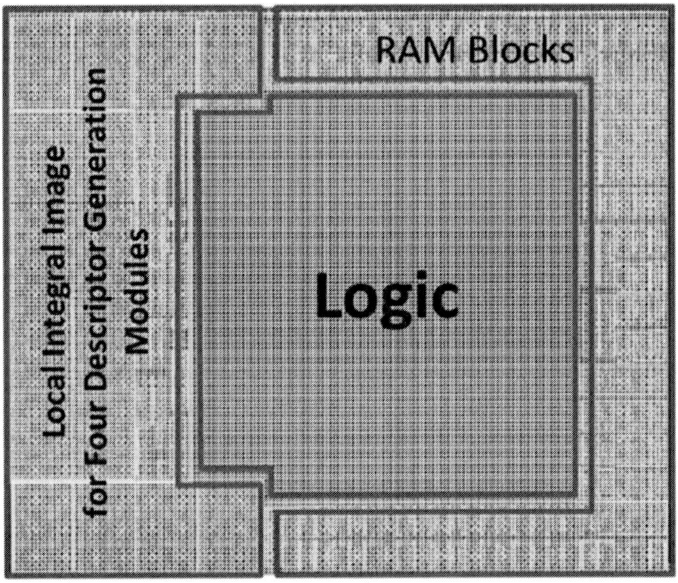

Figure 13. Layout of the proposed AKAZE accelerator.

Table 4. Characteristics of the proposed accelerator.

AKAZE Accelerator	
Process	TSMC 65 nm1p10m
Frequency	200 MHz
Gate	0.95 M
Memory	2.12 M bit
Size	4 mm × 3.3 mm
Speed	127 fps (1920 × 1080)

Table 5. Operations in each stage.

Stage	Average Operations (Operations/Pixel)
Nonlinear Pyramid Build	290
Key Point Location	240
Descriptor Generation	20
Total	550

5. CONCLUSIONS

This paper introduces several algorithm and hardware co-design techniques that can realize real-time acceleration for feature extraction. The proposed system maps AKAZE features to a highly flexible system architecture OSA. This provides elastic pyramid parameters and also considerably reduces hardware costs. Two other techniques, B-DTCNN and RPB, are employed to reduce the memory burden, which is the main bottleneck in the hardware design. Based on the above optimizations, the system achieves significantly high throughput for full HD images while maintaining similar accuracy to the original algorithm.

ACKNOWLEDGMENTS

This work is supported in part by the China National High Technologies Research Program (No. 2012AA012701) and the project from the State Grid Cooperation of China (No. SGRI-WD-71-13-014/008/010/011).

AUTHOR CONTRIBUTIONS

The corresponding author, Leibo Liu, conceived and designed the experiments as well as prepared the manuscript. The first author, Guangli Jiang, performed the experiments, proposed the research ideas and drafted the paper. The third author, Wenping Zhu, contributed the analysis tools. The fourth and fifth authors, Shouyi Yin and Shaojun Wei, both provided general guidance during the research.

CONFLICTS OF INTEREST

The authors declare no conflict of interest.

REFERENCES

1. Chong, C.Y.; Kumar, S.P. Sensor networks: Evolution, opportunities, and challenges. Proc. IEEE **2003**, 91, 1247–1256.
2. Remagnino, P.; Shihab, A.I.; Jones, G.A. Distributed intelligence for multi-camera visual surveillance. Pattern Recognit. **2004**, 37, 675–689.
3. McCurdy, N.J.; Griswold, W.G. A systems architecture for ubiquitous video. In Proceedings of the 3rd International Conference on Mobile Systems, Applications, and Services, Seattle, WA, USA, 6–8 June 2005; pp. 1–14.

4. Kim, G.; Kim, Y.; Lee, K.; Park, S.; Hong, I.; Bong, K.; Shin, D.; Choi, S.; Oh, J.; Yoo, H.-J. A 1.22 TOPS and 1.52 mW/MHz augmented reality multi-core processor with neural network NoC for HMD applications. In Proceedings of the 2014 IEEE International Solid-State Circuits Conference Digest of Technical Papers, San Francisco, CA, USA, 9–13 February 2014; pp. 182–183.

5. Geronimo, D.; Lopez, A.M.; Sappa, A.D.; Graf, T. Survey of Pedestrian Detection for Advanced Driver Assistance Systems. IEEE Trans. Pattern Anal. Mach. Intell. **2010**, 32, 1239–1258.

6. Brown, M.; Lowe, D.G. Recognising panoramas. In Proceedings of the Ninth IEEE International Conference on Computer Vision, Nice, France, 13–16 October 2003; Volume 2, pp. 1218–1225.

7. Besbes, B.; Apatean, A.; Rogozan, A.; Bensrhair, A. Combining SURF-based local and global features for road obstacle recognition in far infrared images. In Proceedings of the 2010 13th International IEEE Conference on Intelligent Transportation Systems, Funchal, Portugal, 19–22 September 2010; pp. 1869–1874.

8. Park, S.; Kim, S.; Park, M.; Park, S.K. Vision-based global localization for mobile robots with hybrid maps of objects and spatial layouts. Inf. Sci. **2009**, 179, 4174–4198.

9. Jain, A.K.; Klare, B.; Park, U. Face Matching and Retrieval in Forensics Applications. IEEE Multimed. **2012**, 19, 20–28.

10. Soro, S.; Heinzelman, W. A Survey of Visual Sensor Networks. Adv. Multimed. **2009**, 2009.

11. Wang, J.; Zhong, S.; Yan, L.; Cao, Z. An Embedded System-on-Chip Architecture for Real-time Visual Detection and Matching. IEEE Trans. Circuits Syst. Video Technol. **2014**, 24, 525–538.

12. Lowe, D.G. Distinctive image features from scale-invariant keypoints. Int. J. Comput. Vis. **2004**, 60, 91–110.

13. Huang, F.C.; Huang, S.Y.; Ker, J.W.; Chen, Y.C. High-Performance SIFT Hardware Accelerator for Real-Time Image Feature Extraction. IEEE Trans. Circuits Syst. Video Technol. **2012**, 22, 340–351.

14. Chiu, L.C.; Chang, T.S.; Chen, J.Y.; Chang, N.Y.C. Fast SIFT Design for Real-Time Visual Feature Extraction. IEEE Trans. Image Process. **2013**, 22, 3158–3167.

15. Jiang, J.; Li, X.Y.; Zhang, G.J. SIFT Hardware Implementation for Real-Time Image Feature Extraction. IEEE Trans. Circuits Syst. Video Technol. **2014**, 24, 1209–1220.

16. Calonder, M.; Lepetit, V.; Strecha, C.; Fua, P. BRIEF: Binary Robust Independent Elementary Features. In Proceedings of the Computer Vision—ECCV, Heraklion, Greece, 5–11 September 2010; Volume 6314, pp. 778–792.

17. Yang, A.Y.; Maji, S.; Christoudias, C.M.; Darrell, T.; Malik, J.; Sastry, S.S. Multiple-view object recognition in band-limited distributed camera networks. In Proceedings of the ICDSC 2009 Third ACM/IEEE International Conference on Distributed Smart Cameras, Como, Italy, 30 August–2 September 2009; pp. 1–8.

18. Park, J.S.; Kim, H.E.; Kim, L.S. A 182 mW 94.3 f/s in Full HD Pattern-Matching Based Image Recognition Accelerator for an Embedded Vision System in 0.13 μm-CMOS Technology. IEEE Trans. Circuits Syst. Video Technol. **2013**, 23, 832–845.

19. Alcantarilla, P.F.; Bartoli, A.; Davison, A.J. KAZE Features. In Proceedings of the Computer Vision—ECCV 2012, Florence, Italy, 7–13 October 2012; Volume 7577, pp. 214–227.

20. Grewenig, S.; Weickert, J.; Bruhn, A. From Box Filtering to Fast Explicit Diffusion. In Pattern Recognition, Proceedings of the 32nd DAGM Symposium, Darmstadt, Germany, 22–24 September 2010; Volume 6376, pp. 533–542.

21. Grewenig, S.; Weickert, J.; Schroers, C.; Bruhn, A. Cyclic Schemes for PDE-Based Image Analysis; Technical Report 327; Department of Mathematics, Saarland University: Saarbrücken, Germany, 2013.

22. Mikolajczyk, K.; Schmid, C. A performance evaluation of local descriptors. IEEE Trans. Pattern Anal. Mach. Intell. **2005**, 27, 1615–1630.

23. Liu, L.L.; Zhang, W.L.; Deng, C.C.; Yin, S.Y.; Cai, S.S.; Wei, S.J. SURFEX: A 57 fps 1080 P resolution 220 mW silicon implementation for simplified speeded-up robust feature with 65 nm process. In Proceedings of the IEEE 2013 Custom Integrated Circuits Conference, San Jose, CA, USA, 22–25 September 2013; pp. 1–4.

24. Bonato, V.; Marques, E.; Constantinides, G.A. A Parallel Hardware Architecture for Scale and Rotation Invariant Feature Detection. IEEE Trans. Circuits Syst. Video Technol. **2008**, 18, 1703–1712.

25. Tola, E.; Lepetit, V.; Fua, P. DAISY: An Efficient Dense Descriptor Applied to Wide-Baseline Stereo. IEEE Trans. Pattern Anal. Mach. Intell. **2010**, 32, 815–830.

26. Leutenegger, S.; Chli, M.; Siegwart, R.Y. BRISK: Binary Robust Invariant Scalable Keypoints. In Proceedings of the 2011 IEEE International Conference on Computer Vision (ICCV), Barcelona, Spain, 6–13 November 2011; pp. 2548–2555.

27. Chua, L.O.; Roska, T. The CNN paradigm. IEEE Trans. Circuits Syst. I Fundam. Theory Appl. **1993**, 40, 147–156.

28. Javier Martinez-Alvarez, J.; Javier Garrigos-Guerrero, F.; Javier Toledo-Moreo, F.; Manuel Ferrandez-Vicente, J. High performance implementation of an FPGA-based sequential DT-CNN. In Nature Inspired Problem-Solving Methods in Knowledge Engineering, Proceedings of the Second International Work-Conference on the Interplay Between Natural and Artificial Computation (IWINAC 2007), La Manga del Mar Menor, Spain, 18–21 June 2007; Volume 4528, pp. 1–9.

29. Belt, H.J.W. Word length reduction for the integral image. In Proceedings of the 15th IEEE International Conference on Image Processing, San Diego, CA, USA, 12–15 October 2008; pp. 805–808.

30. Su, Y.C.; Huang, K.Y.; Chen, T.W.; Tsai, Y.M.; Chien, S.Y.; Chen, L.G. A 52 mW Full HD 160-Degree Object Viewpoint Recognition SoC with Visual Vocabulary Processor for Wearable Vision Applications. IEEE J. Solid-State Circuits **2012**, 47, 797–809.

31. Visual Feature Benchmark. Available online: https://github.com/BloodAxe/OpenCV-Features-Comparison (accessed on 4 April 2015).

32. Deng, W.J.; Zhu, Y.Q.; Feng, H.; Jiang, Z.G. An efficient hardware architecture of the optimised SIFT descriptor generation. In Proceedings of the 2012 22nd International Conference on Field Programmable Logic and Applications, Oslo, Norway, 29–31 August 2012; pp. 345–352.

33. Chen, T.; Du, Z.; Sun, N.; Wang, J.; Wu, C.; Chen, Y.; Temam, O. Diannao: A small-footprint high-throughput accelerator for ubiquitous machine-learning. In Proceedings of the 19th International Conference on Architectural Support for Programming Languages and Operating Systems, Salt Lake City, UT, USA, 1–5 May 2014; pp. 269–284.

34. Jeon, D.; Henry, M.B.; Kim, Y.; Lee, I.; Zhang, Z.; Blaauw, D.; Sylvester, D. An Energy Efficient Full-Frame Feature Extraction Accelerator with Shift-Latch FIFO in 28 nm CMOS. IEEE J. Solid-State Circuits **2014**, 49, 1271–1284.

Chapter 4

Practical Attacks on Mobile Cellular Networks and Possible Countermeasures

Murat Oğul and Selçuk Baktır *

Department of Computer Engineering, Bahçeşehir University, Çırağan Caddesi, No: 4, Beşiktaş 34353, İstanbul, Turkey

ABSTRACT

Due to widespread adoption of mobile communications devices and increasingly high throughput capacity of cellular networks, Third-Generation (3G) and Long Term Evolution (LTE) are becoming ever more popular. Millions of smart phones with 3G capabilities are sold every year and used for mostly browsing the Internet. Hence, mobile operators have been heavily investing in their packet switched networks to meet customer demand and stay ahead in the market. The widespread use of broadband mobile Internet bring along also some IP based threats such as the Denial of Service (DoS) attacks, botnets and malwares. In this paper, we performed DoS and flooding attacks on 3G mobile networks and measured their effect on the most critical elements of a network such as the Radio Network Controller (RNC) and the Serving GPRS Support Node (SGSN) devices. We carried out our experiments on a real mobile network, not just a simulation environment, and hence our findings depict a realistic picture of the vulnerabilities existent in 3G mobile networks. We propose alternative solutions to avoid these vulnerabilities and mitigate the issues raised.

Keywords: mobile networks; 3G; security; DoS; denial of service attack; paging

1. INTRODUCTION

Mobile communication is becoming commonplace and playing an irreplaceable role in our lives. It is facilitating not only voice communication but also high speed data communication. Millions of smart phones with 3G capabilities are sold every year and mostly used for browsing the Internet, banking applications and messaging, resulting in the expectations for the data revenue of mobile cellular network operators to exceed their voice revenue [1]. Especially, with the availability of the high data throughput in 3G, Long Term Evolution (LTE) and LTE-Advanced [2] mobile networks, many subscribers communicate more data than voice. The trend is that data communication, compared to voice communication, is dominating the mobile market. It is expected that mobile revenue will double by 2016, but traffic will increase tenfold due to increasing demand for mobile data communication [3]. As a result, in mobile cellular networks, data communication revenue is forecasted to exceed voice communication revenue [4]. Therefore, mobile operators capitalize on their data services and, in addition to being mobile voice carriers, mobile core networks tend to be used as Internet Service Providers (ISP). The investments made by mobile carriers on their IP based infrastructures and services are motivated mostly by their subscribers' increasing demand for mobile data communication. The acceptance of mobile networks as a venue for Internet access brings along new IP based threats on mobile core networks [5].

Previous studies investigated several vulnerabilities in mobile 3G networks. Ricciato [6] discussed some attacks on 3G networks and showed how unwanted traffic results in resource wastage. He also discussed four different attack models specific for cellular networks and evaluated their results for a more robust network design perspective [7]. Serror et al. [8] investigated attacks on the paging channel and presented their findings through simulations. Bavosa [9] made valuable contributions to the field by discussing the vulnerabilities on cellular networks in detail and proposing countermeasures. Wu et al. [10] studied the randomization method to prevent DoS attacks on 3G cellular networks, again using simulations. Several DoS attacks on 3G networks were realized [11,12,13]. In this work, we particularly focus on measuring the effect of DoS and flooding attacks, which exploit paging [14] vulnerabilities, and realize our experiments in a real production mobile cellular network. Furthermore, we propose solutions to mitigate the existent vulnerabilities.

The rest of the paper is organized as follows. Section 2 discusses the basic architecture of a 3G network and Section 3 analyzes possible threats in 3G as well as LTE networks. In Section 4, we conducted some attacks exploiting paging vulnerabilities on 3G networks and observed their negative effects. Section 5 discusses some countermeasures against the security threats mentioned. Finally, Section 6 is the conclusion.

2. MOBILE CELLULAR NETWORK BASICS

Mobile Cellular Networks are made up of two main components, namely circuit switched (CS) and packet switched (PS) networks [15]. The CS network is used mostly for voice traffic while the PS network is used for data traffic. In other words, a CS network could be called as a Radio Network and a PS network as a Packet Network, as shown in Figure 1 below.

GGSN (Gateway GPRS Support Node) assigns the IP and Domain Name System (DNS) addresses to User Equipment (UE) such as mobile phones, smartphones, GPRS devices, etc. When the UE connects to the network, the GGSN handles packet data protocol (PDP) contexts [16] and also performs packet forwarding between the Packet Network (PN) and the Internet. Serving GPRS Support Nodes (SGSN) are placed in the most critical location in cellular networks. They have direct connection to almost all elements in a cellular network and have responsibilities in managing data channels, such as the PS domain originated paging channels and providing Call Data Register (CDR) information to billing systems. The Radio Network Controller (RNC) is mostly responsible for managing radio resources. User Equipment firstly communicates with NodeB via the air interface. The NodeB performs tasks similar to those performed by a base station in a 2G network. In addition, all the UEs use different timeslots and signaling channels for voice and data communication.

In Cellular Networks, different kind of protocols, such as the GPRS Tunneling Protocol (GTP) tunnels, the User Datagram Protocol (UDP) and the IP, are used between devices. Mostly, the IP protocol is used but other upper layer protocols are also used on different layers of the OSI model [17].

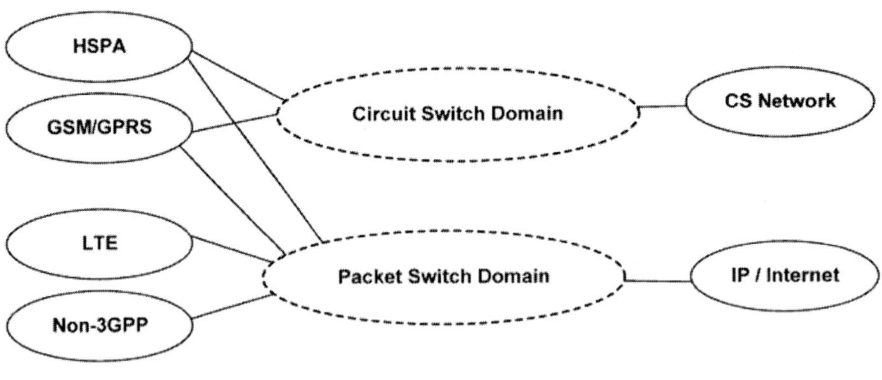

Figure 1. Mobile Cellular Network.

2.1. 3G Mobility Management for Data Communication

In mobile networks, the mobile end point may change its location and hence the systems serving this mobile end point may change. For instance, as seen in Figure 2, when a mobile user connects to a mobile PS network and moves from one point to another without closing the connection, the serving system for the end point, such as the Node B, can change from NodeB-A to NodeB-B. In Global System for Mobile (GSM) networks, knowing the end point location is one of the most critical issues for the network devices of an operator. Therefore, in order to give better service, the mobile operator must know the exact location of the mobile end point.

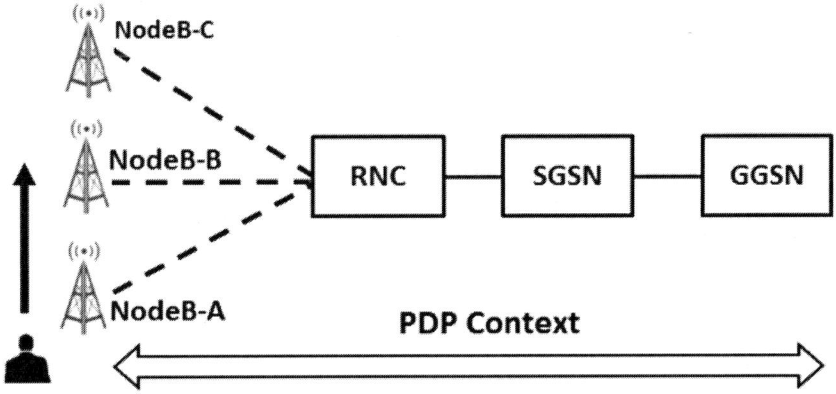

Figure 2. Serving NodeB changes.

Packet Mobility Management (PMM) context is used to track a mobile end point's location and its serving systems such as the NodeB, the RNC and the SGSN. In the PS domain, the SGSN and the mobile end point perform this task by managing PMM contexts. However, GGSN is not directly related to location tracking and does not need to know the mobile end point's PMM context. A PMM context may be different on a mobile end point and on a SGSN.

Mobility Management (MM) is necessary for tracking the physical location of the UE and serving devices such as NodeB. After the establishment of the connection between the UE and cellular network equipments, all PS user data (from or to the UE) is first sent to the GGSN. This means the UE connects to a mobile network and uses the NodeB, the RNC and the SGSN for different session establishments. However, end to end IP level communication takes place between the UE and GGSN. Later, the GGSN forwards the UE's data towards its destination according to per host specific routing on the GGSN.

The connection between the UE and the GGSN is referred to as an active PDP context. Each PDP context has its own IP address that is assigned by a GGSN. As needed, each UE can have up to 16 activated PDP contexts and each of them is uniquely identified by a combination of the subscriber's International Mobile Subscriber Identity (IMSI) and a Network Service Access Point Identifier (NSAPI) [18]. In order to route the UE data correctly, a PDP context should exist from the UE to the GGSN, as shown in Figure 3. A PDP context is necessary only for data communication and not voice.

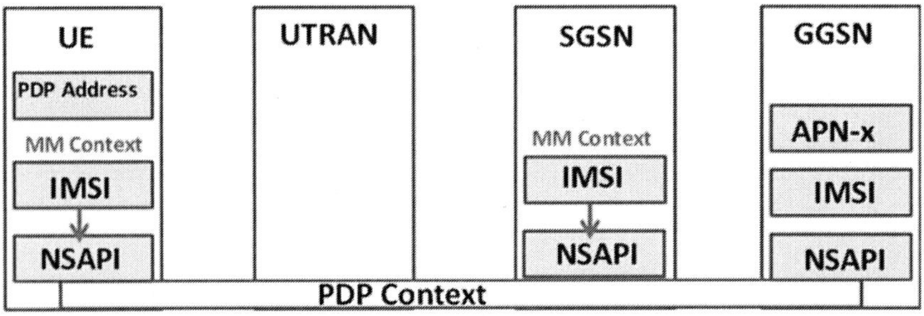

Figure 3. End to End packet data protocol (PDP) Context.

MM is referred to as PMM (Packet Mobility Management) on GPRS 3G networks. In order to track a UE's location easily, mobile networks use a hierarchy of areas, such as Location Area (LA) and Routing Area (RA). Cells form an RA and RAs form an LA, as shown in Figure 4.

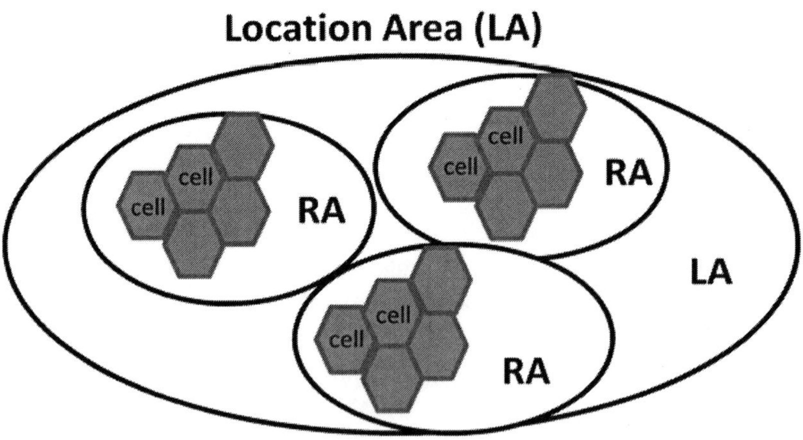

Figure 4. Mobile network hierarchy.

An LA is a set of base stations that are grouped together to optimize signaling. An RNC manages radio channel allocation and handover for all cells in the same LA. Many nodeBs in the Universal Mobile Telecommunications System (UMTS) share a single RNC in UMTS. The size of an LA is an issue since an LA too big or too small could be a problem from the optimization point of view. The size of an LA may also affect paging on the RNC [19].

An RA is used in the PS domain by UEs that are attached to GPRS and have active PDP contexts. A change from an RA to another RA is done in same way as it is done during a change from one LA to another LA. The only difference is that an LA is controlled by an RNC whereas an RA is controlled by an SGSN.

In a PS network, a UE can be in one of the three different PMM states: PMM-Detached, PMM-Idle and PMM-Connected, as shown in Figure 5.

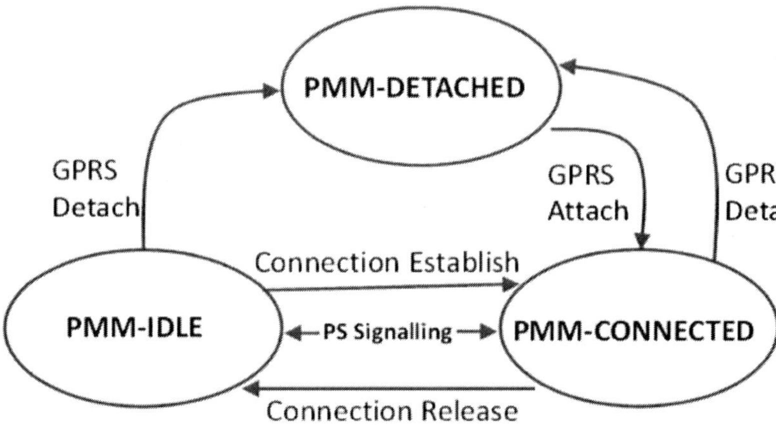

Figure 5. A User equipment (UE)'s packet mobility management (PMM) state transition.

PMM-Idle State: Once a UE is attached to a GPRS network, in addition to IP assignment, end-to-end resources are dedicated for the virtual channel between the UE and the GGSN, and also a timeout timer is started to track the UE activity. If the UE is idle for longer than a certain amount of time, the related systems such the GGSN and the SGSN will release their dedicated resources. After this release, although the UE is still known by an SGSN, a significant signaling overhead is required to reinitiate the data transfer. Connection re-establishment can be initiated by either the UE or the SGSN. The SGSN triggered connection re-establishment occurs only if an SGSN receives data destined for a certain UE, from a GGSN. In order to realize this, the SGSN uses the paging procedure. In that case, the UE's location is known by the SGSN with an accuracy of an RA but not a cell. It is important to note that if there is a PDP context active, the given UE's cell is already known, so a paging message is no longer required. So, the UE and the SGSN have valid PMM contexts. The UE and SGSN should enter the PMM-CONNECTED state when the PS signaling connection is established between the UE and the SGSN.

PMM-Detached state: This state means that the UE is switched off, outside the service area or not attached to a PS network, i.e., the UE and SGSN contexts don't have any location and routing information for the UE, and hence the UE is not reachable by an SGSN. In this state, there is no active PDP context and assigned IP address to the UE. A GPRS attach procedure should be performed in order to transfer to the PMM-Connected state. If any traffic destined to the UE comes to the GGSN, it will be discarded and no paging procedure will be needed.

PMM-Connected State: In this state, the UE's location is known in the SGSN with an accuracy of a serving RNC. Also, the serving RNC tracks the UE with an accuracy of a serving cell. Therefore, if any traffic destined to the UE comes to the GGSN, it will be forwarded to the UE through a PDP context and no paging procedure will be needed.

2.2. Paging

Paging can be triggered by a request from a PS or CS network. Paging triggered by a CS network is not in our concern. On a PS network triggered paging, the paging procedure is used only for a UE in the PMM-idle state. Packets that are destined to a UE may come to the GGSN from the Internet or 3G packet networks, such as from another 3G subscriber. If a PDP context exists, packets will be sent to the SGSN via a GTP tunnel. For these packets to reach to the UE, the SGSN should know about the target UE's cell information, i.e., the cell that is serving the UE. Therefore, if the SGSN does not have the cell information of the destined UE, it will initiate a paging procedure by sending some packets to the RNC. As mentioned before, this means that paging is realized only if the UE is in the PMM-idle state when packets are sent to it. Figure 6 illustrates the signal flow activity for the PS triggered paging.

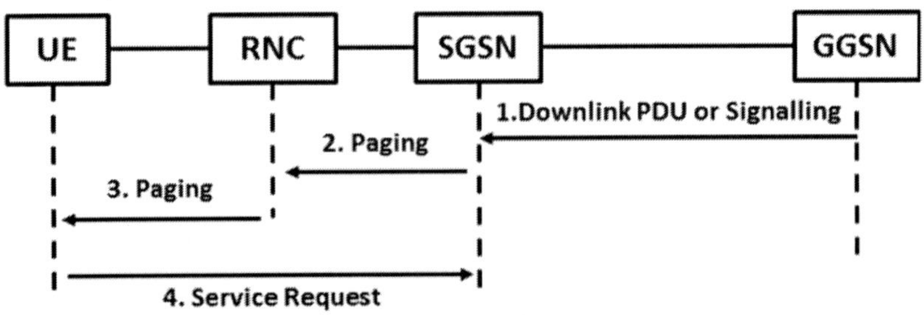

Figure 6. Packet switched (PS) triggered paging procedure.

The paging procedure from the GGSN to the UE, given in Figure 6, is described in detail below.

1. The GGSN sends a packet, destined for the UE in PPM-idle state, to the SGSN;
2. Upon getting packets from the GGSN, the SGSN sends a Paging Request message that includes the IMSI number, RA, etc., to the RNC belonging to the RA where the UE is located;
3. The RNC passes the paging requests to the UE;
4. Upon getting the paging request message, the UE's state changes from PMM-idle to PMM-connected. Also, the UE sends paging request packets back to the SGSN.

After completing the paging procedure, the SGSN knows about the cell, RA and all the other location information of the destination UE. This means that any packet destined from the GGSN to the UE in PMM-idle state can be delivered to the UE.

While a paging procedure is taking place, mostly the RNC's and SGSN's loads rise. This can be exploited by some DoS or Distributed Denial of Service (DDoS) type of attacks, which is the major focus of this study.

3. THREATS ON CELLULAR NETWORKS

Unlike wired networks, mobile cellular networks have a complex structure with many components and protocols resulting in vulnerabilities and threats [20]. However, it is possible to take the necessary precautions and thus mitigate the risks of these threats to an acceptable level. Mobile operators should take the required actions to mitigate the security risks to their subscribers and also to their core networks. It is up to an operator to possibly charge for these security actions.

As discussed in the previous section, in mobile cellular networks, paging effects are important security issues. Paging attacks could be most damaging on the border of a mobile operator's network. One border of cellular networks is where Internet access starts and another one is where subscribers attach to a 3G network. Mobile operators can take most of the countermeasures on these border points.

Thanks to the release of the 3G and LTE technologies, mobile GSM operators have overtaken ISP roles. Some argue that, like traditional ISPs, mobile operators have no obligation to protect their customers, while others argue they should protect their subscribers, as well as their core network infrastructure, against security threats [21]. Although traditional wired network ISPs and mobile network operators both act as ISPs, their core network infrastructure or underlying networks are very different. The security requirement on the border of the Internet and on the subscriber border is higher for mobile operators than for traditional wired ISPs. This is due to the existence of more fragile components and protocols used in mobile networks. In our study, we've focused on these border points. In an actual production mobile network environment, we have investigated the effect of DoS kind of attacks from the Internet against 3G users. We discuss these attacks and possible countermeasures in the next section.

4. MEASURING THE EFFECT OF THE DOS ATTACK ON 3G NETWORKS

We think paging is a big threat in mobile cellular networks. So, we performed some attacks by exploiting vulnerabilities in GSM network components due to paging, and examined the effect of these attacks. A network or company is as strong as its weakest point in terms of security. Therefore, a network should be assessed as a whole when a security solution is put in place. All pieces of a network should be evaluated carefully in terms of security. We performed DoS and flooding attacks in a production mobile cellular network that supports 120,000 concurrent users and investigated their effect on core network equipment such as the RNC and SGSN. We suggest some countermeasures, taking our findings into consideration.

As discussed earlier, some mobile cellular operators think that no security solution is necessary on the Internet border of their network that serves their subscribers during Internet access. So, they typically construct their networks as shown in Figure 7. In this setting, any packet coming from the Internet, destined to a subscriber of the mobile operator, could reach its target.

A subscriber accesses core network devices for authentication and authorization, after passing through the air interface and the transport network of the mobile operator. An IP address is assigned to a subscriber by a GGSN if it is successfully authenticated and authorized. The first point on the mobile cellular network where a subscriber can do IP-based communication is the GGSN. The GGSN acts as a remote access server. A subscriber's data traffic is carried via tunnels until the GGSN. For example, the subscriber traffic from the GGSN to the SGSN is carried by GTP tunnels. Briefly, behind the GGSN, the subscriber data traffic is tunneled and carried to the subscriber as shown in Figure 8. Therefore, when any customer tries to learn the traffic path by using trace applications (e.g., tracert for Windows or traceroute for Linux and Unix), they notice that they can see just the GGSN's IP address, and the devices after the GGSN only if the operator allows. Direct access to core network equipment is not allowed by nature of mobile cellular networks. Even then, we think that the traffic from a subscriber to the Internet, and from the Internet to a subscriber, could affect core network devices dramatically. It is a serious threat and a countermeasure needs to be taken. The most practical way to decrease the risks here to an acceptable level, without monetary investment, is to enable the IP filter feature of the used Internet router without a firewall feature. With this feature, a router examines the source and/or destination IP address fields, and possibly also the TCP/UDP port numbers, in a packet's IP header. By controlling the TCP/UDP port number in the IP header, it might achieve more detailed filtering. On the other hand, Access Control Lists (ACLs) consist of IP/TCP/UDP based filtering entries, and they are typically used to indicate the allowed/blocked IP addresses whose traffic is allowed/blocked. This simple countermeasure will protect the core network devices and also the UEs. However, the basic ACLs on the router supports only basic IP/TCP/UDP restriction and does not provide session state tracking or stateful inspection [22] for all the used protocols such as TCP and UDP. Without a stateful firewall feature set, a router can only support TCP SYN attack protection. However, all other attacks with TCP flags, such as ACK, RST/ACK, FIN, FIN/ACK, etc., which are set to 1, can pass through a router's ACLs without being blocked. Therefore, a router recognizes any packet coming from the Internet with its SYN bit set to 1. If the packet is not given the access right on the ACL, it will be dropped. Thus, a router can block only the unwanted packets with their TCP SYN bit set to 1. And all other TCP or UDP packets will be allowed in order not to block the response to legal customer initiated traffic. In our study, we call this kind of an ACL as TCP ACL. Hence, by using this kind of a solution, we protect customers and mobile core network infrastructure against only TCP SYN based attacks. Although our routers provide only TCP SYN based attack protection, we notice that the number of paging on RNC decreases after the TCP based Access Control List (ACL) is applied. To ensure the positive effect, we temporarily disabled TCP based ACL on our Internet router and observed the paging activity and resource usage of our core network devices. After we applied TCP based ACL on our Internet routers again, the paging activity and CPU usage was influenced positively as shown with the graphs in Figure 9 and Figure 10. In these graphs, the horizontal axis shows time and date, and the vertical axis shows the CPU usage. These graphs have a sampling period of 1875 s. As seen in these graphs, the CPU usage on the RNC and SGSN decrease by approximately 25%, compared to the values when the ACL is not applied.

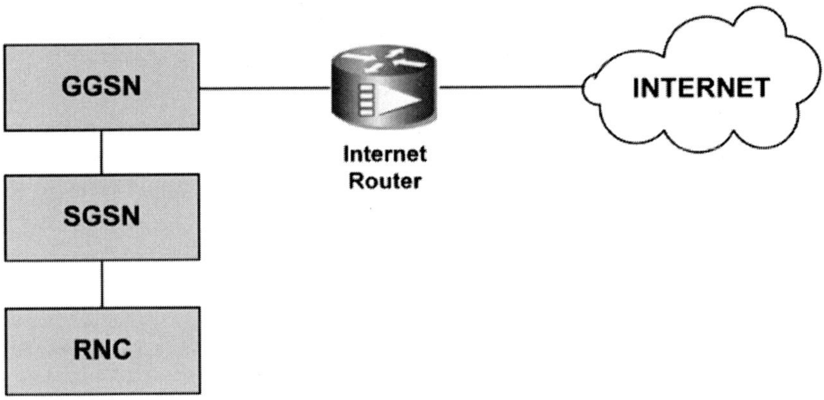

Figure 7. Network designed without a security solution.

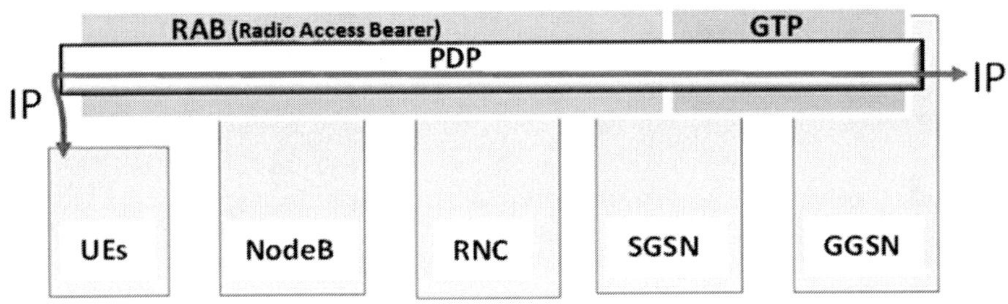

Figure 8. Mobile device's intrusion prevention (IP) connectivity.

These results do not mean that we are under a TCP SYN based attack all the time but there is some TCP based traffic from the Internet to our 3G subscribers' subnet. Therefore, the CPU usage is decreasing after applying the TCP based ACL that blocks the unnecessary traffic from the Internet.

Paging activity on the RNC also decreased from 240,000 to 160,000, i.e., by approximately 33%, after applying the ACL on the Internet router, as shown in Figure 11 where the vertical and horizontal axes show the number of paging activity and date/time, respectively.

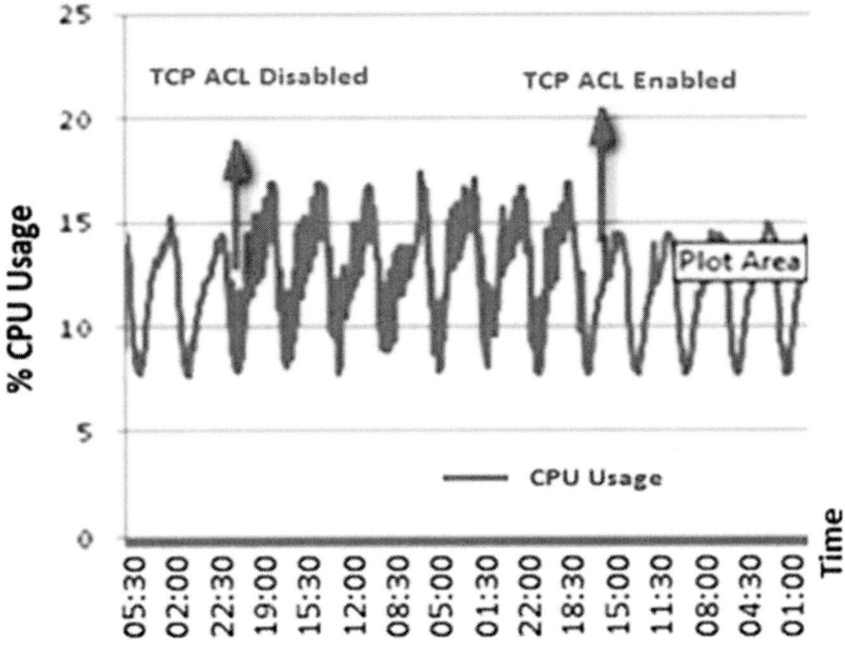

Figure 9. Serving GPRS support node (SGSN) CPU usage with/without TCP access control list (ACL).

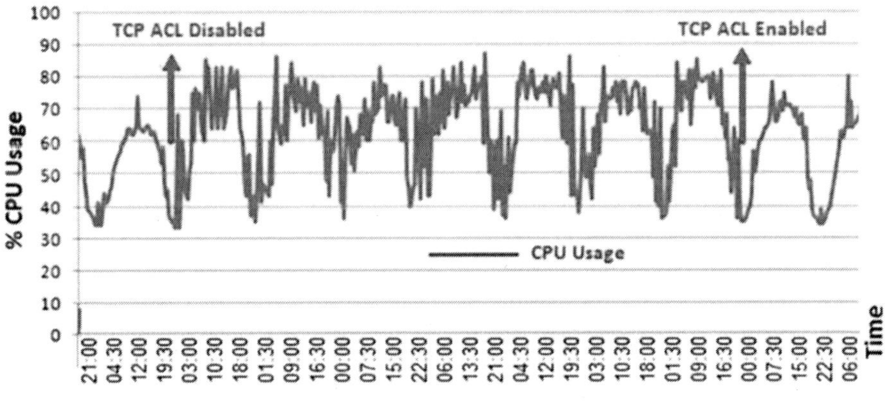

Figure 10. RNC CPU usage with/without TCP ACL.

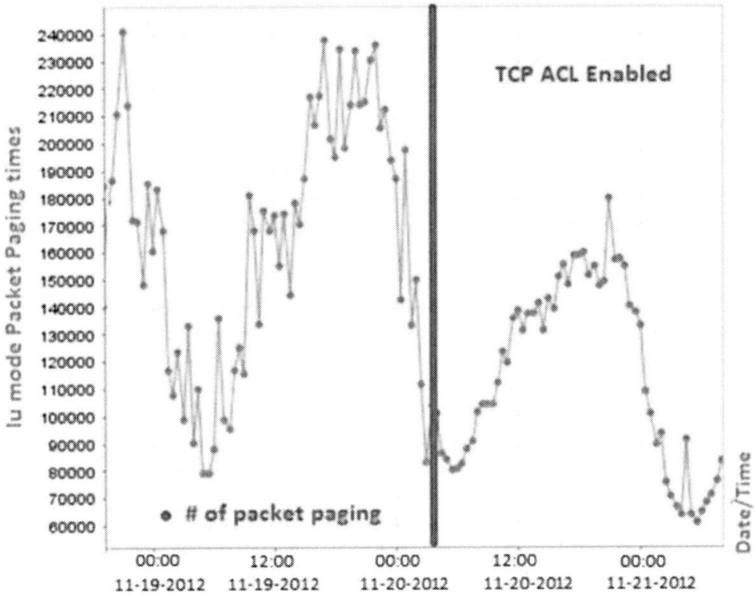

Figure 11. Number of paging activity on an radio network controller (RNC) with/without applying an ACL.

These results made us also think about the risks that could occur due to UDP based attacks. Therefore, using UDP packets, we performed some attacks against our 3G network. By using some free open source tools, we sent UDP packets of size 60 bytes at a rate of 60,000 to 100,000 packets per second from the Internet to our 3G network. We performed our attacks at midnight between 02:30 and 03:00 a.m. to minimize the adverse effects to the customers. During these attacks, the number of paging and amount of CPU usage on our core network devices dramatically increased as shown in Figure 12. As seen in the figure, the RNC's PS paging attempt count was 70,000 just before attack and during the attack it reached 283,000. Hence, the number of pagings increased by approximately 304% due to our attack.

PS paging activity dramatically increased on the RNC and SGSN when we sent bulk data traffic to the subnets of the 3G subscribers. However, CS paging [23] is not affected during that time. Both the PS and CS paging activity decreased after midnight due to the decreasing number of phone calls. Figure 13 shows how the same RNC's CS paging activity changes during the same time interval. This shows that the PS paging and the CS paging are used for different purposes. The CS and PS paging activities are triggered by the MSC and GGSN, respectively. In other words, PS paging is data-oriented and CS paging is voice-oriented. They use different signaling channels until the RNC. However, they use the same control channel from the RNC to NodeB. Therefore, all voice and data transmission can be effected if it is overloaded by an attacker who exploits PS paging.

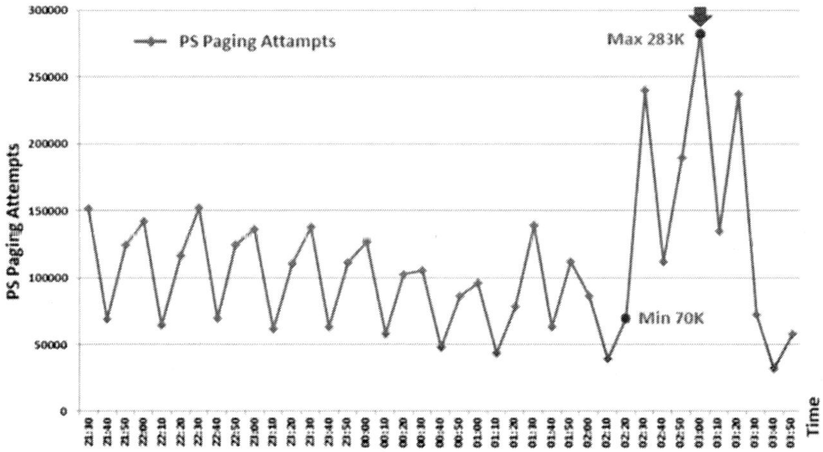

Figure 12. RNC PS paging attempts during the user datagram protocol (UDP) based attacks (sampling period is 30 min).

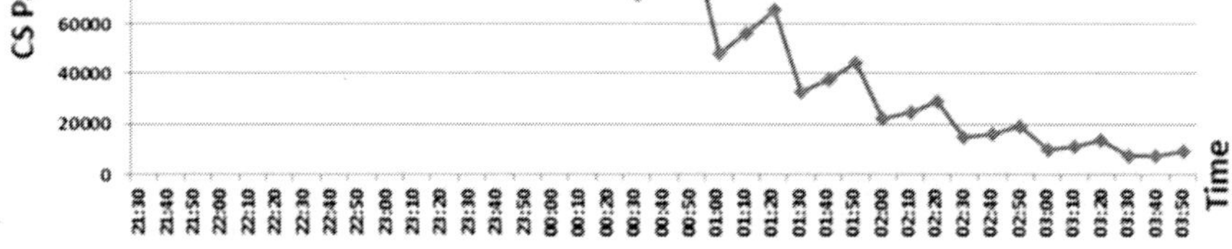

Figure 13. Circuit switched (CS) paging attempts on the RNC during the UDP based attack.

During our attacks, we observed that the CPU usage of an RNC increased. In Figure 14, the RNC CPU usage during the UDP attack is given with a sampling period of 1875 s.

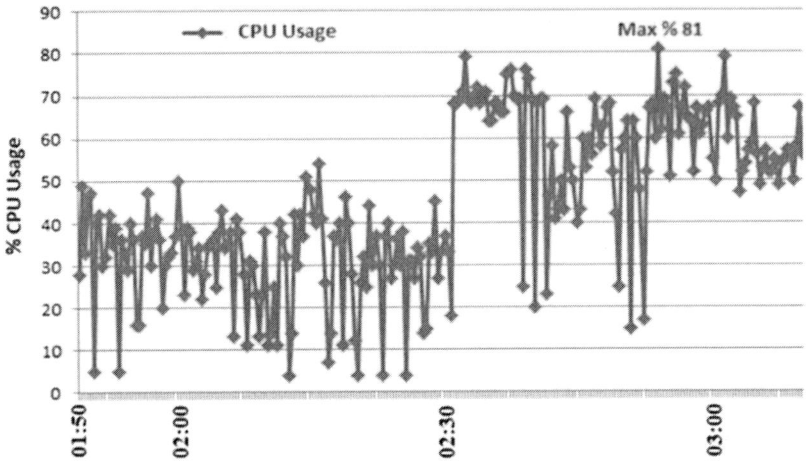

Figure 14. RNCs CPU usage during the UDP based attack.

In this work, we first examined the effect of unwanted data traffic on cellular network devices such as the RNC and SGSN. We realized that blocking unnecessary TCP traffic from the Internet to 3G subscribers' IP subnets decreases not only the resource usage on cellular network devices but also the number of paging attempts on the RNC. However, we did not apply the UDP based access restriction on our Internet router due to the router's limited filtering capability. In order to see the effects of UDP based DoS/DDoS or flooding attacks from the Internet, we performed a UDP based attack at midnight because of the smallest data and voice load on cellular network at that time. The CPU usage and number of PS paging attempts of the RNC was dramatically affected, although the attack was not sophisticated and it was not performed during peak hours. All these results show that mobile cellular networks are more vulnerable against unwanted bulk traffic caused by DoS/DDoS or flooding attacks.

As we discussed before, the RNC is one of the most critical device in a mobile network. It is related to both data and voice services. Hence, both voice and data subscribers will be affected if the RNC is out of service due to a CPU overload caused by high traffic resulting from attacks or malicious software [24] such as viruses [25], worms, etc.

5. COUNTERMEASURES AGAINST 3/4G SECURITY THREATS

The mobile communication industry has been getting a rise during the last three years. Hence, the security issues in mobile communications have become increasingly more important in 2013 [26]. Therefore, not only mobile operators but also mobile device users should be careful when using services over a mobile network. We focused especially on the mobile operators' security vulnerabilities when they don't protect their customers and networks against cyber threats. So, we will mention only the countermeasures that could be taken on the mobile operator's side. Security countermeasures against the security threats on the 3G border and core network devices could be summarized as follows:

1. Stateful IP packet filtering could be applied on the Internet border of the mobile network. This task could be achieved by an Internet router with firewall feature with both TCP and UDP stateful packet filtering functionality. A second option is to use a firewall device between the GGSN and the Internet router. The packet filtering device must be stateful for TCP and UDP, and drop all unnecessary traffic from the Internet to the mobile operator or to the customers. The drawback of the stateful firewall solution is the limitations in a firewall's session state table;
2. Carrier-grade Network Address Translation (NAT [27], CGN [28]) is used both to mitigate IPv4 exhaustion and to block unnecessary traffic from the Internet. But when a CGN is used, it is not possible to allow the traffic from the Internet to the subscribers, which is a drawback in using the CGN;
3. Configuration should be done against IP address spoofing [29] on the IP packet filter device or the Internet router;
4. The flooding and the DoS/DDoS type of attacks are the most dangerous and effective ways to destroy target devices. They could be performed individually or organized, e.g., using a botnet [30,31]. A high number of connection requests could put most stateful firewalls out of service due to limitations in their state tables. Therefore, a DoS/DDoS protection service or product should be used against attacks coming from the Internet. Generally, it is recommended that DDoS protection is placed in the ISP's side [32];
5. Mobile to mobile traffic could cause spreading of worms and viruses. This can be dangerous for both mobile devices and mobile operators, especially when worms/viruses act in an organized manner forming a

botnet. Therefore, mobile to mobile traffic should be filtered by the GGSN or an IP filtering device;

6. Malware [33] or botnet members can generate flooding traffic to their victims such as mobile devices or operators, causing service interruptions on core network devices. Hence, rate limiting [34] should be applied to mobile user traffic on the GGSN or an IP packet filtering device should be used to avoid this threat;
7. Directerd broadcast [35] traffic must be denied on the GGSN to protect against DDoS attacks;
8. RA and LA scope optimization should be done according to the best practices to decrease paging effects;
9. Parameter optimization on RNC and SGSN should be done according to the best practices to decrease the PS paging effect;
10. Mobile operators could locate the GTP aware firewall products between the SGSN and the GGSN to block attacks coming through the GTP protocol;
11. The PS and the CS paging control channels should be isolated to prevent the CS paging from being overloaded by the PS paging channel under attack. Hence, at least voice communication service would not be affected in the case of an attack against the PS paging channel.

A pictorial summary of all the countermeasures to be taken on the mobile operator's side can be seen in Figure 15. The advantages and disadvantages of using the above mentioned countermeasures are listed in the Table 1.

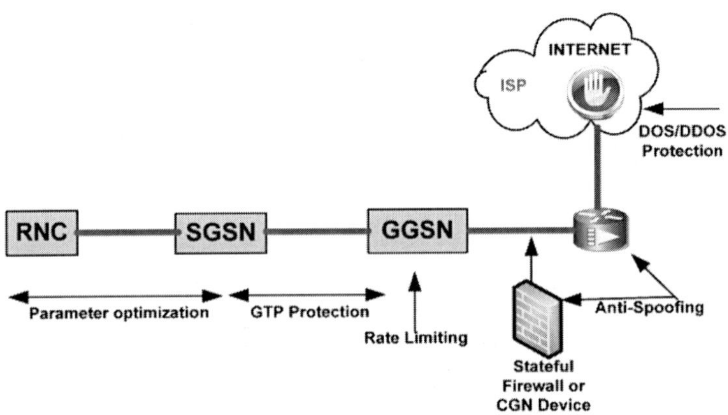

Figure 15. Application points for possible countermeasures on an ISP and a mobile operator's network.

Table 1. Advantages and disadvantages of possible countermeasures against DoS and DDoS attacks on 3G mobile networks.

Protections	PROS	CONS
DDoS Protection on ISP	Effective DDoS protection	Extra cost
Stateful Firewall	Effective and flexible IP filtering	Session state table limitations, Point of failure
CGN	Blocking unwanted traffic from the Internet	NAT table limitations, Lack of IP filtering, Point of failure
GTP Firewall	GTP attack protection	Extra cost, Point of failure
Parameter Optimization	No-cost and protection	N/A
User Isolation	Preventing organized internal attacks	Unforeseen UE-to-UE communication
Rete Limiting	Decreases the effect of attacks	N/A

Security solutions could be applied most easily and efficiently on the border of the Internet. We have shown that cellular mobile networks should be effectively protected by using firewalls, Intrusion Prevention Devices (IPS) and/or DoS/DDoS protection products on the border of the Internet. In addition to these solutions, DoS/DDoS protection on the ISP side is very important, due to the limitations in the session tables of stateful firewalls. DDoS protection service on the ISP side is expected to filter all flooding and DDoS attacks before they overload an Internet connection link and fill up the session table of a stateful firewall. Another security solution is to isolate the customer traffic in order to prevent spreading of malicious software which may otherwise result in subscriber dissatisfaction due to additional usage charges and also harm core network equipment.

6. CONCLUSIONS

In the IP world, most effective and easy to realize attacks are DoS and DDoS attacks. Especially, mobile cellular networks could be affected by these attacks unless necessary countermeasures are taken. We observed these threats on mobile networks and took some precautions to mitigate them. We realized some DoS attacks on a production mobile cellular network and measured their destructive effects on core network devices. Our study shows that security solutions can be applied most easily and effectively on the Internet border of mobile cellular networks and user access devices, before any unwanted traffic ever arrives at the core network. In addition to this, all devices throughout the mobile network should be hardened in terms of security against attacks. Furthermore, DDoS protection service located on the ISP side is becoming necessary for all telecom and enterprise companies.

As future work, we plan to change the Location Area Code (LAC) and Routing Area Code (RAC) parameters on the RNC, and observe resource usage and paging effects on an RNC under attack. Furthermore, by using the CGN and a stateful firewall as a border filtering device, the robustness of a mobile cellular network can be improved.

ACKNOWLEDGMENTS

This work is supported in part by the grant EU FP7 Marie Curie IRG 256544.

CONFLICTS OF INTEREST

The authors declare no conflict of interest.

REFERENCES

1. CBS Interactive Web Page. Japan First Country Where Data Revenues Exceed Voice. Available online: http://www.zdnet.com/japan-first-country-where-data-revenues-exceed-voice-7000011817 (accessed on 28 July 2013).
2. Ghosh, A.; Ratasuk, R.; Mondal, B.; Mangalvedhe, N.; Thomas, T. LTE-advanced: Next-generation wireless broadband technology. IEEE Wirel. Commun. **2010**, 17, 10–22.
3. Technology Marketing Corporation Web Page. Mobile Revenue will Double by 2016, but Traffic will Increase Tenfold. Available online: http://www.tmcnet.com/topics/articles/2012/05/16/290655-mobile-revenue-will-double-2016-but-traffic-will.htm (accessed on 28 July 2013).
4. Itwire Web Page. Asia Pacific Mobile Data Revenues Tipped to Exceed Voice in 2016. Available online: http://www.itwire.com/your-it-news/mobility/49878-asia-pacific-mobile-data-revenues-tipped-to-exceed-voice-in-2016 (accessed on 28 July 2013).
5. Nagy, M.; Kotosová, M. An IP Based Security Threat in Mobile Networks. In Proceedings of the 35th International Convention on Information and Communication Technology, Electronics and Microelectronics (MIPRO), Opatija, Croatia, 21–25 May 2012.
6. Ricciato, F. Unwanted traffic in 3G networks. ACM SIGCOMM Comput. Commun. Rev. **2006**, 36, 53–56.
7. Ricciato, F.; Coluccia, A.; D'Alconzo, A. A review of DoS attack models for 3G cellular networks from a system-design perspective. Comput. Commun. **2010**, 33, 551–558.

8. Serror, J.; Zang, H.; Bolot, J.C. Impact of Paging Channel Overloads or Attacks on a Cellular Network. In Proceedings of the ACM Workshop on Wireless Security (WiSe 06), Los Angeles, CA, USA, 29 September 2006.

9. Whitehouse, O.; Murphy, G. Attacks and Counter Measures in 2.5G and 3G Cellular IP Networks; White Paper; Juniper Networks: Sunnyvale, CA, USA, 2004. [

10. Wu, Z.; Zhou, X.; Yang, F. Defending against DoS Attacks on 3G Cellular Networks via Randomization Method. In Proceedings of the 2010 International Conference on Educational and Information Technology (ICEIT 2010), Chongqing, China, 17–19 September 2010.

11. Lee, P.P.C.; Bu, T.; Woo, T. On the Detection of Signaling DoS Attacks on 3G Wireless Networks. In Proceedings of the 26th IEEE International Conference on Computer Communications (INFOCOM 2007), Anchorage, AK, USA, 6–12 May 2007.

12. Zhao, B.; Chi, C.; Gao, W.; Zhu, S.; Cao, G. A Chain Reaction DoS Attack on 3G Networks: Analysis and Defenses. In Proceedings of the 28th IEEE International Conference on Computer Communications (INFOCOM 2009), Rio de Janeiro, Brazil, 19–25 April 2009.

13. Cheng, C.M.; Kung, H.T.; Tan, K.S. Use of Spectral Analysis in Defense against DoS Attacks. In Proceedings of the Global Telecommunications Conference 2002, Taipei, Taiwan, 17–21 November 2002.

14. The European Telecommunications Standards Institute (ETSI). Technical Specification: 3GPP TS 25.331; version 8.1.0. ETSI: Valbonne, France, 2008. Available online: http://www.etsi.org/deliver/etsi_ts /125300_125399/125331/08.01.00_60/ts_125331v080100p.pdf (accessed on 28 July 2013).

15. Eng, K.Y.; Ali, A.M.; Baradello, C.; Turner, J.; Vlack, D.; Walters, S.M. Packet Switching vs. Circuit Switching in Future Integrated Services Digital Networks. In Proceedings of the 7th IEEE International Conference on Computer Communications; New Orleans, LA, USA: 27–31 March 1988.

16. Killalea, T. Recommended Internet Service Provider Security Services and Procedures; BCP 46 and RFC 3013; Internet Engineering Task Force (IETF): Fremont, CA, USA, 2000.

17. Lyman, C.A. Computer communication standarts. ACM SIGCOMM Comput. Commun. Rev. **1984**, 14, 46–52.

18. Ericsson Inc, GPRS System Survey. In Student Book; LZT 123 5347 R4A; Ericsson: Stockholm, Sweden, 2005.

19. Ozugur, T. Multiobjective Hierarchical Location and Routing Area Optimization in GPRS and UMTS Networks. In Proceedings of the 2002 IEEE International Conference on Communications (ICC 2002), New York, NY, USA, 28 April–2 May 2002.

20. Leavitt, N. Mobile phones: The next frontier for hackers? Computer **2005**, 38, 20–23.

21. Becher, M.; Freiling, F.C.; Hoffmann, J.; Holz, T. Mobile Security Catching Up? Revealing the Nuts and Bolts of the Security of Mobile Devices. In Proceedings of the 2011 IEEE Symposium on Security and Privacy, Berkeley, CA, USA, 22–25 May 2011.

22. Newman, D. Benchmarking Terminology for Firewall Performance; RFC 2647; Internet Engineering Task Force (IETF): Fremont, CA, USA, 1999.

23. The 3rd Generation Partnership Project (3GPP) Technical Specification Group. In Technical Specification; 23.060 V6.11.0; 3GPP: Valbonne, France, 2005.

24. Leavitt, N. Malicious code moves to mobile devices. IEEE Comput. **2000**, 33, 16–19.

25. Dagon, D. Mobile phones as computing devices: The viruses are coming! IEEE Pervasive Comput. **2004**, 3, 11–15.

26. Mobile Marketer Web Page. Top Mobile Security Threats for 2013. Available online: http://www.mobilemarketer.com/cms/news/strategy/14518.html (accessed on 28 July 2013).

27. Egevang, K.; Francis, P. The IP Network Address Translator (NAT); RFC 1631; Internet Engineering Task Force (IETF): Fremont, CA, USA, 1994.

28. Jiang, S.; Carpenter, B. An Incremental Carrier-Grade NAT (CGN) for IPv6 Transition; Internet Engineering Task Force (IETF): Fremont, CA, USA, 2011.

29. Ferguson, P.; Senie, D. Network Ingress Filtering: Defeating Denial of Service Attacks which Employ IP Source Address Spoofing. BCP 38 and RFC 2827; Internet Engineering Task Force (IETF): Fremont, CA, USA, 2000.

30. Tyagi, A.K.; Aghila, G. A wide scale survey on botnet. Int. J. Comput. Appl. **2011**, 34, 9–22.

31. Sharma, R.K.; Chandel, G.S. Botnet detection and resolution challenges: A survey paper. Int. J. Comput. Inf. Technol. Bioinforma. **2009**, 1, 10–15.

32. Douligeris, C.; Mitrokotsa, A. DDoS attacks and defense mechanisms: Classification and state-of-the-art. Comput. Netw. **2004**, 44, 643–666.

33. Lawton, G. Is it finally time to worry about mobile malware? IEEE Comput. **2008**, 41, 12–14.

34. Morrow, C.; Jones, G. Filtering and Rate Limiting Capabilities for IP Network Infrastructure. Available online: https://tools.ietf.org/html/draft-ietf-opsec-filter-caps-09 (accessed on 28 July 2013).

35. The European Telecommunications Standards Institute (ETSI). Technical Specification; 3GPP TS 29.060 V9.3.0 (2010-06); ETSI: Valbonne, France. Available online: http://www.etsi.org/deliver/etsi_ts/129000_129099/129060/09.03.00_60/ts_129060v090300p.pdf (accessed on 28 July 2013).

Chapter 5

A Comparison between Theoretical and Experimental Coverage Analysis in 3G Cellular Networks

Jazmín Ponce-Rojas, Mariana Eslava-Gutierrez, Montserrat Jiménez-Licea, Sergio Vidal-Beltrán

Sección de Estudios de Posgrado e Investigación, ESIME Zac.-IPN, México D.F., México

ABSTRACT

This paper shows a comparison between theoretical and experimental coverage analysis. Theoretical work is based on Okumura-Hata propagation model, which are compared with the measured data obtained through an experimental analysis en México City. It is important because all the network designers and managers have to take in account how the signals will arrive to the mobile devices. If they know it, they can install Nodes B (base stations) in the better place in the area in order to take advantage of the power radiated by the antenna.

Keywords: WCDMA; Okumura-Hata; Propagation Analysis

1. INTRODUCTION

A propagation model is a set of mathematical expressions, diagrams and algorithms used to represent the radio characteristics of a given environment. Generally, prediction models can be classified as empirical or statistical, theoretical or deterministic or a combination of these two (semi-empirical). While empirical models are based on measurements, theoretical models are based on the fundamental principles of the phenomena of radio wave propagation. Propagation models predict the path loss for an RF signal can be between a base station and a mobile or fixed receiver. The advantage of doing a radio channel model taking into account the characteristics of the path between transmitter (Tx) and receiver (Rx), is to determine the feasibility of projects that wish to plan in certain areas, so you can make a estimate of the need, costs and required equipment capacity (technical specifications). The performance of the propagation models is measured by the accuracy of the results compared with actual field measurements. The model described in this article has a good correlation in terms of the comparisons mentioned both in simulation and field measurements. The applicability of a model depends on this specification requires such as: the type of terrain (mountainous, hilly or quasi-smooth), the characteristics of the propagation environment (urban, suburban, and open), characteristics of the atmosphere (refraction index, rainfall intensity), soil electrical properties (conductivity ground), type of urban construction material etc. [1].

Several existing path loss models such as the Free Space Path Loss Model, Okumura-Hata's Model, and Egli's Model are chosen as reference for optimized path loss model development. The best existing path loss model with the smallest mean relative error to the measured path loss will be chosen as a reference for the development of the optimized path loss model. The regression fitting method is used to develop a new empirical linear line by combining the best existing path loss model with the measurement data which is collected from the WCDMA network. The new empirical linear line is used as a reference during optimization to develop an optimized path loss model [2]. For a very irregular and rugged topography, the best performing models are the diffraction losses estimated using the classical knife-edge and its different variations for the extension to multiple knife edges (i.e., Epstein-Peterson, Deygout, Bullington etc.), these models analyze point by point between Tx-Rx path, identifying the losses caused by the main obstacles and added to the solution Friis equation.

The Okumura-Hata's model is an empirical model based on extensive measurements made in Japan at several frequencies within the range of 150 to 1920 MHz. Okumura-Hata's model is developed for macro cells with cell diameters of 1 to 100 km [3]. The height of the BS antenna is between 30 - 1000 meters and the height of Mobil station antenna is around 3 meters. The OkumuraHata's model takes into account some of the propagation parameters such as the type of environment and the terrain irregularity. This model is used to predict de power in a receiver located in an urban area for Mobil communications. Okumura developed a set of curves giving the median attenuation relative to free space in an urban area over a quasi-smooth terrain with a base station effective antenna height of 200 m and mobile antenna height of 3 m. These curves were developed from extensive measurements using vertical Omni-directional antennas at both the base and mobile [3].

The Okumura's Model is expressed as:

$$L50(\mathrm{dB}) = LF + Amu(f,d) - G(ht) - G(hr) - G_{\mathrm{area}}$$

where:

L50 (dB): is the 50th percentile value of propagation path loss;

LF: is free space path loss;

Amu: is free space attenuation;

G(ht): is base station antenna height gain factor;

G(hr): is mobile antenna height gain factor, and G_{area}: is gain corresponding to specific environment.

This paper describes how Okumura's model can be chosen for urban outdoor coverage in the Wide Code Division Multiple Access (WCDMA) system. This optimized path loss model is based on the empirical measurements collected in the WCDMA network focusing on Mexico City.

Wide Code Division Multiple Access is a channel access method that allows several simultaneous information transmissions over a single communication channel, with different distinguishing code patterns. WCDMA employs both the spread-spectrum technology and a special coding scheme to allow multiple users to be multiplexed over the same physical channel. In the WCDMA system, duplex channel is made of two 1.25 MHz-wide bands of spectrum. WCDMA is a technology of the direct sequence of Spread Spectrum; this technology expands the signals over a bandwidth of 5 MHz, buried in the noise in the channel, and has the capacity to carry voice and data simultaneously. WCDMA provides high transfer rates, efficient support of asymmetric traffic, transmission using packet switching through the radio interface and high efficiency in spectrum use. The Base Station (BS) also known as a Node B; is part of the UMTS Terrestrial Radio Access Network (UTRAN). The Node B is to perform fundamental tasks of transmission and reception of radio, filtering of the signal, amplification, modulation and demodulation of the signal and be an interface to the Radio Network Controller (RNC) [5].

The Common Pilot Channel (CPICH) transmits a carrier used to estimate the cannel parameters. It is the physical reference for other channels. It is used to control power, coherent transmission and detection, measurement of adjacent cells and obtaining the Scrambling Code (SC) [6]. Power is divided in different paths which, depending on the quality of the Rake Receiver included in the mobile phone, provide greater or lesser number of signal components. Depending on the fingers that the Rake receiver receives, the orthogonality of the codes transmitted by a Base Station is lost, causing that the received signal fades. Those causes in all the signals sent by the Base Station and particularly in the signals from other subscribers, strong interference on mobile phones that support multipath reception [7].

2. TEST SCENARIO

Measurements were done, along the test area, (Figure 1) it includes several buildings which are representative for their height, such as the 9 buildings with 3 floors and elongated structure, approximately 13 meters high, located from east to west, an adjacent building, perpendicular to these 9, with a north-south orientation with the same height. The area includes an auditorium with an approximate height of 13 meters. There is a library, a planetarium, and the sports area, which there is a stadium with a height of 6 meters. Measurements were made using a BTS Master MT8222A, which is a spectrum analyzer with advanced Software profiles to analyze WCDMA/ HSDPA signals. The measurement capabilities of this unit include Return Loss, Cable Loss, Distance to Fault, two port gain/loss, one port and two ports magnitude and phase, interference analysis, spectrogram, and power meter. The BTS Master can measure Node B performance Over the Air by connecting an antenna or by connecting Node B equipment directly to the BTS Master MT8222A. Measuring a WCDMA signal over the air, the antenna must be connected appropriately to the frequency band to be measured [8].

3. MEASUREMENTS

The measurement procedure was performed in the band V (UMTS Channels), which has a center frequency of 887.5 MHz; which is used in Mexico for WCDMA services. The measurements were made with a spectrum analyzer, BTS Master MT8222A with frequency range of 9 KHz to 7.1 GHz, which has applications software for the analysis of WCDMA signal. The BTS Master can measure the performance of Node B by connecting directly to the Node B equipment or through the air by the connection of an omni-directional antenna, which operates in the frequency band of 870 to 960 MHz. To obtain the location information of each metering requires a GPS (Global Positioning System) antenna.

The first step for a correct analysis of the performance of a Node B is to delineate the area of interest where the measurements will be made; this area is shown in the Figure 1. 1519 measurement were obtained, the distance between them was about 3 meters, in the area shown in Figure 1, whose surface is about 0.7 Km^2. In each measurement the spectrum analyzer was placed at a height between 1.10 and 1.30 meters, since it is the average height to which the user carries his mobile equipment.

Figure 1. Analysis area.

For this work, the latitude and longitude were selected as a geographical reference of the system, and the CPICH power level because through the measurement of this power level, the user terminal is able to establish a connection with the Node B near, which provides the best service. This will allow that the user station know which is the dominant pilot that would define the coverage area.

4. DATA PROCESSING

The measurement equipment is capable of taking grab samples of the power levels at certain points. For adequate analysis requires the proper spacing between measurements, making it easy to apply statistical techniques such as Krige Method [9]; which is an interpolation technique based in a sample regression, with several samples irregularly spaced, it was used to predict unknown values from known values. The interpolation method of Krige facilitates the task of analyzing the distribution of power radiated by an antenna, it is sufficient to take a series of grab samples with the proper equipment (longitude and latitude parameters are needed) to create an experimental variogram, and based on it make the most appropriate approach to the theoretical variogram to perform interpolation and move from discrete samples to a continuous measurement map.

The Method of Krige is a family of generalized algorithms for least squares that from a set of observations provides the optimal linear predictor for the variable in any position. Daniel G. Krige developed the technique in an attempt to predict more accurately the mineral reserves. In recent decades the Krige Method has become a fundamental tool in the field of the geostatistics because founded the basis of linear geostatistical initially developed this method. It is a geostatistical method for estimating points that using a variogram model to obtain data. It is based on the premise that the spatial variation continues with the same pattern.

The variogram or semivariogram is a tool that allows analyzes the spatial behavior of a variable on a defined area, resulting in the influence of data at different distances. From the data provided by the theoretical variogram the estimation is performed by the method of Krige. The semivariogram is a measure of how similar the points are in the space when they are farther apart. To develop a variogram first requires creating an experimental variogram based on the selected simple, and choose a theoretical variogram that fit the experimental, because this is not a function where we can make interpolations. The semivariogram provides information of the spatial behavior of a variable. To quantify the degree and scale of spatial variation is necessary adjusting a function that describes the behavior of the variable [9].

The experimental semi-variogram (see Figure 2) is constructed depending on the variation of the mathematical expectation of the values obtained in the measurements (see Figures 3 and 4), to determine the influence of data at different distances. There are different space-temporal statistical models to obtain the experimental variogram, as: the exponential, Gaussian, Bessel, linear, to name a few. The most appropriate model for the results of the analysis of performance of a Node B is the empirical model "exponential Bessel" since the Bessel function of exponential correlation is one of the most flexible for general use [9].

To minimize the variance of the mean square error of estimate is necessary to have a function that describes the behavior of the phenomenon discussed, as an interpolation between the points of the experimental variogram does not guarantee the existence and uniqueness of the solution of the Kriging system, that is why, based on the experimental variogram, we construct a theoretical variogram that resembles the actual behavior of the variable through the least squares approximation as is showing in the red line in the Figure 5 [9].

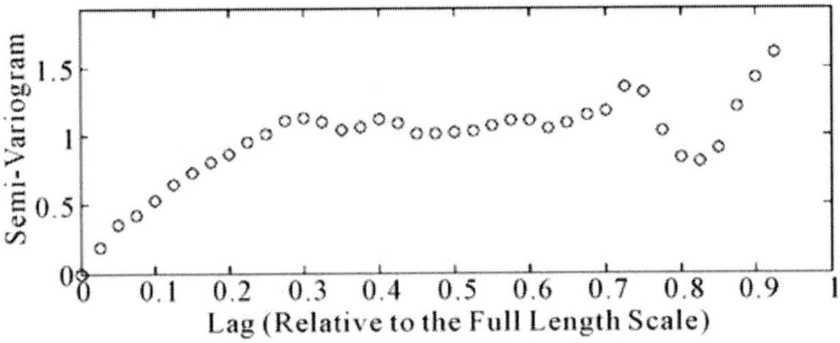

Figure 2. Experimental semi-variogram [9].

The Figure 6 is obtained using Ordinary Krige, because the mean value is not known, and we know that the value is not constant throughout the study area, but locally can be considered constant; because measurements are made at a distance will be very similar to those made in the vicinity of that point [9].

The validation step shown in Figure 7, is for to be sure that the mean square error is within the acceptance region to eliminate any doubt with respect to reliability, efficiency, consistency, sufficiency and unbiasedness of the power level estimator for proper analysis of the performance of a Node B [9].

The number of measurements needed to ensure a correct prediction of the measured power levels, depend on the range of variability of this power level. That is to say, if the power level is very variable, a greater number of measurements must be taken that when the power does not change quickly. Knowing the function most appropriate to the behavior of the measured power level the information is plotted so that geographical coordinates define the axes, and the power level determines the color, which represents the sample [9].

For complete analysis requires the use of different models of Geographic Information Systems (GIS, Geographic Information System) which is an organized integration of hardware, software and geographic data, designed to capture, store, manipulate, analyze and display

LONGITUDE	LATITUDE	CPICH
-99.13472222	19.49777778	-82.553
-99.13472222	19.49777778	-80.167
-99.13472222	19.49777778	-83.676
-99.13472222	19.49805556	-88.068
-99.13472222	19.49833333	-84.118
-99.13472222	19.49833333	-85.757
-99.13472222	19.49833333	-88.932
-99.13472222	19.49861111	-86.631
-99.13472222	19.49861111	-83.616
-99.13472222	19.49888889	-79.743
-99.13472222	19.49916667	-82.326
-99.13472222	19.49944444	-86.039
-99.13472222	19.49972222	-84.787
-99.13472222	19.49972222	-85.6
-99.13472222	19.49972222	-87.987
-99.13472222	19.50027778	-81.967
-99.13472222	19.50027778	-81.457
-99.13472222	19.50083333	-86.443
-99.13472222	19.50083333	-88.63
-99.13472222	19.50111111	-89.686
-99.13472222	19.50111111	-91.257
-99.13472222	19.50166667	-89.88
-99.13472222	19.50166667	-93.867
-99.13472222	19.50166667	-91.075
-99.13472222	19.50166667	-88.946
-99.13472222	19.50194444	-92.528
-99.13444444	19.49583333	-59.089
-99.13444444	19.49583333	-64.404
-99.13444444	19.49583333	-61.553
-99.13444444	19.49583333	-61.553
-99.13444444	19.49583333	-59.87

Figure 3. Data file format [9].

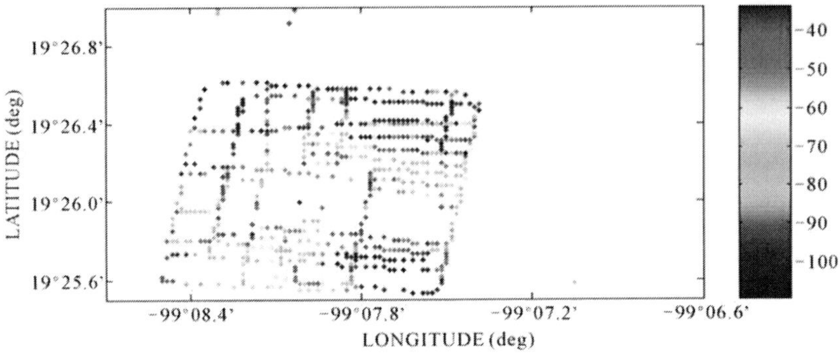

Figure 4. Location of point measurements [9].

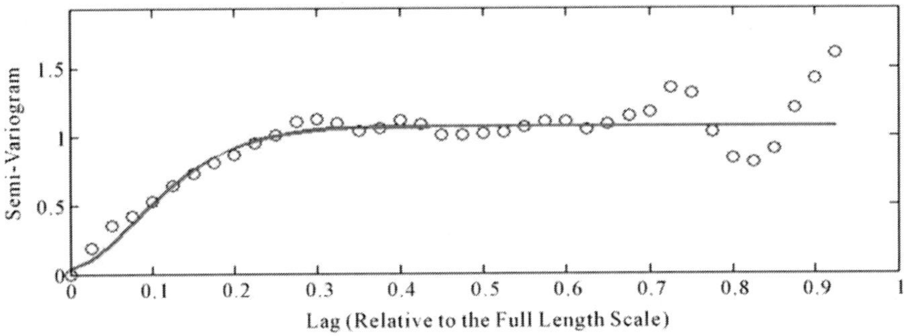

Figure 5. Theoretical variogram from LSQ [9].

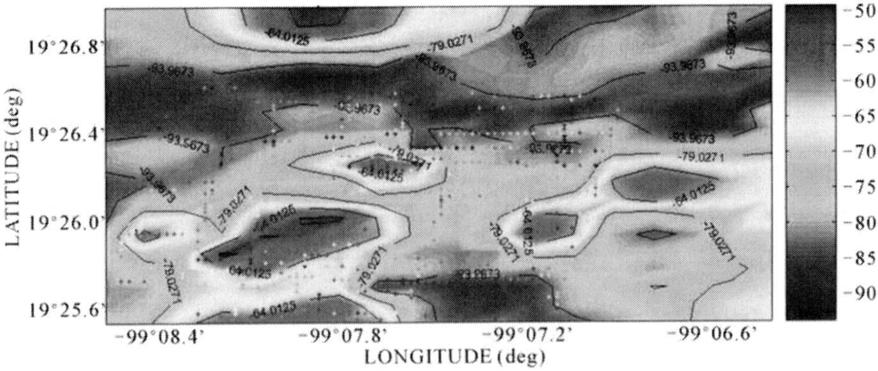

Figure 6. Coverage map [9].

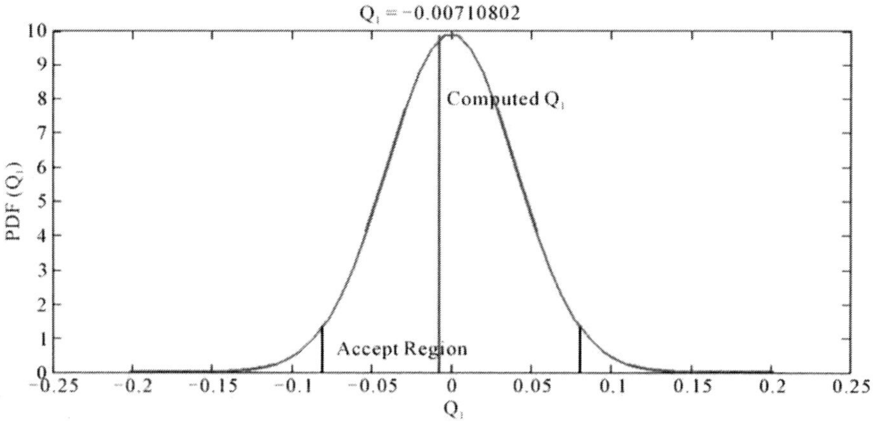

Figure 7. Graphical validation process Krige Method using the software tool "EasyKrig" [9].

all forms of geographically referenced information, to solve complex problems of planning and management to meet specific information necessary for a general vision of the area of interest [10]. Because with this is possible to overlay the power map created by Krige method, in the satellite photography of the interest area. For this work we made a program which works in MatLab. This makes calculations based on Okumura-Hata Method to obtain the approximate value of CPICH power level in the same points where measurements were made. In order to compare the theoretical levels created by Okumura-Hata equation with practical values measured, the program takes position information of each measure to calculate the distance between the measured points (as mobile station position) and the position of the Base Station, and height of both stations. After know all calculated values, those values with position information of measurements are used for created a theoretical CPICH power level.

5. RESULTS

Figure 8 shows the power distribution levels for a particular Base Station, which is the dominant pilot in most of the area of interest. Taking into account the sensitivity of most mobile devices on the market is –80 dBm [11]. The power levels founded are acceptable, because in the boundaries of the coverage area, shown in Figure 8, the user station is able for start a call.

According to Okumura-Hata's Model the power level of CPICH is plotted in Figure 9.

Figure 8. CPICH power distribution for the SC 226 in dBm.

According to Table 1 the bigger errors are in the area around the Base Station. And lower errors are in the nearly area of the Base Station. It is because OkmuraHata model takes in to account the earth curvature. Then, if the distance between the Base Station and the Mobile Station is very short the calculation error is bigger that the error in more large distances. But in general, the Okumura-Hata model is appropriate for model the behavior of the signal in 3G cellular networks, because as we can see in Table 1, although the distance is greater that a kilometer the relative error is small, Because while the distance is into the range of 1 Km from 10 Km the Okumura-Hata model is applicable.

Also, we can realize that the difference between the calculate value and the measurement is bigger in the nearby area to the Base Station, but it does not happen around 0.5 Km as is shown in Figure 10. Where the red points represent the practical values, which are more similar to theoretical values. And the blue points the more different.

In the range of –90 dBm from –70 dBm the average different is of 6.7 dBm. This is a good approximation because for choose a Base Station the Mobile Station needs a difference approximate of 9 dB which is bigger that the difference between practical and theoretical values. Is important to say that the 90.2% of the compared values are between the range of 0 dBm from 10 dBm.

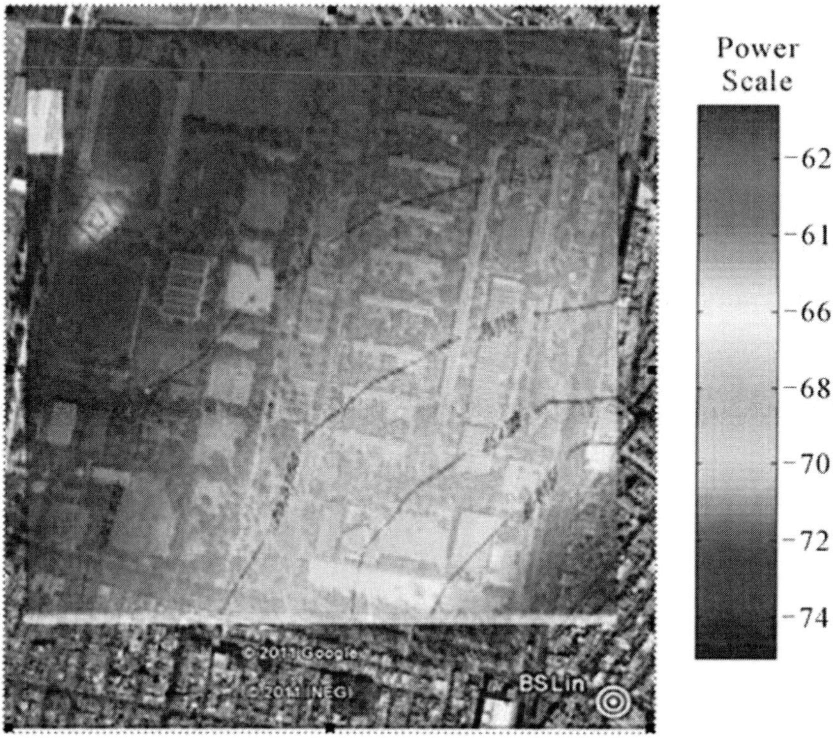

Figure 9. CPICH power distribution for the SC 226 in dBm according to Okumura-Hata model.

Figure 10. Analysis of error in Okumura-Hata model approximation.

Table 1. Analysis of error vs distance.

Measured CPICH (dBm)	Calculated CPICH (dBm)	Distance (BS-MS) (m)	Absolute Error (dBm)	Relative Error
–69.9349	–69.932	454.51	0.0029	0
–63.182	–63.1986	383.99	0.0166	0.0003
–72.008	–71.9887	642.99	0.0193	0.0003
–76.135	–76.1983	831.32	0.0633	0.0008
–72.235	–72.3017	660.24	0.0667	0.0009
–79.518	–79.9258	1089.75	0.4078	0.0051
–65.933	–27.7028	33.08	38.2302	1.38
–61.564	–22.1074	44.28	39.4566	1.7848
–68.267	–22.1062	36.96	46.1608	2.0881

6. CONCLUSIONS

Due to the Okumura-Hata is a model of propagation which was created based on several measurements made in Japan is necessary to ensure that is appropriate use this model in Mexico City due to that those are places with very different conditions for the propagation of the signal. There are a lot of propagation models, but specifically the Okumura-Hata model can be used because it offers an accuracy around of 90% as is shown in this article. It makes of Okumura-Hata a reliable propagation model for this particular case. As this work shows if you use this propagation model very near from the Base Station, is to say, with very short distances, the prediction is bad because the curvature of the earth introduces an important calculation error. Despite the fact that the Okumura-Hata model is good, the result could be better if we make some adjusts about the height of the Base Station and the mobile station. And take into account the specific characteristics of weather, kind of terrain among many others. Finally we can say that, in this particular case, the Okumura-Hata model could be appropriate to predict the behavior of the signal into a 3G Cellular Network in Mexico City, but an extended analysis must be deployed in order to validate the effectiveness of this propagation model.

When we use the Okumura-Hata Model the biggest mistakes are in the vicinity of the base station. And mistakes are lower in more remote areas of middle and base station measurement. This is because the model takes into account the curvature of the earth.

Although the range of distance between the base station and mobile station for the implementation of Okumura-Hata model is 1 to 10 Km the best predictions are given about 0.5 km.

REFERENCES

1. Freely Adapted from a Portion of Jean-Paul M. G. Linmartz's, "Wireless Communication, the Interactive Multimedia CD-ROM," Baltzer Science Publishers, Amsterdam, Vol. 1, No. 1, 1996. http://people.seas.harvard.edu/~jones/es151/prop_models/propagation.html

2. European Journal of Scientific Research, Vol. 45, No. 4, 2010, pp. 508-528. http://www.eurojournals.com/ejsr.htm

3. Helsinki University of Technology Radio Laboratory Publications, "Teknillisen Korkeakoulun Radiolaboratorion Julkaisuja," Espoo, 2002.

4. T. Keji, "WCDMA Mobile Communications System," John Wiley & Sons, New York, 2002.

5. L. Jaana, W. Achim and N. Tomás, "Radio Network Planning and Optimization for UMTS," 2nd Edition, Editorial John Wiley and Sons, New York, 2006

6. Electronic Communications Committee (ECC) within the European Conference of Postal and Telecommunications Administrations (CEPT), "UMTS Coverage Measurements," Ecc Report 103, 2007.

7. QUALCOMM, "WCDMA Network Planning and Optimization," Revision B, 2006.

8. 3rd Generation Partnership Project, "Base Station (BS) Conformance Testing (FDD)," Technical Specification Group Radio Access Network, Release 10, 2011.

9. B. J. Ponce and S. Vidal, "Krige Method Application for the Coverage Analysis of a Node-B in a WCDMA Network," International Journal of Communications, Network and System Sciences, Vol. 4, No. 3, 2011, pp. 180- 188.

10. J. Ponce, S. Vidal, I. Zamudio and M. Eslava, "Performance Experimental Analysis of a WCDMA Node B Based on CPICH, Multipath, EC/Io and RSCP Measurements," 12th National Congress of Electromechanic and Systems Engineering, Mexico City, 2010.

11. M. Ibáñe and A. Simón, "Space-Temporal Models Statistics in Perimetry," Ph.D. Dissertation, Department of Mathematics, School of Technology and Experimental Sciences, Jaume 1 University, Castellón de la Plana, 2003.

12. H. Acevedo, "Geographic Information Systems," 2003. http://www .inbio.ac.cr/es/inbio/inb_sig.htm

13. J. Ponce, S. Vidal, M. Acevedo and M. Jimenez, "A Geographic Information System Applied to Coverage Maps of 3G Cellular Communications Networks," Journal of Geographic Information System, Vol. 3, No. 2, 2011, pp. 140-144.

14. M. Licea, S. Vidal and C. Barroeta, "Characterization of the Performance Parameters of Data Transmission in Mobile Communication Terminals of Broadband," B.S Dissertation, Department of Communications, National Polytechnic Institute, Mexico City, 2010.

15. J. Ponce, S. Vidal and J. Sosa, "Analysis of the Impact of the Multipath Phenomena and Size of Cell in WCDMA Third-Generation Networks," B.S Dissertation, Department of Communications, National Polytechnic Institute, Mexico City, 2010.

Chapter 6

Adaptive Terminal-Modality-Based Joint Call Admission Control for Heterogeneous Cellular Networks

Mahmoud M. Badawy, Salman A. AlQahtani

Computer Engineering Department, King Saud University, Riyadh, Saudi Arabia

ABSTRACT

The coexistence of different Radio Access Technologies (RATs) requires a need for Common Radio Resource Management (CRRM) to support the provision of Quality of Service (QoS) and the efficient utilization of radio resources. The provision of QoS is an important and challenging issue in the design of integrated services packet networks. Call admission control (CAC) is an integral part of the problem. Clearly, without CAC, providing QoS guarantees will be impossible. There is unfairness in allocation of radio resources among heterogeneous mobile terminals in heterogeneous wireless networks. In this paper, an Adaptive-Terminal Modality-Based Joint Call Admission Control (ATJCAC) algorithm is proposed to enhance connection-level QoS and reduce call blocking/dropping probability. The proposed ATJCAC algorithm makes call admission decisions based on mobile terminal modality (capability), network load, adaptive the bandwidth of ongoing call and radio access technology (RAT) terminal support index. Simulation results show that the proposed ATJCAC scheme reduces call blocking/dropping probability.

Keywords: Call Admission Control; Call Blocking; Call Dropping; Next Generation Wireless Network (NGWN); RAT Selection Approaches

1. INTRODUCTION

Network heterogeneity refers to a combination of multiple wireless networks based on different access technologies (e.g. UMTS, EV-DO, LTE, WiMAX, etc.) coexisting in the same geographical area. Due to the coexistence of different Radio Access Technologies (RATs), Next Generation Wireless Networks (NGWN) are predicted to be heterogeneous in nature. The coexistence of different RATs requires a need for Common Radio Resource Management (CRRM) to support the provision of Quality of Service (QoS) and the efficient utilization of radio resources. With joint radio resource management in NGWN, mobile users will be able to communicate through any of the available radio access technologies (RATs) and roam from one RAT to another, using multimode terminals (MTs) as shown in Figure 1 [1-3].

Next generation wireless cellular networks, including 3G and 4G technologies are envisaged to support more mobile users and variety of high-speed Wireless Multimedia Services (WMSs). A WMS enables the simultaneous transmission of voice, data, text and images through radio links by means of the new wireless technologies. Different WMSs have diverse bandwidth and Quality of Service (QoS) requirements from their users that need to be guaranteed by wireless cellular networks. In wireless cellular networks, user's QoS requirements can be quantitatively expressed in terms of probabilistic connection-level QoS parameters such as new call blocking probability (NCBP) and handoff call dropping probability (HCDP) [4]. The NCBP is the probability of a new arriving call being rejected while the HCDP is the probability that an accepted call is terminated before the completion of its service, i.e., the probability that a handoff attempt fails [4].

Provisioning connection-level QoS in wireless cellular networks becomes complex due to 1) The limited radio link bandwidth, and 2) The high rate of handoff events as the next generation of wireless cellular networks will use micro/pico cellular architectures in order to provide higher capacity. Therefore, one of the most important connection-level QoS issues is how to reduce/control handoff drops due to lack of available resources in the new cell, since mobile users should be able to continue their ongoing connections. Since it is practically impossible to completely eliminate handoff drops, the best one can do is to provide some forms of probabilistic QoS guarantees by keeping HCDP below a predetermined value [5].

In the 3G and beyond wireless systems, multimedia services such as voice, video, data, and audio are to be offered with various quality-of-service (QoS) profiles. Hence, more sophisticated call admission control (CAC) schemes are developed to cope with these changes. Traffic of admitted calls is then controlled by other RRM techniques such as scheduling, handoff, power, and rate control schemes.

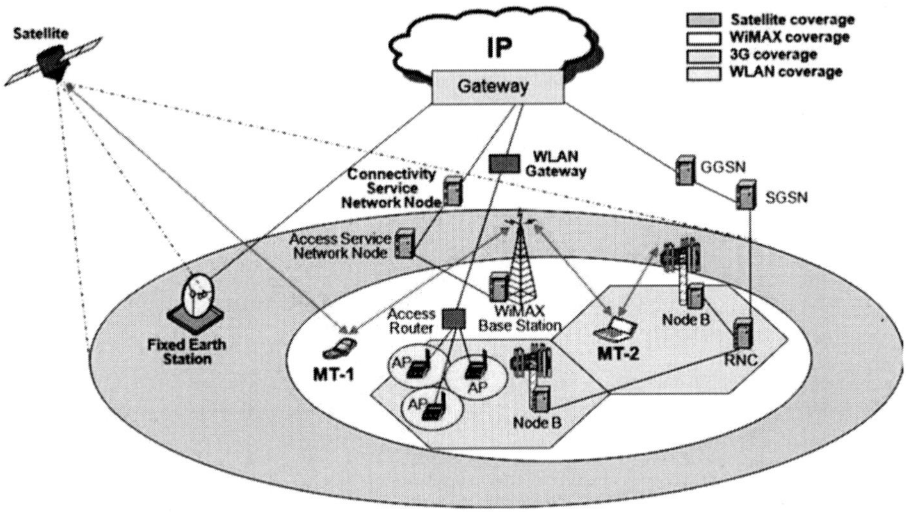

Figure 1. The coexistence of different RATs.

RAT selection algorithms are part of the CRRM algorithms. Simply, their role is to verify if an incoming call will be suitable to fit into a heterogeneous wireless network, and to decide which of the available RATs is most suitable to fit the need of the incoming call and admit it. Guaranteeing the requirements of QoS for all accepted calls and at the same time being able to provide the most efficient utilization of the available radio resources is the goal of RAT selection algorithm. Call admission control is a key element in the provision of guaranteed quality of service in wireless networks. The design of call admission control algorithms for mobile cellular networks is especially challenging given the limited and highly variable resources, and the mobility of users encountered in such networks.

Generally, CAC algorithms are triggered by any of the following events: New call arrival and handoff call arrival. The normal call admission control algorithms do not provide a solution to fit a heterogeneous wireless network. Therefore, there is a need to develop RAT selection algorithm in addition to Call admission control. This guarantees a term called Joint call admission control (JCAC) algorithm.

In this paper, an Adaptive-Terminal Modality-Based Joint Call Admission Control (ATJCAC) algorithm is proposed to enhance connection-level QoS and reduce call blocking/dropping probability. The ATJCAC scheme is designed to simultaneously achieve the following objectives in heterogeneous cellular networks:

1) Ensure fairness in allocation of radio resources among heterogeneous mobile terminals;

2) Adapt the bandwidth of ongoing calls to improve connection-level QoS;

3) Guarantee the QoS requirement of all admitted calls;

4) Prioritize handoff calls over new calls.

The rest of this paper is organized as follows. The related work is presented in the next section. In Section 3, the system model is described. The proposed adaptiveTJCAC scheme is presented in Section 4. In Section 5, result discussions of the proposed scheme are provided. Finally, the conclusion of this research is presented in Section 6.

2. LITERATURE REVIEW

A number of RAT selection algorithms including initial RAT selection and vertical handover have been proposed in the literature for heterogeneous wireless networks [1,2, 6-14]. Reference [14] presents a good revision on these algorithms. Each one has its benefits and limitations. O. E. Falowo et al. in paper [1] review the recent call admission control algorithms for heterogeneous wireless networks. The benefits and requirements of JCAC algorithms are discussed. The authors examine eight different approaches for selecting the most appropriate RAT for incoming calls in HWN and classify the JCAC algorithms based on these approaches. The advantages and disadvantages of each approach are discussed. The same authors in [2] propose a JCAC algorithm which considers the users preference in making an admission decision and a specific case where the user prefers to be served by the RAT which has the least service cost is modeled and evaluated. In [6] a JCAC scheme for multimedia traffic that maximizes the overall network revenue with QoS constraints over coupled WLAN and CDMA cellular network is considered. X. G. Wang et al. in [7] proposed an adaptive call admission control for integrated cellular and WLAN network. In this proposed scheme, call admission decisions are based on requested QoS and availability of radio resources in the considered RATs. D. Karabudak et al. in [8] proposed a call admission control scheme for the heterogeneous network using genetic algorithm. The objectives of the scheme are to achieve maximum wireless network utilization and meet QoS requirements. A network capacity policy based joint admission controller is presented by K. Murray et al. in [9,10]. D. Qiang et al. in [11] proposed a joint admission control scheme for multimedia traffic that exploits vertical handoffs as an effective tool to enhance radio resource management while guaranteeing handoff users QoS requirements. The network resources utilized by the vertical handoff user are captured by a link utility function. X. Li et al. in [12] proposed an efficient joint session admission control scheme that maximizes overall network revenue with QoS constraints over both the WLAN and the TD-SCDMA cellular networks. In [13], the authors proposed a call admission control reservation algorithm that takes resource fluctuations into consideration. They considered two types of applications denoted by wide-band and narrow band. The performance of the algorithm was modeled through a queuing theory approach and its main performance measures are compared with a conventional algorithm through simulation. The authors in [14] proposed an algorithm, which incorporates traditional Admission Control (AC) and Wiener Process (WP)-based prediction algorithms to determine when to carry out access service network gateway relocation.

Gelabert et al. in [15] presented a Markovian approach to RAT selection in heterogeneous wireless networks. They developed an analytical model for RAT selection algorithms in a heterogeneous wireless network comprising GSM/EGDE and UMTS. The proposed algorithm selects just one RAT for each call. In [16], a service-class based JCAC algorithm was proposed. it admits calls into a particular RAT based on the class of service, such as voice, video streaming, real-time video, web browsing, etc. in [17], a terminal-modality-based JCAC scheme was proposed. It consists of two main components: joint call admission controller and band-width reservation unit.

3. SYSTEM MODEL AND ASSUMPTIONS

We consider a heterogeneous cellular network which consists of J number of RATs with co-located cells. A typical example of a heterogeneous wireless network, adapted from [16] is shown in Figure 2. In the heterogeneous network, radio resources are jointly managed. Cellular networks such as GSM, UMTS (3G) and LTE can have the same and fully overlapped coverage, which is technically feasible, and may also save on installation costs [18,19]. Let H denote the set of all available RATs in the heterogeneous wireless network.

Then, H is given as follows:

$$H = \{\mathrm{RAT1}, \mathrm{RAT2}, \cdots \mathrm{RAT}j\}$$

where J is the total number of RATs in the heterogeneous cellular network. The heterogeneous cellular network supports k-classes of calls, and each RAT in set H is optimized to support certain classes of calls. Let H_i ($H_i \subseteq H$) denote the set of RATs which can support class-i calls in the heterogeneous cellular network, and let h_i ($h_i \subseteq h$) denote the set of indices of all RATj which belong to H_i, where h = {1, 2, ⋯, J}. Furthermore, let J_i ($J_i \leq J$) denote the total number of RATs that can support class-i calls. Let D_j ($D_j \subseteq D$) denotes the set of call classes that can be supported by RATj (j = 1, 2, ⋯, J) where D = {class-1, class-2,

···, class-k}. Note that the idea that different networks support different classes of calls is true in reality. For example, LTE and UMTS network can support video streaming whereas GSM network cannot support video streaming.

Each cell in RATj (j = 1, ···, J) has a total of Bj basic bandwidth units (bbu). The physical meaning of a unit of radio resources (such as time slots, code sequence, etc.) is dependent on the specific technological implementation of the radio interface [20]. However, no matter which multiple access technology (FDMA, TDMA, WCDMA or OFDMA) is used, system capacity could be interpreted in terms of effective or equivalent bandwidth [21-22]. Therefore, this research refers to the bandwidth of a call as the number of bbu that is adequate for guaranteeing the desired QoS for the call, which is similar to the approach used for homogeneous networks in [22,23].

It is assumed that packet-level QoS is stochastically assured by allocating at least the minimum effective bandwidth required to guarantee a given maximum probability on packet drop, delay, and jitter. The approach used is to decompose a heterogeneous cellular network into groups of co-located cells as shown in Figure 3.

For example, cell 1a and cell 2a form a group of colocated cells. Similarly, cell 1b and cell 2b form another group of co-located cells, and so on. When a mobile user with an ongoing call is moving outside the coverage area of a group of co-located cells, the call must be handed over to one of the cells that can support the call in the neighboring group of co-located cells. For example, in the two-class three-RAT heterogeneous cellular network illustrated in Figure 3, an ongoing class-1 call can be handed over from cell 2a to cell 2b, or from cell 2a to cell 1b. Note that hand-off comprises both horizontal and vertical handoffs.

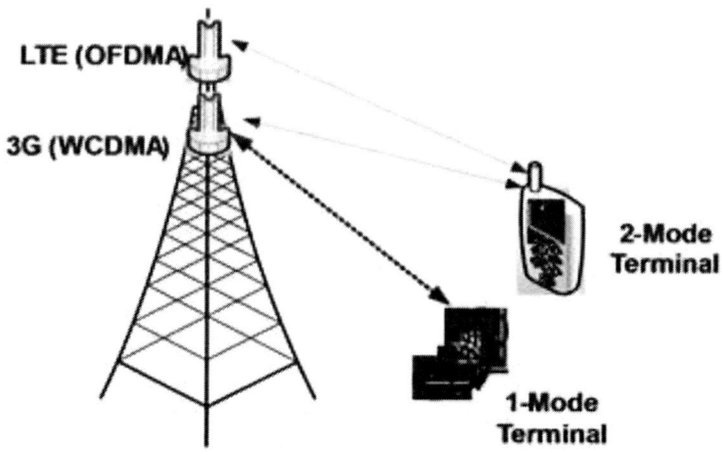

Figure 2. A typical two-RAT heterogeneous cellular network with co-located.

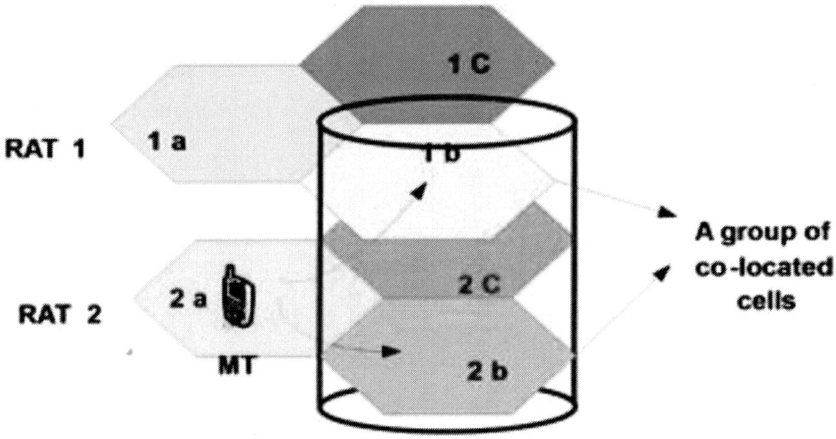

Figure 3. Two-RAT heterogeneous cellular networks with co-located cells.

The correlation between the groups of co-located cells results from handoff connections between the cells of corresponding groups. Under this formulation, each group of co-located cells can be modeled and analyzed individually. Therefore, a single group of co-located cells is considered in this research. The heterogeneous network supports K classes of calls. Each class is characterized by bandwidth requirement, arrival distribution, and channel holding time. Each class-i call requires a discrete bandwidth value, $b_{i,w}$ where $b_{i,w}$ belongs to the set $B_i = \{b_{i,w}\}$ for $i = 1, 2,\cdots, K$ and $w = 1, 2,\cdots, W_i$. W_i is the number of different bandwidth values that a class-i call can be allocated. $b_{i,1}$ (also denoted as $b_{i,min}$) and $b_{i,Wi}$ (also denoted as $b_{i,max}$) are, respectively, the minimum and maximum bandwidth that can be allocated to a class-i call. Note that $b_{i,w} < b_{i,(w+1)}$ for $i = 1, 2\cdots K$ and $w = 1, 2\cdots (W_i - 1)$.

The requested bandwidth of class-i call is denoted by $b_{i,req}$, where $b_{i,req} \in B_i$. Let $m_{i,j}$ and $n_{i,j}$ denote, respectively, the number of new call of class-i and handoff call of class-i, in RATj. with $1 \le c \le m_{i,j}$ (for new calls) and $1 \le c \le n_{i,j}$ (for handoff calls). Let $b_{i,\text{assigned } c}$ denote the bandwidth assigned to call c of class-i in RAT-j in the group of co-located cells, where $b_{i,\text{assigned } c} \in B_i$. A call c of class-i is degraded if $b_{i,\text{assigned } c} < b_{i,req}$ whereas the call is upgraded if $b_{i,\text{assigned } c} > b_{i,req}$. If a class of calls (i.e., class-i calls) requires a fixed number of bbu (i.e. constant bit-rate service), it becomes a special case in our model in which $b_{i,min} = b_{i,max}$ and the set B_i has only one element. However, it will not be possible to upgrade or degrade this class of calls.

We define the following terms commonly used in the literature to be used throughout this paper.

1) Call holding time: It is duration of the requested call connection. This is a random variable which depends on the user behavior (call characteristics).

2) Cell residency time: It is amount of time during which a mobile terminal stays in a cell during a single visit. Cell residency is a random variable which depends on the user behavior and system parameters, e.g. cell geometry.

3) Channel holding time: How long a call which is accepted in a cell and is assigned a channel will use this channel before completion or handoff to another cell? This is a random variable which can be computed from the call holding time and cell residency time and generally is different for new calls and handoff calls.

Following are the general assumptions in the studied cellular networks. The New call arrival of class-i arrive is assumed to follow Poisson process with rate λ_i^n, n denoted to new call. Handoff call of class-i arrive according to Poisson process with rate λ_i^h, h denoted to handoff call. Call holding time (CHT) of class-i is assumed to exponential distribution with mean μ_{ci}^{-1}. Cell residence time (CRT) is assumed to follow an exponential distribution with mean μ_{hi}^{-1}, h denoted to handoff rate. Channel holding time for call of class-i is assumed to exponential distribution with mean μ_i^{-1} where $\mu_i = \mu_{ci} + \mu_{hi}$.

4. PROPOSED ADAPTIVE TJCAC SCHEME

This section describes the proposed adaptive terminalmodality-based JCAC scheme. In fact, the joint call admission control (JCAC) algorithm is one of the RRM algorithms. The basic function of JCAC algorithms is to decide whether an incoming call can be accepted or not. They also decide which of the available radio access technology is most suitable to accommodate the incoming call. Figure 4 shows call admission control procedure in heterogeneous cellular networks.

When these mobile terminals make a call, then they will send a service request to the JCAC algorithm. The JCAC scheme, which executes the JCAC algorithm, will then select the most suitable RAT for the incoming call.

Figure 5 illustrates the problem of unfairness in radio resource allocation in a three-RAT heterogeneous wireless network when terminal modality is not considered in making RAT selection decisions. Assume that 1) all the three RATs have equal capacity; 2) all the arriving calls belong to the same class; and 3) each RAT can support only two calls. Figure 5 shows six consecutively arriving calls (1 to 6) in the heterogeneous wireless network. A load-balancing JCAC scheme, for instance, will admit the

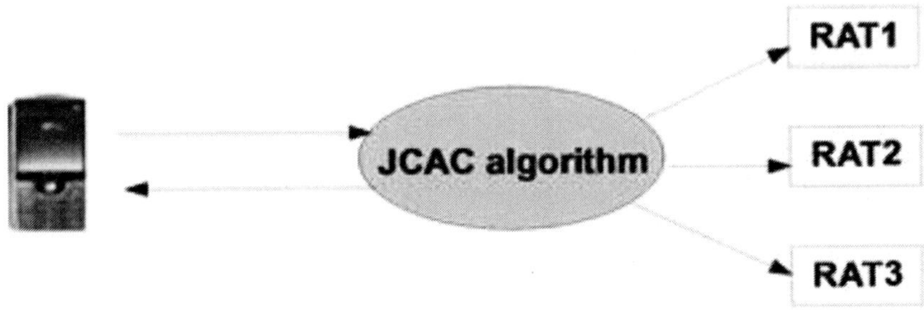

Figure 4. JCAC procedure in heterogeneous cellular networks.

Figure 5. Unfair allocation of radio resources among heterogeneous mobile terminals.

first incoming call (call 1 from a triple-mode terminal) into RAT 1. It will admit call 2 (from a dual-mode terminal) into RAT 2, admit call 3 (from a triple-mode terminal) into RAT 3, and admit call 4 (from a dual-mode terminal into RAT1).

The JCAC scheme cannot admit the fifth incoming call (call 5) into RAT 2 or RAT 3 be-cause call 5 is from a single-mode terminal that is sup-ported only by RAT 1. The JCAC scheme will then try to admit call 5 into RAT 1. Call 5 will be blocked in RAT 1 because it is already fully loaded (maximum of two calls). In a similar manner, the sixth incoming call (call 6) from a single-mode terminal that is supported by RAT 1 only, will be blocked.

In order to reduce this problem of unfairness in allocation of radio resources and also enhance QoS and system utilization among heterogeneous terminals in heterogeneous wireless networks, adaptive-terminal modalitybased JCAC (ATJCAC) scheme is proposed for heterogeneous wireless networks. ATJCAC consists of the following three components: joint call admission controller, band-width reservation unit and bandwidth adaptation controller. The main component of the proposed scheme is shown in Figure 6.

The following are the descriptions of these main components.

4.1. The Joint Call Admission Controller

The joint call admission controller implements the JCAC algorithm. The basic function of the JCAC algorithm is to make call admission decisions and guarantee fairness among the different heterogeneous terminals in the heterogeneous wireless network. The proposed JCAC algorithm makes RAT selection decisions based on the modality of the mobile terminal initiating a call, the terminal support index of each RAT that can support the call, and the current load in the available RATs.

Terminal support index of a RAT indicates the ratio of terminals supported by the RAT to the total terminals registered in the heterogeneous wireless network. Terminal support index of RATj (R^J) is defined as follows:

$$R^j = T^J / T, 0 < R^j \leq 1 \tag{1}$$

where T^j is the number of mobile terminals that have an interface for RAT-j (i.e. that can be admitted into RAT-j), and T is the total number of mobile terminals registered in the network.

During a call setup, a mobile terminal initiating a call sends a service request to the joint call admission control which implements the JCAC algorithm. The JCAC algorithm is part of the joint resource management entity of the heterogeneous wireless network. The location of the joint resource management entity depends on the specific network architecture deployed by the network operator. The service request contains the call type, capability of terminal (number and types of network supported by the terminal), service class, and bandwidth requirements. Figure 7 is the flow chart of the proposed JCAC algorithm. As shown in Figure 7, x_{ij} and y_{ij} represent the residual bbu available for new and handoff class-i calls, respectively, in RAT-j. L_j, and T^j represent the current load and Terminal support index of RAT-j, respectively. H_i is the set of RATs that can support an incoming class-i call, and the mobile terminal (based on terminal modality) initiating the call. h_i is the set of indexes of RATs that belong to set H_i.

As shown in Figure 7, whenever a class-i call arrives from a y-mode terminal in the heterogeneous wireless network, the JCAC algorithm determines the set H_i of RAT-j that can support the class-i call and that can support the mobile terminal initiating the call (based on terminal modality). The JCAC algorithm then sorts the RATs in set H_i in increasing order of their terminalsupport index (R_j). Starting from the RAT with the lowest terminal-support index, the JCAC algorithm attempts to allocate the maximum bbu for this call (i.e., set $b_{i,min}$ =

$b_{i,max}$) provided that the available bbu in the selected RAT is greater than or equal to $b_{i,max}$. If the available bbu in the selected RAT is less than $b_{i,max}$ but greater than or equal to $b_{i,req}$, the call will be assigned a bandwidth between $b_{i,req}$ and $b_{i,max}$. If the available bbu is less than $b_{i,req}$ but greater than or equal to $b_{i,1}$ ($b_{i,min}$), the call will be assigned a bandwidth between $b_{i,1}$ and $b_{i,req}$.

If the available bbu in RATs with the lowest terminal-support index is less than $b_{i,1}$, BA algorithm (BAA) will be invoked to reduce the bandwidth of some ongoing call(s) in the chosen RAT. If the available bbu is still less than $b_{i,1}$, the next available RAT with the second-lowest terminal support index will be selected for the call, and so on. If two or more RATs that can support the incoming call have the same low terminal-support index, the least loaded of the two or more RATs will be selected for the incoming call.

The class-i call is rejected if none of the RATs in set H_i has enough basic bandwidth units (bbu) to accommodate the call. By attempting to admit an incoming call into the RAT with the lowest terminal-support index that can support the class of the call, the proposed ATJCAC scheme reserves other RATs with a higher terminalsupport index for calls from low-capability terminals, thereby reducing the blocking probability of such calls.

Figure 8 illustrates how the proposed JCAC scheme reduces the problem of unfairness in allocation of radio resources among heterogeneous mobile terminals. As show in Figure 8, and using Equation (1), the terminal support index of RAT 1, RAT 2, and RAT 3 are 6/6, 4/6, and 2/6 respectively. Note that terminal support index is calculated based on the number of terminals registered in the heterogeneous wireless network, and it is assumed that there only six terminals in this example.

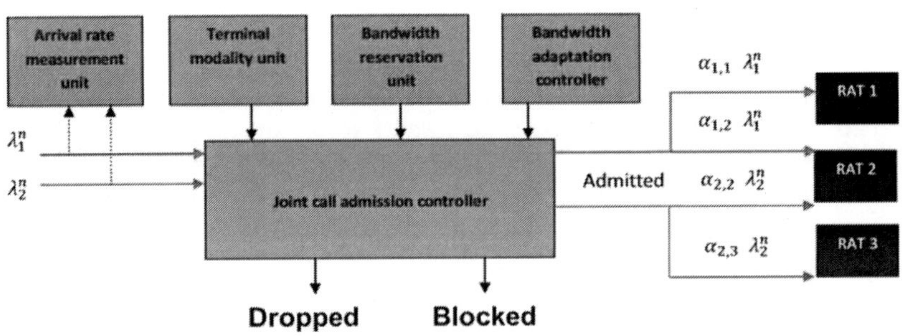

Figure 6. Component of the proposed ATJCAC scheme.

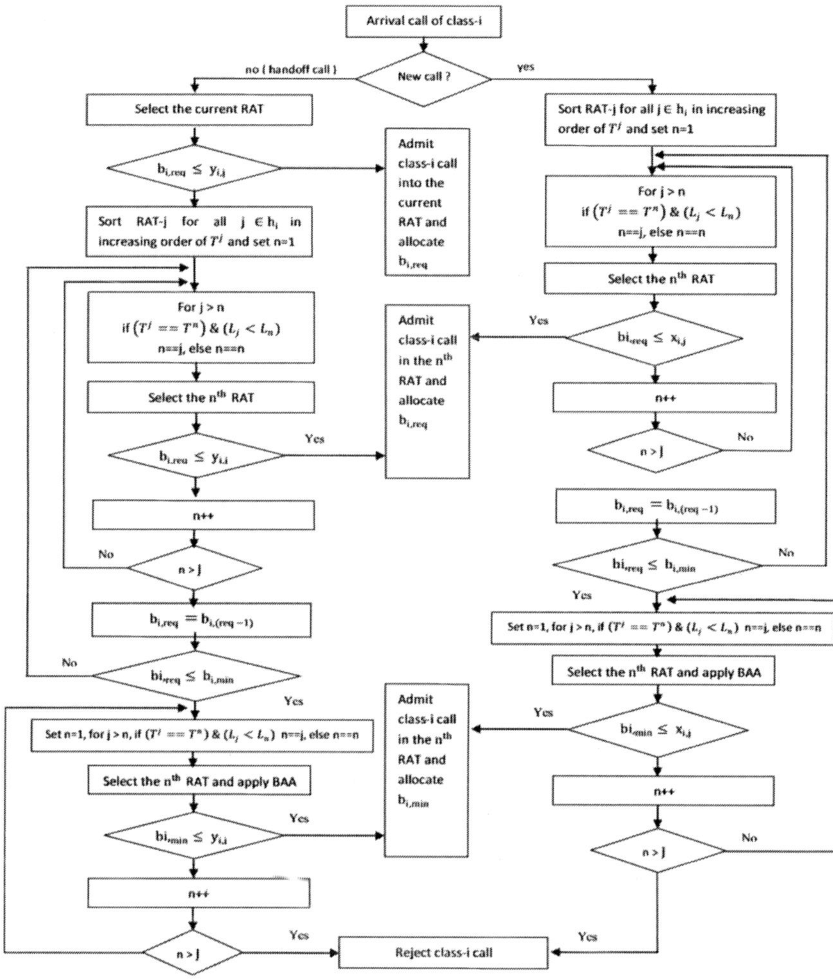

Figure 7. Flow chart of the proposed ATJCAC algorithm

As shown in Figure 8, the first incoming call (call 1 from a triple-mode terminal) can be admitted into any of the three RATs. The proposed ATJCAC will select the RAT with the lowest terminal support index (RAT 3), thereby reserving RAT 1 and RAT 2 for calls from lowcapability mobile terminals. The second incoming call (call 2 from a dual-mode terminal) can be admitted into RAT 1 or RAT 2, The ATJCAC scheme selects RAT 2, which has the lower terminal support index. The third incoming call (call 3 is from a triple mode terminal) can be admitted into any of the three RATs. The ATJCAC selects RAT 3 with the lowest terminal support index. The fourth call is admitted into RAT 2. The fifth and sixth calls are then admitted into RAT 1, which has the highest terminal support index. Thus the problem of unfairness in allocation of radio resources is reduced.

4.2. Bandwidth Reservation Unit

In order to maintain lower handoff calls over new calls, we use different a threshold-based bandwidth reservation unit. The policy reserves bandwidth for aggregate handoff calls, thus gives them priority over new calls. The policy also prioritizes among different classes of handoff calls according to their QoS constraints by assigning a series of bandwidth thresholds $t_{0,j}$, $t_{1,j}$, ⋯, ⋯, $t_{k,j}$ for handoff calls such that

$$t_{0,j} \leq t_{1,j} \leq \cdots \leq t_{i,j} \leq t_{(i+1),j} \leq t_{k,j} = B_{j,\forall j} \tag{2}$$

where $t_{0,j}$: the total bbu available in RAT j for new call (is the threshold after which new calls will be rejected in RAT-j) and $t_{i,j}$ denotes the total number of bbu available for handoff class-i calls in RATj. B_j denotes the total number of bbu available in RATj. When given any val ues of threshold (ex. $t_{0,j}$), there exist optimal values of $a_{i,j}$ and b_{ij} (j = 1, ⋯, J, i Î $d_{i,j}$ where $d_{i,j}$ denotes the set of indices of all class-i calls that can be supported by RAT-j) that minimizes the overall blocking probability in the heterogeneous wireless network. Figure 9 shows the band-width reservation unit for a two-class two-RAT heterogeneous cellular network j noting that $C_j = C_{hij}$ and $t_{0,J} = C_{nij}$.

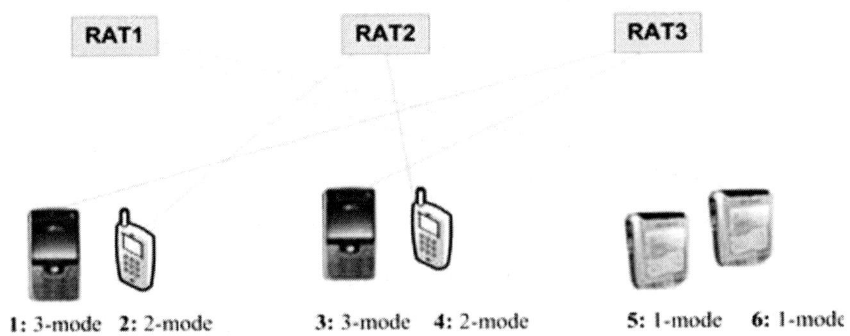

Figure 8. Fair allocation of radio resources among heterogeneous mobile terminals.

4.3. Bandwidth Adaptation/Reallocation Controller

The bandwidth adaptation controller executes the BAA which is triggered when a new call arrives or when a call is completed. Most multimedia applications are adaptive. For example, voice can be encoded at 16 kbps, 32 kbps, 64 kbps, and 128 kbps by choosing appropriate encoding mechanisms. Similarly, video applications can be made rate adaptive by using, for instance, a layered coding method. In layer coding method, the lowest layer (i.e., the base layer) contains the critical information for decoding the image sequence at its minimum visual quality. Additional layers provide increasing quality. All these encoded layers may be transmitted when the network is underutilized. However, when the network resources are being fully utilized, only based layer(s) which contain critical information may be transmitted.

As an illustration, if one would watch a 30-minute video clip encoded at 256 kbps and 64 kbps respectively. At 256 kbps, one will see better pictures with better resolution than at 64 kbps. Therefore, the bandwidth adaptation affects the quality of the real-time applications rather than the transmission time. However, the minimum requested QoS is maintained by ensuring that the bbu of the calls are not degraded below the required minimum.

Under conditions of heavy traffic load, i.e. the sum of the requested bandwidth exceeds the unused bandwidth capacity so that not all the requests can be completely served, the role of bandwidth adaptation technique are essential. These algorithms are needed to reduce the requested or already connected call bandwidth allocation. In designing the algorithm, we assume that a service with degraded QoS is better than an outright rejection of service requests. The quality grade is determined by the amount of Acceptable Bandwidth level (ABL).

The Bandwidth Reallocation (i.e. degradation or upgrading of resource allocation) module is deployed to reallocate the bandwidth capacity. The reallocation forms the crust of the scheme/algorithm. The strategy implemented is by cohesively coupling lower and priority calls in a trade-off manner. In the event of a new call or handoff from a higher priority traffic encountering insufficient bandwidth level, sustaining the call is compulsory in the presence of hybrid traffic. Bandwidth of lower priority connections (i.e. non-real-time and real-time VBR traffic) are decreased to the level of streams of the lowest priority or the highest ABL. In the event of a vise versa conditions, bandwidth of lower priority traffic are able to be increased to a maximum level of the highest priority level. The process of reallocating may be involve either an upgrade or degrade of the bandwidth allocation based on the ABL.

The ABL is obtained by subtracting the maximum required bandwidth with the minimum required bandwidth. The difference is called the degradable range/spectrum. The degradable spectrum is further divided into N (i.e. where N = 1, 2 ··· n) levels, called micro-ABLs as shown in Figure 10.

The concept of bandwidth allocation as a discrete component is applied into the structuring and derivation of the micro-ABLs. Subsequent to this theory, the bandwidth allocation for the micro-ABLs form the discrete set B = {BW_{min},

$BW_{(min+1)}$, $BW_{(min+2)}$, ⋯, BW_{avg}, ⋯, $BW_{max}\}$ where BW_{min} is the minimum bound, BW_{avg} is the average bound and BW_{max} is the maximum bound for bandwidth allocation.

The computation and deployment of bandwidth reallocation consumes an amount of time capable of jeopardizing the probability of a mobile to continue is con nection. Thus, to avoid this time delay, the process of bandwidth reallocation is carried out in a distributed manner. Each Base Station (BS) does the computation process independent from other BSs. An important prerequisite is to ensure that the computation process of bandwidth reallocation algorithm should be completed before the system does the real bandwidth reallocation of the ongoing connections in the network.

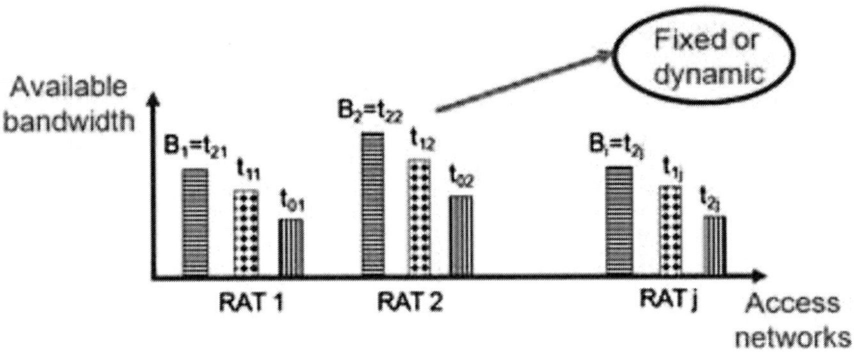

Figure 9. Fair allocation of radio resources among heterogeneous mobile terminals.

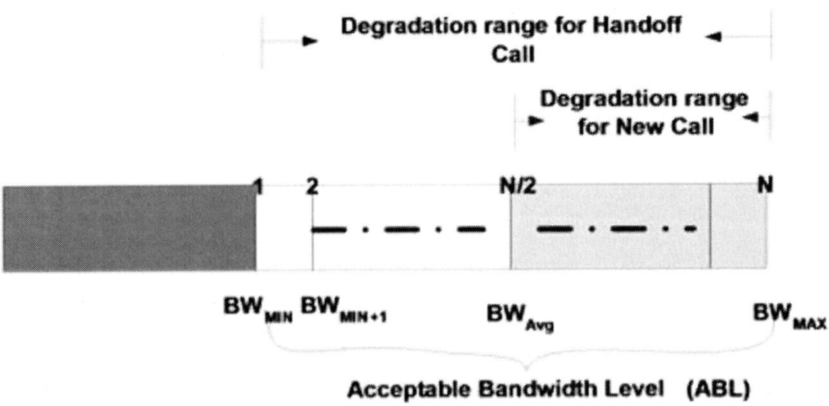

Figure 10. ABL and Micro-ABLs range/spectrum.

5. RESULTS AND DISCUSSION

In this section, the performance of the proposed Adaptive-terminal-modality based JCAC scheme is evaluated with respect to New Call Blocking Probability (NCBP) and Handoff Call Dropping Probability (HCDP), using a one-class three-RAT heterogeneous wireless network supporting heterogeneous mobile terminals. A new call from i-mode terminal is blocked in the group of co-located cells if none of the available RATs supported by i-mode terminal has enough bbu to accommodate the new call. NCBP is the average number of blocked new call over all arrived new call. A handoff call from i-mode terminal is dropped in the group of co-located cells if none of the available RATs supported by i-mode terminal has enough bbu to accommodate the handoff call. HCDP is the average number of dropped new call over all arrived handoff call is known as HCDP.

A numerical simulation is conducted using C++ program and MatLab. The following system parameters are used: $C_1 = C_2 = C_3 = 10$, $t_{01} = t_{02} = t_{03} = 5$, $b_1 = \{1, 2, 3\}$, $\mu_1 = 0.5$, $\lambda_1^n = [1,5]$, $\lambda_1^h = 0.5\lambda_1^n$. In order to investigate the performance of the proposed ATJCAC scheme, two scenarios with different combinations of heterogeneous terminals are examined.

- First scenario (Terminals Dominated by 1-Mode Terminals) 1M:2M:3M = 50:25:25: is 1-mode = 50 terminal, 2-mode = 25 terminal, and 3-mode = 25 terminals.
- Second scenario (Terminals Dominated by 2-Mode Terminals) 1M:2M:3M = 25:50:25: 1-mode = 25 terminal, 2-mode = 50 terminal, and 3-mode = 25 terminals.

For each of the two scenarios, the performance of the proposed ATJCAC scheme is compared with the performance of two other JCAC schemes namely, the Terminal modality-based JCAC scheme ("NATJCAC" scheme) and the service-class-based JCAC scheme ("SJCAC" scheme). Results obtained from the two scenarios are discussed in the following subsections. In all figures we use letter "A" to indicate ATJCAC, "N" to indicated NATJCAC and letter "S" to indicate SJCAC.

5.1. First Scenario: 1M:2M:3M = 50:25:25

Figure 11 shows the effect of varying the call arrival rate on the NCBP (Pb) of 1-mode, 2-mode, and 3-mode terminals for SJCAC, NATJCAC and the proposed ATJCAC. As shown in Figure 11, Pb for the three JCAC schemes and three classes of terminals increases with an increase in arrival rate for three JCAC schemes. This is expected. However, for 1-mode terminals, Pb of the NATJCAC scheme is lower than the corresponding Pb1 of the SJCAC schemes. The NATJCAC scheme is able to reduce the Pb by admitting most of the calls from 3-mode terminals into RAT 3, which has the lowest terminal support index, thereby reserving RAT 1 and RAT 2 for calls from 1-mode and 2-mode terminals, respectively. However, the reduction in 1-mode terminals Pb for NATJCAC scheme is at the expense of the Pb1 of calls from 3-mode terminals. It can be seen that for 3-mode terminals, the Pb of the NATJCAC scheme is higher than the corresponding Pb of the SJCAC schemes.

The NCBP for the ATJCAC scheme is always less than the corresponding NCBP for the NATJCAC scheme. Note that lower NCBP of the ATJCAC scheme implies that its connection-level QoS is better than that of the NATJCAC scheme. The reason why the NCBP of the ATJCAC scheme is less than the NATJCAC scheme is as follows. When the total bbu allocated to new calls is being fully utilized, incoming new calls are rejected by the NATJCAC scheme whereas the ATJCAC scheme adapts (degrades) the bandwidth of some of the ongoing adaptive calls to free just enough bbu to accommodate the incoming new calls. Consequently, the NCBP of the ATJCAC is less than that of the NATJCAC. However, an adaptive class-i call is never degraded below the minimum bbu necessary to guarantee its minimum QoS requirements. For SJCAC the Pb for one-mode terminals is very high whereas Pb for two-mode and three-mode terminals is very low. Thus, one-mode terminals are treated unfairly by the SJCAC scheme.

Figure 12 shows the effect of varying the call arrival rate on the HCDP (Pd) of 1-mode, 2-mode, and 3-mode terminals for SJCAC, NATJCAC and the proposed ATJCAC.

As shown in Figure 12, Pd for the three JCAC schemes and three classes of terminals increases with an increase in arrival rate for three JCAC schemes. This is expected. However, for 1-mode terminals, Pd of the NATJCAC scheme is lower than the corresponding Pd of the SJCAC schemes. The NATJCAC scheme is able to reduce the Pd by admitting most of the calls from 3-mode terminals into RAT 3, which has the lowest terminal support index, thereby reserving RAT 1 and RAT 2 for calls from 1-mode and 2-mode terminals, respectively. However, the reduction in 1-mode terminals Pd for NATJCAC scheme is at the expense of the Pd of calls from 3-mode terminals. It can be seen that for 3-mode terminals, the Pd of the NATJCAC scheme is higher than the corresponding Pd of the SJCAC schemes.

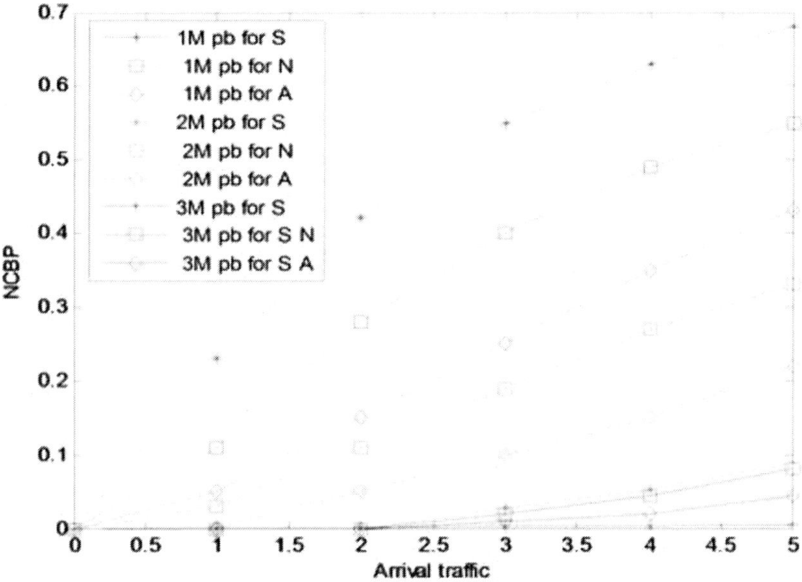

Figure 11. NCBP (Pb) for class-1 calls with 1M:2M:3M = 50:25:25.

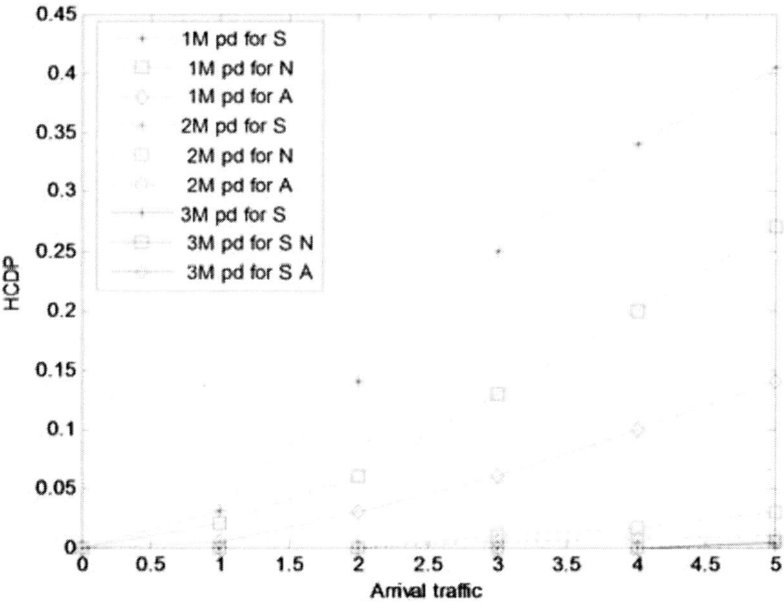

Figure 12. HCDP (Pd) for class-1 calls with 1M:2M:3M = 50:25:25.

The HCDP for the ATJCAC scheme is always less than the corresponding HCDP for the NATJCAC scheme. Note that lower HCDP of the ATJCAC scheme implies that its connection-level QoS is better than that of the NATJCAC scheme. The reason why the HCDP of the ATJCAC scheme is less than the NATJCAC scheme is as follows. When the total bbu allocated to handoff calls is being fully utilized, incoming handoff calls are rejected by the NATJCAC scheme whereas the ATJCAC scheme adapts (degrades) the bandwidth of some of the ongoing adaptive calls to free just enough bbu to accommodate the incoming handoff calls. Consequently, the HCDP of the ATJCAC is less than that of the NATJCAC. However, an adaptive class-i call is never degraded below the minimum bbu necessary to guarantee its minimum QoS requirements. For SJCAC the Pd for one-mode terminals is very high whereas Pd for two-mode and three-mode terminals are very low. Thus, one-mode terminals are treated unfairly by the SJCAC scheme.

5.2. Second Scenario: 1M:2M:3M = 25:50:25

Figure 13 shows the effect of varying the call arrival rate on the NCBP (Pb) of 1-mode, 2-mode, and 3-mode terminals for SJCAC, NATJCAC and the proposed ATJCAC. As shown in Figure 13, Pb for the three JCAC schemes and three classes of terminals increases with an increase in arrival rate for three JCAC schemes. This is expected. However, for 1-mode terminals, Pb of the NATJCAC scheme is lower than the corresponding Pb of the SJCAC schemes. The NATJCAC scheme is able to reduce the Pb by admitting most of the calls from 3-mode terminals into RAT 3, which has the lowest terminal support index, thereby reserving RAT 1 and RAT 2 for calls from 1-mode and 2-mode terminals, respectively. However, the reduction in 1-mode terminals Pb for NATJCAC scheme is at the expense of the Pb1 of calls from 3-mode terminals. It can be seen that for 3-mode terminals, the Pb of the NATJCAC scheme is higher than the corresponding Pb of the SJCAC schemes.

The NCBP for the ATJCAC scheme is always less than the corresponding NCBP for the NATJCAC scheme. Note that lower NCBP of the ATJCAC scheme implies that its connection-level QoS is better than that of the NATJCAC scheme. The reason why the NCBP of the ATJCAC scheme is less than the NATJCAC scheme is as follows. When the total bbu allocated to new calls is being fully utilized, incoming new calls are rejected by the NATJCAC scheme whereas the ATJCAC scheme adapts (degrades) the bandwidth of some of the ongoing adaptive calls to free just enough bbu to accommodate the incoming new calls. Consequently, the NCBP of the ATJCAC is less than that of the NATJCAC. However, an adaptive class-i call is never degraded below the minimum bbu necessary to guarantee its minimum QoS requirements.

For SJCAC the Pb for one-mode terminals is very high whereas Pb for two-mode and three-mode terminals are very low. Thus, one-mode terminals are treated unfairly by the SJCAC scheme.

Figure 14 shows the effect of varying the call arrival rate on the HCDP (Pd) of 1-mode, 2-mode, and 3-mode terminals for SJCAC, NATJCAC and the proposed ATJCAC. As shown in Figure 14, Pd for the three JCAC schemes and three classes of terminals increases with an increase in arrival rate for three JCAC schemes. This is expected.

However, for 1-mode terminals, Pd of the NATJCAC scheme is lower than the corresponding Pd of the SJCAC schemes. The NATJCAC scheme is able to reduce the Pd by admitting most of the calls from 3-mode terminals into RAT 3, which has the lowest terminal support index, thereby reserving RAT 1 and RAT 2 for calls from 1-mode and 2-mode terminals, respectively. However, the reduction in 1-mode terminals Pd for NATJCAC scheme is at the expense of the Pd1 of calls from 3-mode termi nals. It can be seen that for 3-mode terminals, the Pd of the NATJCAC scheme is higher than the corresponding Pd of the SJCAC schemes.

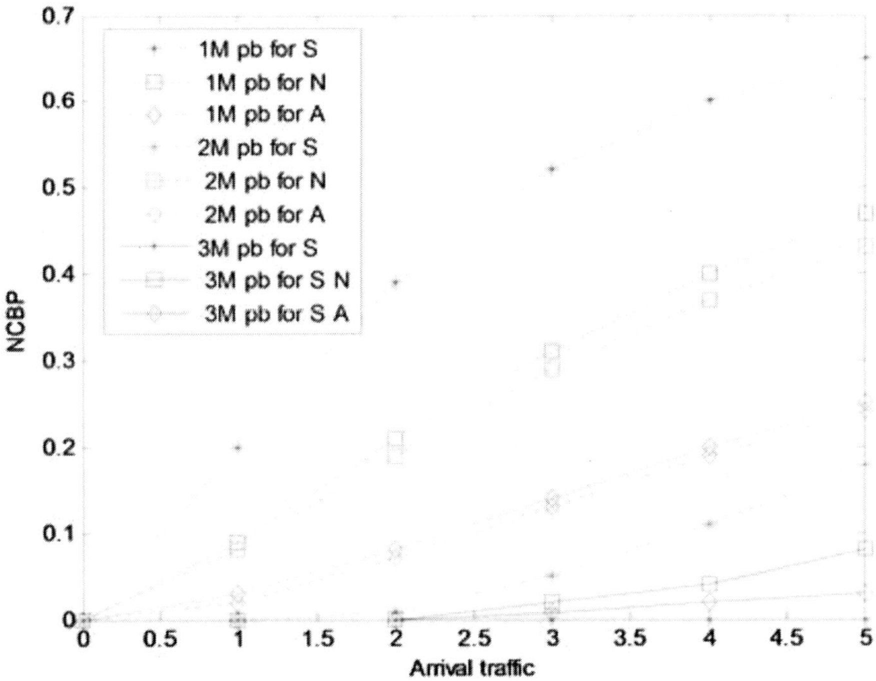

Figure 13. NCBP (Pb) for class-1 calls with 1M:2M:3M = 25:50:25.

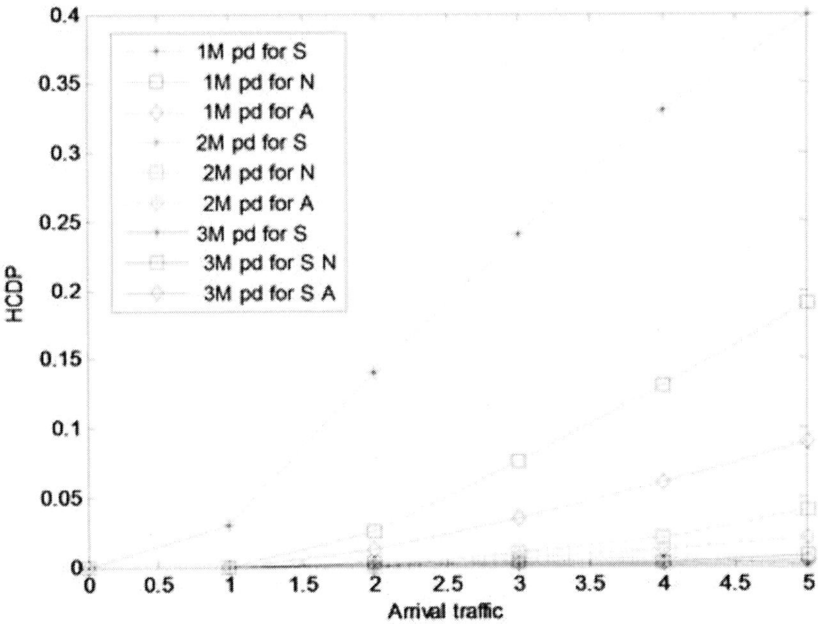

Figure 14. HCDP (Pd) for class-1 calls with 1M:2M:3M = 25:50:25.

The HCDP for the ATJCAC scheme is always less than the corresponding HCDP for the NATJCAC scheme. Note that lower HCDP of the ATJCAC scheme implies that its connection-level QoS is better than that of the NATJCAC scheme. The reason why the HCDP of the ATJCAC scheme is less than the NATJCAC scheme is as follows. When the total bbu allocated to handoff calls is being fully utilized, incoming handoff calls are rejected by the NATJCAC scheme whereas the ATJCAC scheme adapts (degrades) the bandwidth of some of the ongoing adaptive calls to free just enough bbu to accommodate the incoming handoff calls. Consequently, the HCDP of the ATJCAC is less than that of the NATJCAC. However, an adaptive class-i call is never degraded below the minimum bbu necessary to guarantee its minimum QoS requirements. For SJCAC the Pd for one-mode terminals is very high whereas Pd for two-mode and three-mode terminals are very low. Thus, one-mode terminals are treated unfairly by the SJCAC scheme.

6. CONCLUSION

In heterogeneous wireless network, There is a need to make RAT selection decisions in addition to call admission decisions. In future wireless networks multimedia, traffic will have different QoS requirements. The adaptive-TJCAC scheme fair radio resource allocation, guarantees the QoS requirements of all accepted call and reduces both new call blocking probability and handoff call dropping probability in the heterogeneous wireless networks. It prioritizes handoff calls over new calls by making the handoff call rejection thresholds higher than the new call rejection thresholds. The proposed ATJCAC algorithm makes call admission decisions based on mobile terminal modality (capability), network load, adaptive the bandwidth of ongoing call and radio access technology (RAT) terminal support index. Performance of the adaptive-TJCAC scheme is compared with nonadaptive TJCAC scheme and SJCAC scheme. Simulation results show that the proposed ATJCAC scheme reduces call blocking/dropping probability.

REFERENCES

1. O. E. Falowo and H. A. Chan, "Joint Call Admission Control Algorithms: Requirements, Approaches, and Design Considerations," Computer Communications, Vol. 31, No. 6, 2008, pp. 1200-1217.
2. O. E. Falowo and H. A. Chan, "Joint Call Admission Control for Next Generation Wireless Network," IEEE Conference on Electrical and Computer Engineering, Ottawa, May 2006, pp. 1151-1154.
3. S. Lee, K. Sriram, K. Kim, Y. Kim and N. Golmie, "Vertical Handoff Decision Algorithms for Providing Optimized Performance in Heterogeneous Wireless Networks," IEEE Transactions on Vehicular Technology, Vol. 58, No. 2, 2009, pp. 865-881.
4. S. S. Rappaport, "The Multiple-Call Hand-Off Problem in High-Capacity Cellular Communications Systems," IEEE Transactions on Vehicular Technology, Vol. 40, No. 3, 1991, pp. 546-557.
5. M. Naghshineh and M. Schwartz, "Distributed Call Admission Control in Mobile/Wireless Networks," IEEE Journal on SAC, Vol. 14 No. 4, 1996, pp. 711-717.
6. Y. Fei and V. Krishnamurthy, "Optimal Joint Session Admission Control in Integrated WLAN and CDMA Cellular Networks with Vertical Handoff," IEEE Transactions on Mobile Computing, Vol. 6, No. 1, 2007, pp. 126- 139.
7. X. G. Wang, G. Min and J. E. Mellor, "Adaptive QoS Control in Cellular and WLAN Interworking Networks," Second International Working Conference on the Performance Modeling and Evaluation of Heterogeneous Networks (HET-NETs'04), July 2004.

8. D. Karabudak, C. Hung and B. Bing, "A Call Admission Control Scheme Using Genetic Algorithms," Symposium on Applied Computing (SAC 04), March 2004, pp. 1151- 1158.

9. K. Murray and D. Pesch, "Policy Based Access Management and Handover Control in Heterogeneous Wireless Networks," The IEEE Vehicular Technology Conference, Vol. 5, 2004, pp. 3319-3323.

10. K. Murray and D. Pesch, "Intelligent Network Access and Inter-System Handover Control in Heterogeneous Wireless Networks for Smart Space Environments," 1st International Symposium on Wireless Communication Systems, 2004, pp. 66-70.

11. Q. Deng, D.-L. Xie, B. Hu, Y. Shi and S.-Z. Chen, "Joint Admission Control through Vertical Handoffs in Heterogeneous Wireless Network," Global Mobile Congress (GMC), Shanghai, 18-19 October 2010, pp. 1-5.

12. X. Li, H. Ji, P. Si and L. Zhang, "Joint Session Admission Control Scheme in Integrated WLAN and 3G Networks," 5th International ICST Conference on Communications and Networking in China (CHINACOM 2010), August 2010, pp. 1-5.

13. M. Khabazian, O. Kubbar and H. Hassanein, "Call Admission Control with Resource Reservation for MultiService OFDM Networks," 2012 International Conference on Computing, Networking and Communications (ICNC), Maui, 30 January-2 February 2012, pp. 781-785.

14. Z.-H. Liu and J.-C. Chen, "Design and Analysis of the Gateway Relocation and Admission Control Algorithm in Mobile WiMAX Networks," IEEE Transactions on Mobile Computing, Vol. 11, No. 1, 2012, pp. 5-18.

15. X. Gelabert, J. Perez-Romero, O. Sallent and R. Agustı, "A Markovian Approach to Radio Access Technology Selection in Heterogeneous Multiaccess/Multiservice Wireless Networks," IEEE Transactions on Mobile Computing, Vol. 7, No. 10, 2008, pp. 1257-1270.

16. O. E. Falowo, "Terminal-Modality-Based Joint Call Admission Control Algorithm for Fair Radio Resource Allocation in Heterogeneous Cellular Networks," International Journal of Communications, Network and System Sciences, Vol. 5, No. 7, 2012, pp. 392-404.

17. O. E. Falowo and H. A. Chan, "Joint Call Admission Control Algorithm for Fair Radio Resource Allocation in Heterogeneous Wireless Networks Supporting Heterogeneous Mobile Terminals," 7th Annual IEEE Consumer and Communication & Networking Conference (IEEE CCNC), Las Vegas, 9-12 January 2010, pp. 1-5.

18. W. Zhang, "Performance of Real-Time and Data Traffic in Heterogeneous Overlay Wireless Networks," Proceedings of the 19th International Teletraffic Congress, 29 August-2 September 2005.

19. H. Holma and A. Toskala, "WCDMA for UMTS," 2nd Edition, John Wiley & Sons, New York, 2001.

20. V. Pla, J. M. Gimenez-Guzmany, J. Martinez and V. Casares-Giner, "Optimal Admission Control Using Handover Prediction in Mobile Cellular Networks," Proceedings of the 2nd International Working Conference on Performance Modelling and Evaluation of Heterogeneous Networks (HET-NETs '04), 26-28 July 2004.

21. G. Kesidis, J. Walrand and C.-S. Chang, "Effective Bandwidths for Multi-Class Markov Fluids and Other ATM Sources," IEEE/ACM Transactions on Networking, Vol. 1, No. 4, 1993, pp. 424-428.

22. N. Nasser and H. Hassanein, "Dynamic Threshold-Based Call Admission Framework for Prioritized Multimedia Traffic in Wireless Cellular Networks," Proceedings of the IEEE Global Telecommunications Conference (GLOBECOM'04), 29 November-3 December, 2004, pp. 644- 649.

23. S.-P. Chung and J.-C. Lee, "Performance Analysis and Overflowed Traffic Characterization in Multi-Service Hierarchical Wireless Networks," IEEE Transactions on Wireless Communications, Vol. 4, No. 3, 2005, pp. 904- 918.

Chapter 7

Best Signal Quality in Cellular Networks: Asymptotic Properties and Applications to Mobility Management in Small Cell Networks

Van Minh Nguyen,[1] Franc¸ois Baccelli,[2] Laurent Thomas,[1] and Chung Shue Chen[2]

[1]*Network and Networking Department, Bell Labs, Alcatel-Lucent, 91620 Nozay, France*

[2]*TREC (Research Group on Network Theory and Communications), INRIA-ENS, (´ Ecole Normale Sup´erieure), 75214 Paris, France*

ABSTRACT

The quickly increasing data traffic and the user demand for a full coverage of mobile services anywhere and anytime are leading mobile networking into a future of small cell networks. However, due to the high-density and randomness of small cell networks, there are several technical challenges. In this paper, we investigate two critical issues: *best signal quality* and *mobility management*. Under the assumptions that base stations are uniformly distributed in a ring-shaped region and that shadowings are lognormal, independent, and identically distributed, we prove that when the number of sites in the ring tends to infinity, then (i) the maximum signal strength received at the center of the ring tends in distribution to a Gumbel distribution when properly renormalized, and (ii) it is asymptotically independent of the interference. Using these properties, we derive the distribution of the best signal quality. Furthermore, an optimized random cell scanning scheme is proposed, based on the evaluation of the optimal number of sites to be scanned for maximizing the user data throughput.

1. INTRODUCTION

Mobile cellular networks were initially designed for voice service. Nowadays, broadband multimedia services (e.g., video streaming) and data communications have been introduced into mobile wireless networks. These new applications have led to increasing traffic demand. To enhance network capacity and satisfy user demand of broadband services, it is known that reducing the cell size is one of the most effective approaches [1–4] to improve the spatial reuse of radio resources.

Besides, from the viewpoint of end users, full coverage is particularly desirable. Although today's macro- and micro-cellular systems have provided high service coverage, 100%-coverage is not yet reached because operators often have many constraints when installing large base stations and antennas. This generally results in potential coverage holes and dead zones. A promising architecture to cope with this problem is that of small cell networks [4, 5]. A small cell only needs lightweight antennas. It helps to replace bulky roof top base stations by small boxes set on building facade, on public furniture or indoor. Small cells can even be installed by end users (e.g., femtocells). All these greatly enhance network capacity and facilitate network deployment. Pervasive small cell networks have a great potential. For example, Willcom has deployed small cell systems in Japan [6], and Vodafone has recently launched home 3G femtocell networks in the UK [7].

In principle, high-density and randomness are the two basic characteristics of small cell networks. First, reducing cell size to increase the spatial reuse for supporting dense traffic will induce a large number of cells in the same geographical area. Secondly, end users can set up small cells by their own means [2]. This makes small cell locations and coverage areas more random and unpredictable than traditional mobile cellular networks. The above characteristics have introduced technical challenges that require new studies beyond those for macro- and micro-cellular networks. The main issues concern spectrum sharing and interference mitigation, mobility management, capacity analysis, and network self-organization [3, 4]. Among these, the *signal quality*, for example, in terms of signal-to-interference-plus-noise ratio (SINR), and *mobility management* are two critical issues.

In this paper, we first conduct a detailed study on the properties of *best signal quality* in mobile cellular networks. Here, the best signal quality refers to the maximum SINR received from a number of sites. Connecting the mobile to the best base station is one of the key problems. The best base station here means the base station from which the mobile receives the maximum SINR. As the radio propagation experiences random phenomena such as fading and shadowing, the best signal quality is a random quantity. Investigating its stochastic properties is of primary importance for many studies such as capacity analysis, outage analysis, neighbor cell scanning, and base station association. However, to the best of our knowledge, there is no prior art in this area.

In exploring the properties of best signal quality, we focus on cellular networks in which the propagation attenuation of the radio signal is due to the combination of a distance-dependent path-loss and of lognormal shadowing. Consider a ring B of radii $R_{\min}$ and R_B such that $0 < R_{\min} < R_B < \infty$. The randomness of site locations is modeled by a uniform distribution of homogeneous density in B. Using extreme value theory (c.f., [8, 9]), we prove that the maximum signal strength received at the center of B from n sites in B converges in distribution to a Gumbel distribution when properly renormalized and it is asymptotically independent of the total interference, as $n \to \infty$. The distribution of the best signal quality can thus be derived.

The second part of this paper focuses on applying the above results to mobility support in dense small cell networks. Mobility support allows one to maintain service continuity even when users are moving around while keeping efficient use of radio resources. Today's cellular network standards highlight mobile-assisted handover in which the mobile measures the pilot signal quality of neighbor cells and reports the measurement result to the network. If the signal quality from a neighbor cell is better than that of the serving cell by a handover margin, the network will initiate a handover to that cell. The neighbor measurement by mobiles is called *neighbor cell scanning*. Following mobile cellular technologies, it is known that small cell networking will also use mobile-assisted handover for mobility management.

To conduct cell scanning [10–12], today's cellular networks use a *neighbor cell list*. This list contains information about the pilot signal of selected handover candidates and is sent to mobiles. The mobiles then only need to measure the pilot signal quality of sites included in the neighbor cell list of its serving cell. It is known that the neighbor cell list has a significant impact on the performance of mobility management, and this has been a concern for many years in practical operations [13, 14] as well as in scientific research [15–18]. Using neighbor cell list is not effective for the scanning in small cell networks due to the aforementioned characteristics of high density and randomness.

The present paper proposes an optimized *random cell scanning* for small cell networks. This random cell scanning will simplify the network configuration and operation by avoiding maintaining the conventional neighbor cell list while improving user's quality-of-service (QoS). It is also implementable in wideband technologies such as WiMAX and LTE.

In the following, Section 2 describes the system model. Section 3 derives the asymptotic properties and the distribution of the best signal quality. Section 4 presents the optimized random cell scanning and numerical results. Finally, Section 5 contains some concluding remarks.

2. SYSTEM MODEL

The underlying network is composed of cells covered by base stations with omnidirectional antennas. Each base station is also called a *site*. The set of sites is denoted by $\Omega \subset \mathbb{N}$. We now construct a model for studying the maximum signal strength, interference, and the best signal quality, after specifying essential parameters of the radio propagation and the spatial distribution of sites in the network.

As mentioned in the introduction, the location of a small cell site is often not exactly known even to the operator. The spatial distribution of sites seen by a mobile station will hence be treated as completely random [19] and will be modeled by an homogeneous Poisson point process [20] with intensity λ.

In the following, it is assumed that the downlink pilot signal is sent at constant power at all sites. Let $R_{\min}$ be some strictly positive real number. For any mobile user, it is assumed that the distance to his closest site is at least $R_{\min}$ and hence the path loss is in the far field. So, the signal strength of a site i received by a mobile at a position $\mathbf{y} \in \mathbb{R}^2$ is given by

$$P_i(\mathbf{y}) = A(|\mathbf{y} - \mathbf{x}_i|)^{-\beta} X_i, \quad \text{for } |\mathbf{y} - \mathbf{x}_i| \geq R_{\min}, \tag{1}$$

where $\mathbf{x}_i \in \mathbb{R}^2$ is the location of site i, A represents the base station's transmission power and the characteristics of propagation, β is the path loss exponent (here, we consider $2 < \beta \leq 4$), and the random variables $X_i = 10^{X_i^{dB}/10}$, which represent the lognormal shadowing, are defined from $\{X_i^{dB}, i = 1, 2, \ldots\}$, an independent and identically distributed (i.i.d.) sequence of Gaussian random variables with zero mean and standard deviation σ_{dB}. Typically, σ_{dB} is approximately 8 dB [21, 22]. Here, we consider that fast fading is averaged out as it varies much faster than the handover decision process.

Cells sharing a common frequency band interfere one another. Each cell is assumed allocated no more than one frequency band. Denote the set of all the cells sharing frequency band kth by Ω_k, where $k = 1, \ldots, K$. So $\Omega_k \cap \Omega_{k'} = \varnothing$ for $k \neq k'$, and $\bigcup_{k=1}^{K} \Omega_k = \Omega$. The SINR received at $\mathbf{y} \in \mathbb{R}^2$ from site $i \in \Omega_k$ is expressible as

$$\zeta_i(\mathbf{y}) = \frac{P_i(\mathbf{y})}{N_0 + \sum_{j \neq i, j \in \Omega_k} P_j(\mathbf{y})}, \quad \text{for } i \in \Omega_k, \tag{2}$$

where N_0 is the thermal noise average power which is assumed constant. For notational simplicity, let $A := A / N_0$. Then $\zeta_i(\mathbf{y})$ is given by

$$\zeta_i(\mathbf{y}) = \frac{P_i(\mathbf{y})}{1+\sum_{j\neq i, j\in\Omega_k} P_j(\mathbf{y})}, \quad \text{for } i \in \Omega_k . \tag{3}$$

In the following, we will use (3) instead of (2).

3. BEST SIGNAL QUALITY

In this section, we derive the distribution of the best signal quality. Given a set of sites $S \subset \Omega$, the *best signal quality* received from S at a position $\mathbf{y} \in \mathbb{R}^2$, denoted by $Y_S(\mathbf{y})$, is defined as

$$Y_S(\mathbf{y}) = \max_{i \in S} \zeta_i(\mathbf{y}) . \tag{4}$$

Let us first consider a single-frequency network (i.e., $K = 1$).

Lemma 1.

In the cell set S of single-frequency network, the site which provides a mobile the maximum signal strength will also provide this mobile the best signal quality, namely,

$$Y_S(\mathbf{y}) = \frac{M_S(\mathbf{y})}{1+I(\mathbf{y})-M_S(\mathbf{y})}, \quad \forall\, \mathbf{y} \in \mathbb{R}^2, \tag{5}$$

Where

$$M_S(\mathbf{y}) = \max_{i \in S} P_i(\mathbf{y}) \tag{6}$$

.

is the maximum signal strength received at $\mathbf{y}$ from the cell set S, and

$$I(\mathbf{y}) = \sum_{i \in \Omega} P_i(\mathbf{y}) \tag{7}$$

is the total interference received at $\mathbf{y}$.

Proof.

Since $\zeta_i(\mathbf{y}) = P_i(\mathbf{y}) / \{1 + I(\mathbf{y}) - P_i(\mathbf{y})\}$ and $P_i(\mathbf{y}) < I(\mathbf{y})$, (5) follows from the fact that no matter which cell $i \in \Omega$ is considered, $I(\mathbf{y})$ is the same and from the fact that $x / (c - x)$ with c constant is an increasing function of $x < c$.

Let us now consider the case of multiple-frequency networks. Under the assumption that adjacent-channel interference is negligible compared to cochannel interference, cells of different frequency bands do not interfere one another. Thus, for a given network topology $\mathcal{T}$, the SINRs received from cells of different frequency bands are independent. In the context of a random distribution of sites, the SINRs received from cells of different frequency bands are therefore conditionally independent given $\mathcal{T}$. Write cell set S as

$$S = \bigcup_{k=1}^{K} \{S_k : S_k \subset \Omega_k\}, \tag{8}$$

with S_k the subset of S allocated to frequency k. Let

$$Y_{S_k}(\mathbf{y}) = \max_{i \in S_k} \zeta_i(\mathbf{y}) \tag{9}$$

be the best signal quality received at $\mathbf{y}$ from sites which belong to S_k. The random variables $\{Y_{S_k}(\mathbf{y}), k = 1, \dots, K\}$ are conditionally independent given $\mathcal{T}$. As a result,

$$\mathbf{P}\{Y_S(\mathbf{y}) \le y \mid \mathcal{T}\} = \prod_{k=1}^{K} \mathbf{P}\left\{Y_{S_k}(\mathbf{y}) \le y \mid \mathcal{T}\right\}. \tag{10}$$

Remark 1.

For the coming discussions, we define

$$I_S(\mathbf{y}) = \sum_{i \in S} P_i(\mathbf{y}), \tag{11}$$

which is the interference from cells in set S. In the following, for notational simplicity, the location variable $\mathbf{y}$ appearing in $Y_S(\mathbf{y})$, $M_S(\mathbf{y})$, $I_S(\mathbf{y})$, and $I(\mathbf{y})$ will be omitted in case of no ambiguity. We will simply write Y_S, M_S, I_S, and I. Note that $I_S \le I$ since $S \subset \Omega$.

Following Lemma 1, the distribution of Y_S can be determined by the joint distribution of M_S and I, which is given below.

Corollary 1.

The tail distribution of the best signal quality received from cell set S is given by

$$\overline{F}_{Y_S}(y) = \int_{u=0}^{\infty} \int_{v=u}^{((1+y)/y)u-1} f_{(I,M_S)}(v,u) \mathrm{d}v \, \mathrm{d}u, \tag{12}$$

where $f_{(I,M_S)}$ is the joint probability density of I and M_S.

Proof.

By Lemma 1, we have

$$\begin{aligned} \mathbf{P}\{Y_S \geq y\} &= \mathbf{P}\left\{\frac{M_S}{(1+I-M_S)} \geq y\right\} \\ &= \mathbf{P}\left\{I \leq \frac{1+y}{y} M_S - 1\right\} \\ &= \int_{u=0}^{\infty} \int_{v=u}^{((1+y)/y)u-1} f_{(I,M_S)}(v,u) \mathrm{d}v \, \mathrm{d}u . \end{aligned} \tag{13}$$

In view of Corollary 1, we need to study the properties of the maximum signal strength M_S as well as the joint distribution of M_S and I. As described in the introduction, in dense small cell networks, there could be a large number of neighbor cells and a mobile may thus receive from many sites with strong enough signal strength. This justifies the use of extreme value theory within this context.

For some $R_{\min}$ and R_B such that $0 < R_{\min} < R_B < \infty$, let $B \subset \mathbb{R}^2$ be a ring with inner and outer radii $R_{\min}$ and R_B, respectively. In this section, we will establish the following results.

1. The signal strength P_i received at the center of B belongs to the *maximum domain of attraction* (MDA) of the Gumbel distribution (c.f., Theorem 1 in Section 3.1).
2. The maximum signal strength and the interference received at the center of B from n sites therein are *asymptotically independent* as $n \to \infty$ (c.f., Corollary 3 in Section 3.1).
3. The distribution of the best signal quality is derived (c.f., Theorem 2 in Section 3.3).

3.1. Asymptotic Properties

To begin with, some technical details need to be specified. Given a ring B as previously defined, we will study metrics (such as e.g., signal strength, interference, etc.) as seen at the center of B for a set $S \subset \Omega$ of n sites located in B. We will use the notation M_n, Y_n, and I_n instead of M_S, Y_S, and I_S, respectively, with

$$M_n = \max_{i=1, i \in S}^{n} P_i, \qquad I_n = \sum_{i=1, i \in S}^{n} P_i, \qquad Y_n = \max_{i=1, i \in S}^{n} \zeta_i . \tag{14}$$

Lemma 2.

Assume that $0 < R_{\min} < R_B < \infty$, that sites are uniformly distributed in B, and that the shadowing X_i follows a lognormal distribution of parameters $(0, \sigma_X)$. Then the cdf of the signal strength P_i received at the center of B from a site located in B is given by

$$\begin{aligned} F_P(x) &= c\left\{a^{-2/\beta} G_1(x) - b^{-2/\beta} G_2(x) \right. \\ &\quad \left. - e^{\nu} x^{-2/\beta} G_3(x) + e^{\nu} x^{-2/\beta} G_4(x)\right\}, \end{aligned} \tag{15}$$

where $a = AR_B^{-\beta}$, $b = AR_{\min}^{-\beta}$, $c = A^{2/\beta}(R_B^2 - R_{\min}^2)^{-1}$, $\nu = 2\sigma_X^2/\beta^2$, and G_j, $j = 1, \ldots, 4$, refers to the cdf of a lognormal distribution of parameters (μ_j, σ_X), in which

$$\begin{aligned} \mu_1 &= \log a, \qquad \mu_3 = \mu_1 + \frac{2\sigma_X^2}{\beta}, \\ \mu_2 &= \log b, \qquad \mu_4 = \mu_2 + \frac{2\sigma_X^2}{\beta}. \end{aligned} \tag{16}$$

Proof.

See Appendix .

Under the studied system model, $\{P_i, i = 1, 2, \ldots\}$ are independent and identically distributed (i.i.d.), and so the cdf F_{M_n} and probability density function (pdf) f_{M_n} of M_n are directly obtained as follows.

Corollary 2.

Under the conditions of Lemma 2, the cdf and the pdf of M_n are given, respectively, by

$$F_{M_n}(x) = F_P^n(x), \tag{17}$$

$$f_{M_n}(x) = nf_P(x)F_P^{n-1}(x), \tag{18}$$

where $F_P(x)$ is given by (15), and f_P is the pdf of P_i, $f_P(x) = \mathrm{d}F_P(x) / \mathrm{d}x$.

Since M_n is the maximum of i.i.d. random variables, we can also study its asymptotic properties by extreme value theory. Fisher and Tippett [9, Theorem 3.2.3] proved that under appropriate normalization, if the normalized maximum of i.i.d. random variables tends in distribution to a nondegenerate distribution H, then H must have one of the three known forms: Fréchet, Weibull, or Gumbel distribution. In the following, we prove that P_i belongs to the MDA of a Gumbel distribution. First of all, we establish the following result that is required to identify the limiting distribution of M_n.

Lemma 3.

Under the conditions of Lemma 2, the signal strength received at the center of B from a site located in B has the following tail equivalent distribution:

$$\overline{F}_P(x) \sim K \frac{\exp\left(-(\log x - \mu_2)^2 / \left(2\sigma_X^2\right)\right)}{(\log x - \mu_2)^2 / \left(2\sigma_X^2\right)} \tag{19a}$$

$$\sim K \frac{2\sqrt{2\pi}\,\sigma_X \overline{G}_2(x)}{\log x - \mu_2}, \quad \text{as } x \to \infty, \tag{19b}$$

where $\overline{G}_2(x) = 1 - G_2(x)$, and $K = (\sigma_X / \sqrt{2\pi}\beta)(R_{\min}^2 / (R_B^2 - R_{\min}^2))$.

Proof.

See Appendix .

Equation (19b) shows that the tail distribution of the signal strength P_i is close to that of G_2, although it decreases more rapidly. The fact that G_2 determines the tail behavior of F_P is in fact reasonable, since G_2 is the distribution of the signal strength received from the closest possible neighboring site (with $b = AR_{\min}^{-\beta}$ and σ_X). The main result is given below.

Theorem 1.

Assume that $0 < R_{\min} < R_B < \infty$, that sites are uniformly distributed in B, and that shadowings are i.i.d. and follow a lognormal distribution of parameters $(0, \sigma_X)$ with $0 < \sigma_X < \infty$. Then there exists constants $c_n > 0$ and $d_n \in \mathbb{R}$ such that

$$c_n^{-1}(M_n - d_n) \xrightarrow{d} \Lambda \quad \text{as } n \to \infty, \tag{20}$$

where Λ is the standard Gumbel distribution:

$$\Lambda(x) = \exp\{-e^{-x}\}, \quad x \in \mathbb{R}, \tag{21}$$

and $\xrightarrow{d}$ represents the convergence in distribution. A possible choice of c_n and d_n is

$$c_n = \sigma_X(2\log n)^{-1/2} d_n,$$
$$d_n = \exp\left\{\mu_2 + \sigma_X\left(\sqrt{2\log n} + \frac{-\log\log n + \log\kappa}{\sqrt{2\log n}}\right)\right\}, \tag{22}$$

with κ given by Lemma 3.

Proof.

See Appendix .

By Theorem 1, the signal strength belongs to the MDA of the Gumbel distribution, denoted by $\mathrm{MDA}(\Lambda)$. From [23, 24], we have the following corollary of Theorem 1.

Corollary 3.

Let σ_P^2 be the variance and μ_P be the mean of signal strength P_i. Let $\tilde{I}_n = (I_n - n\mu_P)/(\sqrt{n}\sigma_P)$. Let $\tilde{M}_n = (M_n - d_n)/c_n$, where c_n and d_n are given by (22). Under the conditions of Theorem 1,

$$(\tilde{M}_n, \tilde{I}_n) \xrightarrow{d} (\Lambda, \Phi) \quad \text{as } n \to \infty, \tag{23}$$

where Λ is the Gumbel distribution and Φ the standard Gaussian distribution, and where the coordinates are independent.

Proof.

Conditions $0 < R_{\min}$ and $\sigma_X < \infty$ provide $\sigma_P^2 \leq \mathrm{var}\{AR_{\min}^{-\beta} X_i\} < \infty$. Then the result follows by Theorem 1 and [23, 24].

Note that the total interference I can be written as $I = I_n + I_n^c$ where I_n^c denotes the complement of I_n in I. Under the assumptions that the locations of sites are independent and that shadowings are also independent, I_n and I_n^c are independent. The asymptotic independence between M_n and I_n thus induces the asymptotic independence between M_n and I. This observation is stated in the following corollary.

Corollary 4.

Under the conditions of Theorem 1, M_n and I are asymptotically independent as $n \to \infty$.

This asymptotic independence facilitates a wide range of studies involving the total interference and the maximum signal strength. This result will be used in the coming sub-section to derive the distribution of the best signal quality.

Remark 2.

The asymptotic properties given by Theorem 1 and Corollaries 3 and 4 hold when the number of sites in *a bounded area* tends to infinity. This corresponds to a network densification process in which more sites are deployed in a given geographical area in order to satisfy the need for capacity, which is precisely the small cell setting.

3.2. Convergence Speed of Asymptotic Limits

Theorem 1 and Corollaries 3 and 4 provide asymptotic properties when $n \to \infty$. In practice, n is the number of cells to be scanned, and so it can only take moderate values. Thus, it is important to evaluate the convergence speed of (20) and (23). We will do this based on simulations and will measure the discrepancy using a symmetrized version of the Kullback-Leibler divergence (the so-called Jensen-Shannon divergence (JSdiv)).

Let us start with some numerical evaluations of the convergence of $\tilde{M}_n$ to its limiting distribution. Figure 1(a) shows $F_{\tilde{M}_n}$ for different n and compares to empirical simulation results. As expected the analytical distributions obtained by (17) of Corollary 2 match with the empirical distributions for all n. Figure 1(b) plots the analytical distribution and its limiting distribution, that is, the Gumbel distribution Λ. There is a discrepancy in the negative regime (see the circled region in Figure 1(b)). It is worth pointing out that as a maximum of signal strengths, $M_n \geq 0$ and thus $\tilde{M}_n \geq -d_n / c_n = -\sqrt{2\log n} / \sigma_X$ since $\tilde{M}_n = (M_n - d_n) / c_n$. This means that $F_{\tilde{M}_n}(x) = 0$, $\forall x \leq -\sqrt{2\log n} / \sigma_X$, whereas $\Lambda(x) > 0$, $\forall x > -\infty$. This explains the gap observed for small n. This dissimilarity should have limited impact as long as we only deal with positive values of M_n (resp., $\tilde{M}_n \geq -\sqrt{2\log n} / \sigma_X$).

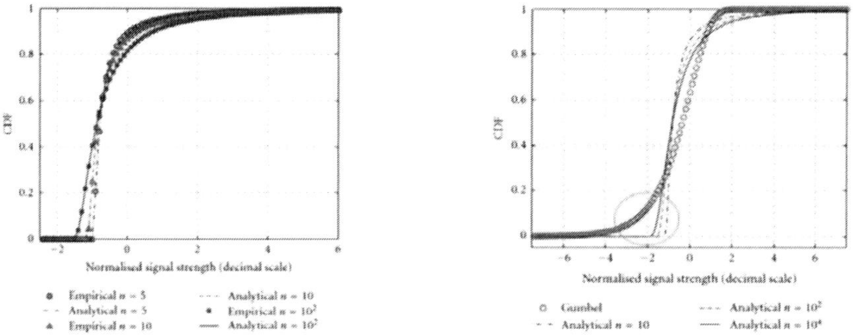

Figure 1.CDF of $\tilde{M}_n$ under different n: $\sigma_{\text{dB}} = 8, \beta = 3$. Comparison between analytical and empirical CDFsComparison between analytical and limiting CDFs

We now study the symmetrized divergence between the analytical and limiting distributions of $\tilde{M}_n$for some moderate values of $n$and under different $\sigma_{\text{dB}}$and β. The convergence is best for σ_{dB}around 10 dB and βaround two to four. For practical systems, $\sigma_{\text{dB}}$is approx. 8 dB and $2 < \beta \leq 4$. We compute the Jensen-Shannon divergence for $\beta = 3$and $\sigma_{\text{dB}} = 8$and plot the results in Figures 2(a) and 2(b), respectively. For these (and other) values (within the range given above) of $\sigma_{\text{dB}}$and β, $\tilde{M}_n$, and Λhave low divergence.

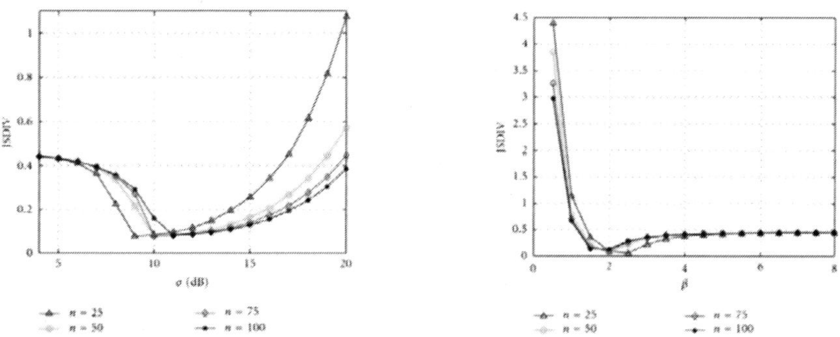

Figure 2 . Jensen-Shannon divergence between $\tilde{M}_n$and Λ. Under different σ_{dB}: $\beta = 3$Under different β: $\sigma_{\text{dB}} = 8$

Let us now measure the (dis)similarity between the empirical joint distribution, $\mathbf{P}\{\tilde{M}_n \leq u, \tilde{I}_n \leq v\}$, and the product of the empirical marginal distributions, $\mathbf{P}\{\tilde{M}_n \leq u\} \times \mathbf{P}\{\tilde{I}_n \leq v\}$. Figure 3 shows an example with $n = 50, \beta = 3$, and $\sigma_{\text{dB}} = 8$. We see that these two density functions are very similar. Figure 4 compares these two density functions for different values of σ_{dB} and β. Within the range defined above, the divergence between the two distributions is again small. Comparing Figure 2 and Figure 4, one can conclude that even if the convergence of $\tilde{M}_n \xrightarrow{d} \Lambda$ remains slow, $\tilde{M}_n$ and $\tilde{I}_n$ quickly become uncorrelated. Thus, the independence between M_n and I_n holds for moderate values of n, that is,

$$f_{(M_n, I_n)}(u, v) \approx f_{M_n}(u) \times f_{I_n}(v), \tag{24}$$

and so the same conclusion holds for the independence between M_n and I.

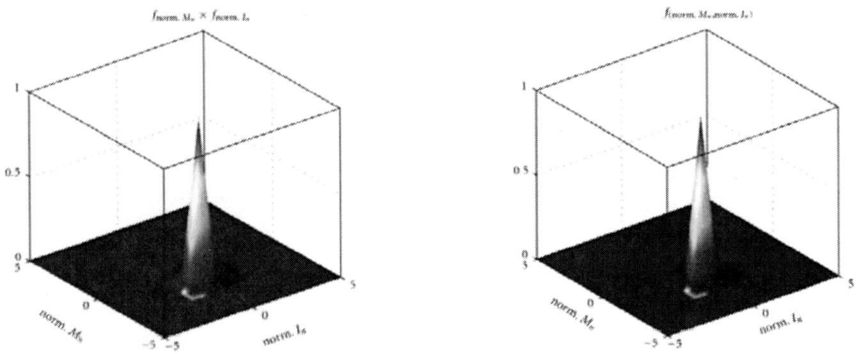

Figure 3 Example of joint densities of $\tilde{M}_n$ and $\tilde{I}_n$: $n = 50$, $\sigma_{dB} = 8$, $\beta = 3$. Here, norm. M_n refers to $\tilde{M}_n$, while norm. I_n refers to $\tilde{I}_n$.

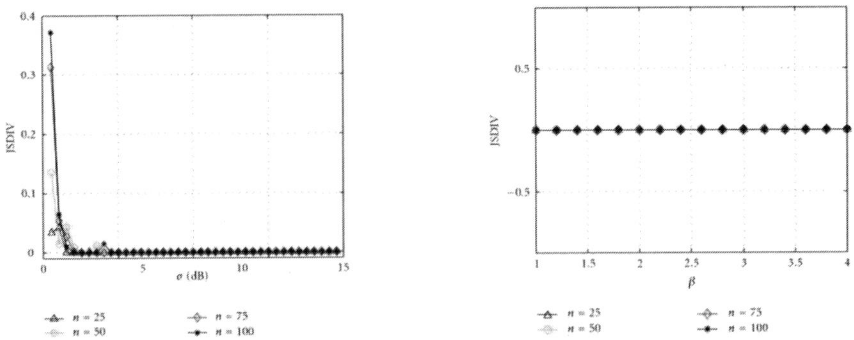

Figure 4 Jensen-Shannon divergence between $f_{(\tilde{M}_n, \tilde{I}_n)}$ and $f_{\tilde{M}_n} \times f_{\tilde{I}_n}$. Under different σ_{dB}: $\beta = 3$ Under different β: $\sigma_{dB} = 8$

3.3. Distribution of the Best Signal Quality

From the above results, we have the distribution of M_n and the asymptotic independence between M_n and I. In order to derive the distribution of the best signal quality, we also need the distribution of the total interference.

Lemma 4.

Assume that shadowings are i.i.d. and follow a lognormal distribution of parameters $(0, \sigma_X)$, $\beta > 2$, and that sites are distributed according to a Poisson point process with intensity λ outside the disk of radius $R_{min} > 0$. Let I be the interference received at the disk center, and ϕ_I be the characteristic function of I. Then:

$$\phi_I(w) = \exp\left\{-\pi\lambda\alpha(A|w|)^{\alpha}\int_0^{A|w|/R_{min}^{\beta}} \frac{1-\phi_X(\text{sign}(w)t)}{t^{\alpha+1}}dt\right\}, \tag{25}$$

where $\alpha = 2/\beta$, and ϕ_X is the characteristic function of X_i;

$|\phi_I(w)|^p \in \mathbf{L}$ for all $p = 1, 2, \dots$, where $\mathbf{L}$ is the space of absolutely integrable functions;

If $AR_{min}^{-\beta}$ is large, then ϕ_I admits the following approximation:

$$\phi_I(w) \approx \exp\left(-\delta|w|^{\alpha}\left[1 - j\,\text{sign}(w)\tan\left(\frac{\pi\alpha}{2}\right)\right]\right), \tag{26}$$

where $\delta = \pi\lambda A^{\alpha}\exp((1/2)\alpha^2\sigma_X^2)\Gamma(1-\alpha)\cos(\pi\alpha/2)$, with $\Gamma(\cdot)$ denoting the gamma function.

Proof.

See Appendix .

Theorem 2.

Under the assumptions of Lemma 4, let B be the ring of inner and outer radii R_{min} and R_B, respectively. Denote the best signal quality received at the center of B from n sites in B by Y_n. Assume that $0 < \sigma_X < \infty$, $0 < R_{min} < R_B < \infty$, that $AR_{min}^{-\beta}$ is large, and that $\pi\lambda(R_B^2 - R_{min}^2) > n$, with high probability, where n is some positive integer. Then the tail distribution of Y_n can be approximated by

$$\overline{F}_{Y_n}(y) \approx \int_{y}^{\infty}\left\{f_{M_n}(u)\int_{0}^{\infty}\frac{2}{\pi w}e^{-\delta w^{\alpha}}\sin\left(w\frac{u-y}{2y}\right)\right.$$
$$\left.\times\cos\left(wu+w\frac{u-y}{2y}-\delta w^{\alpha}\tan\frac{\pi\alpha}{2}\right)dw\right\}du\,. \tag{27}$$

Proof.

See Appendix .

The approximation proposed in Theorem 2 will be used in Section 4 below. It will be validated by simulation in the context considered there.

4. RANDOM CELL SCANNING FOR DATA SERVICES

In this section, the theoretical results developed in Section 3 are applied to random cell scanning.

4.1. Random Cell Scanning

Wideband technologies such as WiMAX, WCDMA, and LTE use a predefined set of codes for the identification of cells at the air interface. For example, 114 pseudonoise sequences are used in WiMAX [25], while 504 physical cell identifiers are used in LTE [26]. When the mobile knows the identification code of a cell, it can synchronize over the air interface and then measure the pilot signal quality of the cell. Therefore, by using a predefined set of codes, these wideband technologies can have more autonomous cell measurement conducted by the mobile. In this paper, this identification code is referred to as cell synchronization identifier (CSID).

In a dense small cell network where a large number of cells are deployed in the same geographical area, the mobile can scan any cell as long as the set of CSIDs used in the network is provided. This capability motivates us to propose *random cell scanning* which is easy to implement and has only very few operation requirements. The scheme is detailed below.

1. When a mobile gets admitted to the network, its (first) serving cell provides him/her the whole set of CSIDs used in the network. The mobile then keeps this information in its memory.
2. To find a handover target, the mobile randomly selects a set of nCSIDs from its memory and conducts the standardized scanning procedure of the underlying cellular technology, for example, scanning specified in IEEE 802.16 [25], or neighbor measurement procedure specified in 3G [27] and LTE [12].

3. The mobile finally selects the cell with the best received signal quality as the handover target.

In the following, we determine the number of cells to be scanned which maximizes the data throughput.

4.2. Problem Formulation

The optimization problem has to take into account the two contrary effects due to the number of cells to be scanned. On one hand, the larger the set of scanned cells, the better the signal quality of the chosen site, and hence the larger the data throughput obtained by the mobile. On the other hand, scanning can have a linear cost in the number of scanned cells, which is detrimental to the throughput obtained by the mobile.

Let us quantify this using the tools of the previous sections.

Let W be the average cell bandwidth available per mobile and assume that it is a constant. Under the assumption of additive white Gaussian noise, the maximum capacity ξ_n that the mobile can have by selecting the best among n randomly scanned cells is

$$\xi_n = W \log(1 + Y_n). \tag{28}$$

Hence

$$\begin{aligned} \mathbf{E}\{\xi_n\} &= W\mathbf{E}\{\log(1+Y_n)\} \\ &= W\int_{y=0}^{\infty} \log(1+y) f_{Y_n}(y)dy, \end{aligned} \tag{29}$$

where f_{Y_n} is the pdf of Y_n. By an integration by parts of $\log(1+y)$ and $f_{Y_n}(y)dy = -d\overline{F}_{Y_n}(y)$, this becomes

$$\mathbf{E}\{\xi_n\} = W\int_{y=0}^{\infty} \frac{\overline{F}_{Y_n}(y)}{1+y} dy. \tag{30}$$

Note that $\mathbf{E}\{\xi_n\}$ is the expected throughput from the best cell. Since Y_n is the maximum signal quality of the n cells, Y_n increases with n and so does ξ_n. Hence, the mobile should scan as many cells as possible. However, on the other hand, if scanning many cells, the mobile will consume much time in scanning and thus have less time for data transmission with the serving cell. A typical situation is that where the scanning time increases proportionally with the number of cells scanned and where the data transmission is suspended. This for instance happens if the underlying cellular technology uses a *compressed mode* scanning, like for example, in IEEE 802.16 [25] and also inter-frequency cell measurements defined by 3GPP [12, 27]. In this mode, scanning intervals, where the mobile temporarily suspends data transmission for scanning neighbor cells, are interleaved with intervals where data transmission with the serving cell is resumed.

Another scenario is that of *parallel scanning-transmission*: here scanning can be performed in parallel to data transmission so that no transmission gap occurs; this is the case in, for example, intrafrequency cell measurements in WCDMA [27] and LTE [12].

Let T be the average time during which the mobile stays in the tagged cell and receives data from it. Let s be the time needed to scan one cell (e.g., in WCDMA, the mobile needs $s = 25$ ms if the cell is in the neighbor cell list and $s = 800$ ms if not [28], whereas in WiMAX, $s = 10$ ms, i.e., two 5-ms frames). Let $L(n)$ be the duration of the suspension of data transmission due to the scanning of the n cells:

$$L(n) = \begin{cases} s \times n & \text{if } \textit{compressed mode} \text{ is used,} \\ 0 & \text{if } \textit{parallel scan.-trans}\text{. is enabled.} \end{cases} \tag{31}$$

Finally, let $\mathbf{E}\{\xi_0\}$ be the average throughput received from the serving cell when no scanning at all is performed (this would be the case if the mobile would pick as serving site one of the sites of set S at random).

The gain of scanning n cells can be quantified by the following metric, that we will call the *acceleration*

$$\begin{aligned} \rho_n &\triangleq \frac{T \cdot \mathbf{E}\{\xi_n\}}{T \cdot \mathbf{E}\{\xi_0\} + L(n) \cdot \mathbf{E}\{\xi_0\}} \\ &= \frac{T}{T + L(n)} \times \frac{\mathbf{E}\{\xi_n\}}{\mathbf{E}\{\xi_0\}}. \end{aligned} \tag{32}$$

In this definition, $T \cdot \mathbf{E}\{\xi_n\}$ (resp., $T \cdot \mathbf{E}\{\xi_0\} + L(n) \cdot \mathbf{E}\{\xi_0\}$) is the expected amount of data transmitted when scanning n cells (resp., doing no scanning at all). We aim at finding the value of n that maximizes the acceleration ρ_n.

It is clear that $T/(T + L(n)) = 1$ when (i) $T \to \infty$, that is, the mobile stays in and receives data from the tagged cell forever, or (ii) $L(n) = 0$, that is, parallel scanning-transmission is enabled. In these cases, ρ_n increases with n and the mobile "should" scan as many cells as possible. However, ρ_n is often concave and the reward of scanning then decreases. To characterize this, we introduce a *growth factor* g defined as follows:

$$g_n \triangleq \frac{\rho_n}{\rho_{n-1}} = \frac{T+L(n-1)}{T+L(n)} \times \frac{\mathbf{E}\{\xi_n\}}{\mathbf{E}\{\xi_{n-1}\}}. \tag{33}$$

Special cases as those considered above can be cast within a general framework which consists in finding the value of n that maximizes ρ_n under the constraint that $g_n \geq 1+\Delta_g$, where $\Delta_g > 0$ is a threshold.

4.3. Numerical Result

In the following, we show how to apply the above results to find the optimal n. We adopt WCDMA as the underlying cellular network technology. 100 omnidirectional small cell base stations are deployed in a square domain of $1\,\text{km} \times 1\,\text{km}$. The network density is thus equal to

$$\lambda = 10^{-4} \text{ small base stations} / \text{m}^2. \tag{34}$$

It is assumed that any cell synchronization identifier can be found in a radius $R_B = 1\,\text{km}$. We take R_{min} equal to 2 meters. The propagation path loss is modeled by the picocell path loss model [29]:

$$PL_{[\text{dB}]}(d) = 37 + 30\log_{10}(d) + 18.3f^{((f+2)/(f+1)-0.46)}, \tag{35}$$

where d is the distance from the base station in meters, f the number of penetrated floors in the propagation path. For indoor office environments, $f = 4$ is the default value [22]; however, here, the small cell network is assumed to be deployed in a general domain including outdoor urban areas where there are less penetrated walls and floors. So, we use $f = 3$ in our numerical study.

It is assumed that the total transmission power including the antenna gain of each small cell base station is $P_{\text{Tx},[\text{dBm}]} = 32\,\text{dBm}$. Shadowing is modeled as a random variable with lognormal distribution with an underlying Gaussian distribution of zero mean and $8\,\text{dB}$ standard deviation. The signal strength received at any distance d from a base station i is expressible as

$$P_{i,[\mathrm{dBm}]}(d) = P_{\mathrm{Tx},[\mathrm{dBm}]} - PL_{[\mathrm{dB}]}(d) + X_i^{\mathrm{dB}}. \tag{36}$$

By (35), we have

$$\begin{aligned} P_{i,[\mathrm{dBm}]}(d) &= P_{\mathrm{Tx},[\mathrm{dBm}]} - 37 - 18.3f^{((f+2)/(f+1)-0.46)} \\ &\quad -30\log_{10}(d) + X_i^{\mathrm{dB}}. \end{aligned} \tag{37}$$

The parameters A and β appearing in $P_i = Ad^{-\beta}X_i$ can be identified from (37) after converting the received signal strength from the dBm scale to the linear scale:

$$\beta = 3, \qquad A_{[\mathrm{mW}]} = 10^{(P_{\mathrm{Tx},[\mathrm{dBm}]} - 37 - 18.3f^{((f+2)/(f+1)-0.46)})/10}. \tag{38}$$\

The received noise power N_0 is given by\

$$N_{0,[\mathrm{mW}]} = k_B T_{\mathrm{Kelvin}} \times NF_{[\mathrm{W}]} \times W_{[\mathrm{Hz}]} \times 10^3, \tag{39}$$

where the effective bandwidth $W_{[\mathrm{Hz}]} = 3.84 \times 10^6$ Hz, k_B is the Boltzmann constant, and T_{Kelvin} is the temperature in Kelvin, $k_B T_{\mathrm{Kelvin}} = 1.3804 \times 10^{-23} \times 290$ W/Hz and $NF_{[\mathrm{dB}]}$ is equal to 7 dB.

It is assumed that the mobile is capable of scanning eight identified cells within 200 ms [28]. So, the average time needed to scan one cell is given by $s = 25\,\mathrm{ms}$.

In order to check the accuracy of the approximations used in the analysis, a simulation was built with the above parameter setting. The interference field was generated according to a Poisson point process of intensity λ in a region between $R_{\min}$ and $R_{\infty} = 100\,\mathrm{km}$. For a number n, the maximum of SINR received from n base stations which are randomly selected from the disk B between radii $R_{\min}$ and R_B was computed. After that the expectation of the maximal capacity $\mathbf{E}\{\xi_n\}$ received from the n selected BSs was evaluated.

In Figure 5(a), the expectation of the maximal throughput $\mathbf{E}\{\xi_n\}$ for different n is plotted, as obtained through the analytical model and simulation. The agreement between model and simulation is quite evident. As shown in Figure 5(a), $\mathbf{E}\{\xi_n\}$ increases with n, though the increasing rate slows down as

n increases. Note that in Figure 5(a), $\mathbf{E}\{\xi_n\}$ is plotted after normalization by $\mathbf{E}\{\xi_{250}\}$.

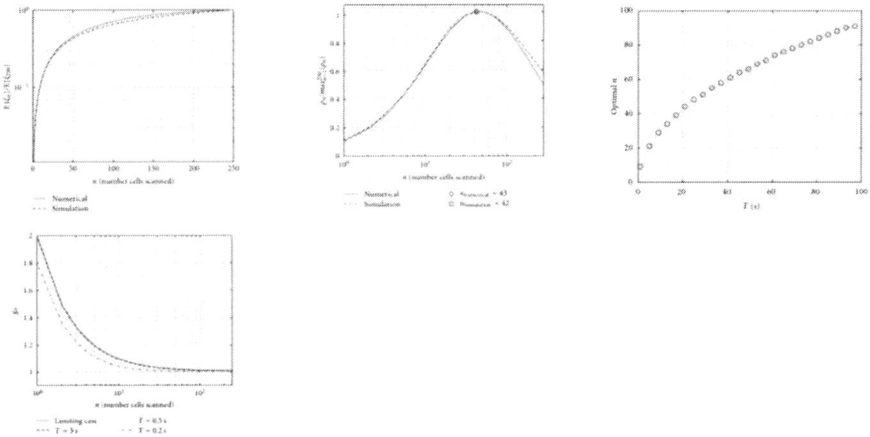

Figure 5 Numerical results in the random cell scanning optimization. Plot of $\mathbf{E}\{\xi_n\} / \max_k \mathbf{E}\{\xi_k\}$ Plot of $\rho_n / \max_k \rho_k$, $T = 0.5$ secondOptimal number of cells to be scannedGrowth factor g_n under different T

Figure 5(b) gives an example of acceleration ρ_n for $T = 0.5$ second and $L(n) = n \times 25\,\text{ms}$. In the plot, ρ_n is normalized by its maximum. Here, an agreement between model and simulation is also obtained. We see that ρ_n first increases rapidly with n, attains its maximum at $n = 42$ by simulation and $n = 43$ by model, and then decays.

Next, using the model we compute the optimal number of cells to be scanned and the growth factor g_n for different T. Note that in (32), the factor $T / (T + L(n))$ can be rewritten as

$$\frac{T}{T+L(n)} = \frac{1}{1+n \times s / T} \quad \text{for} \begin{cases} T < \infty, \\ L(n) = n \times s\,. \end{cases} \tag{40}$$

It is clear that this factor also depends on the ratio T/s. Figure 5(c) plots the optimal n for different values of T/s. Larger T/s will drive the optimal n towards larger values. Since T can be roughly estimated as the mobile residence time in a cell, which is proportional to the cell diameter divided by the user speed, this can be rephrased by stating that the faster the mobile, the smaller T and thus the fewer cells the mobile should scan.

Finally, Figure 5(d) plots the growth factor g_n with different T. In Figure 5(d), the "limiting case" corresponds to the case when $T \to \infty$ or $L(n) = 0$. We see that

g_n is quite stable w.r.t. the variation of T. Besides, g_n flattens out at about 30 cells for a wide range of T. Therefore, in practice this value can be taken as a recommended number of cells to be scanned in the system.

5. CONCLUDING REMARKS

In this paper, we firstly develop asymptotic properties of the signal strength in cellular networks. We have shown that the signal strength received at the center of a ring shaped domain B from a base station located in B belongs to the maximum domain of attraction of a Gumbel distribution. Moreover, the maximum signal strength and the interference received from n cells in B are asymptotically independent as $n \to \infty$. The above properties are proved under the assumption that sites are uniformly distributed in B and that shadowing is lognormal. Secondly, the distribution of the best signal quality is derived. These results are then used to optimize scanning in small cell networks. We determine the number of cells to be scanned for maximizing the mean user throughput within this setting.

REFERENCES

1. Lee WCY: Smaller cells for greater performance. IEEE Communications Magazine 1991, 29(11):19-23. 10.1109/35.109660
2. Claussen H, Ho LTW, Samuel LG: Financial analysis of a pico-cellular home network deployment. Proceedings of the IEEE International Conference on Communications, June 2007 5604-5609.
3. Chandrasekhar V, Andrews JG, Gatherer A: Femtocell networks: a survey. IEEE Communications Magazine 2008, 46(9):59-67.
4. Saunders S, Carlaw S, Giustina A, Bhat RR, Rao VS, Siegberg R: Femtocells: Opportunities and Challenges for Business and Technology. John Wiley & Sons, New York, NY, USA; 2009.
5. Urie A: Keynote: the future of mobile networking will be small cells. Proceedings of the IEEE International Workshop on Indoor and Outdoor Femto Cells (IOFC '09), September 2009, Tokyo, Japan
6. Chika Y: Keynote: true BWA—eXtended Global Platform. Proceedings of the IEEE International Workshop on Indoor and Outdoor Femto Cells (IOFC '09), September 2009, Tokyo, Japan
7. Judge P: Vodafone Launches Home 3G Femtocell in the UK. eWeekEurope, London, UK; 2009.

8. Leadbetter MR, Lindgren G, Rootzén H: Extremes and Related Properties of Random Sequences and Processes. Springer, New York, NY, USA; 1983.

9. Embrechts P, Klüppelberg C, Mikosch T: Modelling Extremal Events for Insurance and Finance. Springer, New York, NY, USA; 1997.

10. Nawrocki M, Aghvami H, Dohler M: Understanding UMTS Radio Network Modelling, Planning and Automated Optimisation: Theory and Practice. John Wiley & Sons, New York, NY, USA; 2006.

11. WiMAX Forum : Mobile System Profile. Approved Spec. Release 1.0, Revision 1.4.0, May 2007

12. 3GPP TS 36.331 : Evolved Universal Terrestrial Radio Access (EUTRA) Radio Resource Control (RRC): protocol Specification (Release 8). Tech. spec. v8.8.0, December 2009

13. NGMN Alliance : Next Generation Mobile Networks Use cases related to self-organising network, Overall description. December 2008., (v2.02):

14. NGMN Alliance : Next Generation Mobile Networks Recommendation on SON and O&M requirements. Req. Spec. v1.23, December 2008

15. Magnusson S, Olofsson H: Dynamic neighbor cell list planning in a micro cellular network. Proceedings of the Annual International Conference on Universal Personal Communications Record, October 1997 1: 223-227.

16. Guerzoni R, Ore I, Valkealahti K, Soldani D: Automatic neighbor cell list optimization for UTRA FDD networks: theoretical approach and experimental validation. Proceedings of the Western Province Motor Club (WPMC '05), 2005, Aalborg, Denmark

17. Soldani D, Ore I: Self-optimizing neighbor cell list for UTRA FDD networks using detected set reporting. Proceedings of the 65th IEEE Vehicular Technology Conference, 2007 694-698.

18. Amirijoo M, Frenger P, Gunnarsson F, Kallin H, Moe J, Zetterberg K: Neighbor cell relation list and measured cell identity management in LTE. Proceedings of the IEEE Network Operations and Management Symposium, April 2008 152-159.

19. Win MZ, Pinto PC, Shepp LA: A mathematical theory of network interference and its applications. Proceedings of the IEEE 2009, 97(2):205-230.

20. Baccelli F, Błaszczyszyn B: Stochastic geometry and wireless networks volume 1: theory. Foundations and Trends in Networking 2009, 3(3-4):249-449. 10.1561/1300000006

21. 3GPP TR 36.942 : Evolved Universal Terrestrial Radio Access (EUTRA): Radio Frequency (RF) system scenarios (Release 8). May 2009., (v8.2.0):

22. WiMAX Forum : WiMAX systems evaluation methodology. Spec. v2.1, July 2008

23. Chow T, Teugels J: The sum and the maximum of i.i.d. random variables. Proceedings of the 2nd Prague Symposium on Asymptotic Statistics, 1978 81-92.

24. Anderson CW, Turkman KF: The joint limiting distribution of sums and maxima of stationary sequences. Journal of Applied Probability 1991, 28(1):33-44. 10.2307/3214738

25. IEEE 802.16 : Air Interface for Broadband Wireless Access Systems. IEEE, Standard Std 802.16-2009, May 2009

26. 3GPP TS 36.300 : Evolved Universal Terrestrial Radio Access (EUTRA) and Evolved Universal Terrestrial Radio Access Network (EUTRAN)—Overall description: stage 2 (Release 8). Tech. Spec. v8.11.0, December 2009

27. 3GPP TS 25.331 : Radio Resource Control (RRC): protocol Specification (Release 8). Tech. Spec. v8.6.0, March 2009

28. 3GPP TS 25.133 : Requirements for support of Radio Resource Management FDD (Release 8). Tech. Spec. v8.9.0, Dec. 2009

29. ETSI TR 101.112 : Selection procedures for the choice of radio transmission technologies of the UMTS. April 1998., (v3.2.0):

30. Abramowitz M, Stegun IA: Handbook of Mathematical Functions. Dover, New York, NY, USA; 1965.

31. Takahashi R: Normalizing constants of a distribution which belongs to the domain of attraction of the Gumbel distribution. Statistics and Probability Letters 1987, 5(3):197-200. 10.1016/0167-7152(87)90039-3

32. Feller W: An Introduction to Probability Theory and Its Applications. Volume 2. 2nd edition. John Wiley & Sons, New York, NY, USA; 1971.

Chapter 8

Energy Efficiency Analysis of Relay-Assisted Cellular Networks

Huan Yu[1,2], Yunzhou Li[2*], Marios Kountouris[3], Xibin Xu[2] and Jing Wang[2]

[1]*Department of Electronic Engineering, Tsinghua University, Beijing 100084, China.*

[2]*Research Institute of Information Technology, State Key Laboratory on Microwave and Digital Communications, Tsinghua National Laboratory for Information Science and Technology(TNList), Tsinghua University, 100084*

Beijing, China.

[3]*SUPELEC, Department of Telecommunications, 3 rue Joliot-Curie, Gif-sur-Yvette 91192, France.*

ABSTRACT

To meet the demand for higher throughput, improved coverage and enhanced reliability, future wireless cellular networks face significant technical challenges. One promising solution is to place relay stations between transmitters and receivers in the cellular network. Meanwhile, as energy consumption reduction has been an important concern for the wireless industry, energy-efficient communications is of prime interest for future networks. In this paper, we study whether and how relays can improve the energy efficiency of cellular networks. Specifically, the energy efficiency of relay-assisted cellular networks is analyzed using tools of stochastic geometry. We first derive the coverage probability for the macro base station (MBS) to user (UE), the MBS to relay station (RS), and the RS to UE links, and then we model the power consumption at the MBS and RS. Based on the analytical model and expressions, the energy efficiency of relay-assisted cellular networks is then evaluated and is shown to be strictly quasi-concave on the transmit power for MBS to UE link or the RS to UE link. Numerical results show that the energy efficiency first improves while it hits a ceiling as the MBS density increases.

1 INTRODUCTION

With the exponential growth of wireless traffic, future cellular networks face huge challenges for catering higher data rate and transmission reliability. One effective and promising way to meet the traffic demands is to deploy relay stations (RSs) in the traditional cellular networks, as a means to increase coverage, throughput, and reliability [1]. Meanwhile, as the energy consumption reduction has become a global trend, the research for energy-efficient wireless networks has increasingly attracted attention recently [2].

There have been some works in the literature about energy-efficient relay-assisted networks from the link layer to the network layer [3–8]. For the link layer, the energy efficiency problem for multi-carrier amplify-and-forward (AF) relay links is considered in [3]. The optimization problem cast in the paper is non-convex and is converted into a quasi-concave problem by two approximate methods, thus providing a suboptimal solution. In [4–6], energy-efficient two-way relay links under different forwarding schemes are studied. The energy efficiency for one-way relay transmission and two-way relay transmission is compared when the AF strategy is adopted in [4], showing that when the bidirectional data amounts are equal, two-way relay transmission performs better; otherwise one-way relay transmission may offer higher energy efficiency. In [5], hybrid relay transmission transmitting partial messages with one-way relay transmission and the remaining messages with two-way relay transmission is proposed. Simulation results show that the performance of hybrid relay transmission is better than that of the two-way relay transmission and one-way relay transmission. Energy efficiency of the two-way relay link using decode-and-forward (DF) strategy is investigated in [6] and an improved bisection algorithm is proposed to find the optimal energy efficiency. For the network layer, in [7, 8], energy-efficient multi-relay networks are studied, in which the source node sends massage to the destination node via the relay nodes and the best one or several relay nodes by a certain criterion are selected to forward the message from source node in order to achieve the highest energy efficiency.

In previous studies on the performance of traditional cellular networks, the cell shape is often supposed to be hexagonal or square, which may not be the case in realistic scenarios (e.g., due to shadowing) or in heterogeneous and uncoordinated networks. Besides, complex time-consuming system-level simulations must be used. In [9], an approach to derive the coverage and rate in cellular networks based on stochastic geometry is proposed, which is more tractable compared to the traditional grid models. Under this approach, in [10], the design of energy-efficient heterogeneous cellular networks through the employment of base station sleeping mode strategies as well as small cells are investigated, in which some methods can be considered in the relay-assisted cellular networks. The relationship among the energy efficiency and the intensity of non-cooperative users (UEs) and cooperative UEs is studied in [11] based on the stochastic geometry approach. In [12], the effect of base station density on the energy efficiency of relay-assisted cellular networks is investigated using the stochastic geometry model. However, the paper only considers the special case where the path loss exponent α=4; while in our work, we derive expressions that are valid for any α>2.

In our work, we analyze the energy efficiency for relay-assisted cellular networks using models based on stochastic geometry [9], where the macro base stations (MBSs) and relay stations(RSs) locations are distributed according to independent homogeneous Poisson point process (HPPP). The mobile users (UEs) in the network are divided into two types, i.e., UEs that communicate directly with MBSs (M-UEs) and UEs that communicate with MBSs via the help of RSs (R-UEs). The main contributions of this paper are summarized as follows:

- Building on [9], a downlink relay-assisted cellular network model is proposed based on the stochastic geometry theory.
- The coverage probabilities for the MBS-RS, the MBS-UE, and the RS-UE links are derived, and the power consumption for MBSs as well as RSs are modeled.
- The energy efficiency expression for relay-assisted cellular networks is analytically deduced under specific power consumption models.
- The energy efficiency is proved to be strictly quasi-concave on the transmit power for MBS to UE link or RS to UE link, and an algorithm based on bisection method is proposed to find the optimal value.

The rest of this paper is organized as follows: in Section 2, the system model is described. In Section 3, the energy efficiency for the downlink relay-assisted cellular networks is derived. Simulation results are shown in Section 4, and conclusions are given in Section 5.

2 SYSTEM MODEL

We consider a relay-assisted cellular network where the MBSs and RSs are located according to independent HPPPs Θ_M and Θ_R with densities λ_M and λ_R, respectively, in the Euclidean plane. Half-duplex relay nodes using DF strategy are considered, where each RS connects to the geographically closest MBS and has coverage of a disk of radius R, i.e., the RSs deployed in the Voronoi cell of an MBS are connected with it (cf. Figure 1). UEs that communicate directly with the MBS (M-UE) are distributed according to some independent stationary point process and connect with the closest MBS. Meanwhile, UEs that communicate with the MBS via the RS (R-UE) are arranged with reference to some independent stationary point process within each RS's circular area. Note that these two different types of UEs are not distinguished in Figure 1.

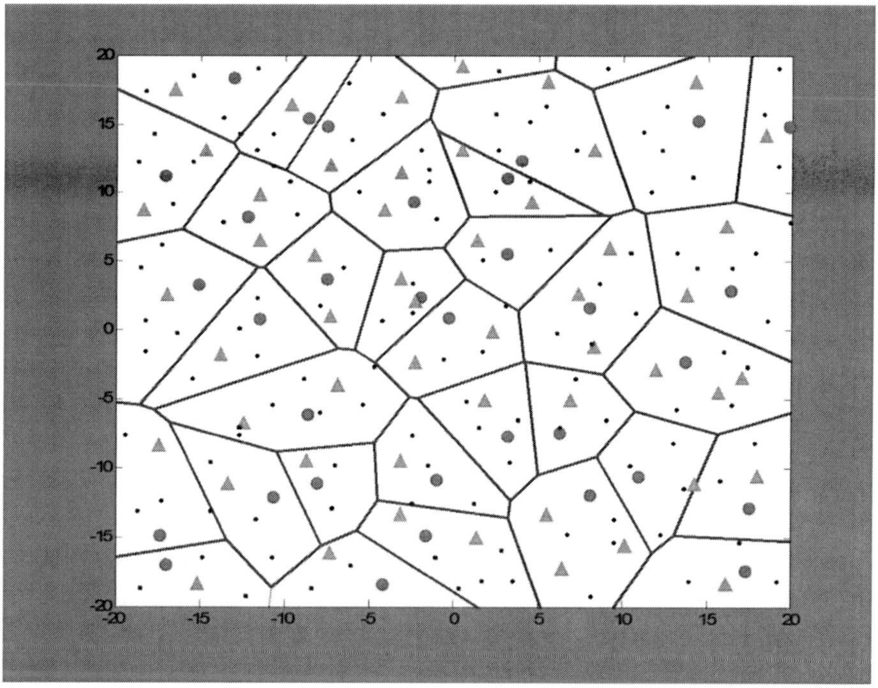

Figure 1 HPPP distributed MBSs and RSs with two types of UEs. Big dots, MBSs; triangles, RSs; little dots, UEs.

The distance between M-UE (or RS) and the target MBS is denoted by r. As Θ_M is a HPPP with density λ_M, the probability density function (pdf) of r is $f_r(r)=2\pi\lambda_M r\exp(-\pi\lambda_M r^2)$ [9]. Furthermore, since the distribution of RSs follows a HPPP and R-UEs are distributed according to some independent stationary point process within each relay's circular area, the distance between the R-UE and its corresponding RS r follows a distribution with pdf $f_R(r)=2r/R^2$.

As shown in Figure 2, in each cell, the MBS-UE link and the MBS-RS-UE link work in non-overlapping frequency bands with bandwidth ω_M and ω_R, respectively. For the MBS-RS-UE link with half-duplex relays, the transmission is divided into two phrases. In the first phrase, the MBS sends messages to the RS, and in the second phrase, the RS decodes the messages and forwards them to the user. The time duration of both phases is equal. We assume that each MBS serves at most one M-UE and RS at any time. If more than one M-UE (RS) are located in the MBS's cell, orthogonal resource sharing such as time division multiple access (TDMA) is performed. Similarly, each RS serves one R-UE at any time and TDMA may be used if multiple R-UEs are distributed in the RS's serving area. Once TDMA is selected, the MBS can schedule the resource and select the RS to work with, the RSs only need to know which one is selected.

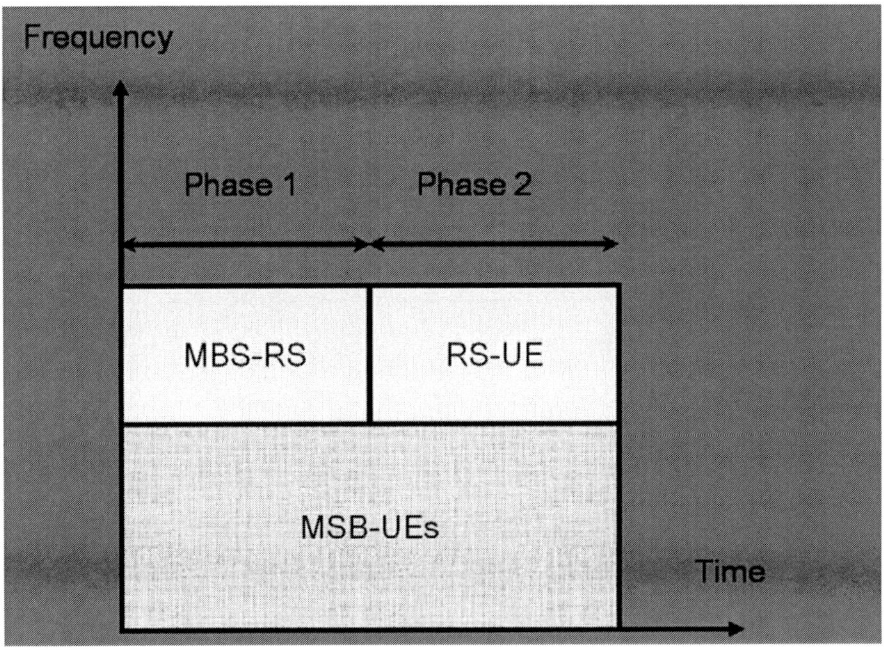

Figure 2 Relay-assisted cellular network frame structure.

3 ENERGY EFFICIENCY ANALYSIS

In this section, the coverage probabilities are first derived, and then, the energy efficiency expression is given for certain power consumption models.

3.1 Coverage probability

The signal-to-interference-plus-noise ratio (SINR) in the downlink from the transmitter (MBS or RS) s to the receiver (RS or UE) d is given by

$$\mathrm{SINR}\,(s \rightarrow d) = \frac{Phr^{-\alpha}}{I + \sigma^2} \tag{1}$$

where P is the transmit power for node s, h is the channel power gain due to the small-scale fading, r is the distance between s and d, α is the path loss exponent, I is the aggregate interference from all the other active transmitters operating in the same frequency band, and σ^2 is the variance of the additive white Gaussian background noise. All channels are assumed to be subject to Rayleigh fading, i.e., $h \sim \exp(1)$.

The coverage probability is defined as the probability that the receive SINR is above a certain threshold Γ and can be written as

$$p_c = \mathbb{P}\,(\mathrm{SINR} \geq \Gamma) \tag{2}$$

3.1.1 Coverage probability for the MBS-UE link

First, we assume that M-UEs that are directly connected to MBS are distributed according to an independent stationary point process. Second, the transmit power from MBS to M-UE is P_{MU}, the receive threshold is denoted as Γ_{MU}, and I_{MU} is the aggregate interference from all the other active MBSs except the target MBS m_0. Then, the coverage probability for a typical MBS-UE link is defined as

$$p_{c\_\mathrm{MU}} = \mathbb{P}\,(\mathrm{SINR}_{\mathrm{MU}} > \Gamma_{\mathrm{MU}}) \tag{3}$$

where $\mathrm{SINR}_{\mathrm{MU}} = \frac{P_{\mathrm{MU}} h_0 r^{-\alpha}}{I_{\mathrm{MU}} + \sigma^2}$ and $I_{\mathrm{MU}} = \sum_{i \in \Theta_M \setminus \{m_0\}} P_{\mathrm{MU}} h_i r_i^{-\alpha}$

and $f_r\,(r) = 2\pi\lambda_M r \exp\left(-\pi\lambda_M r^2\right)$.

Theorem 1 The coverage probability for the MBS-UE link pc_MU

is given by

$$p_{c\_MU} = 2\pi\lambda_M \int_{r>0} \exp\left(-\pi\lambda_M r^2 \left(1 + \rho\left(\Gamma_{MU}, \alpha\right)\right)\right) \times \exp\left(-\Gamma_{MU} r^{\alpha} \sigma^2 / P_{MU}\right) r\mathrm{d}r \quad (4)$$

where $\rho\left(\Gamma_{MU}, \alpha\right) = \Gamma_{MU}^{2/\alpha} \int_{\Gamma_{MU}^{-2/\alpha}}^{\infty} \frac{1}{1+v^{\alpha/2}} \mathrm{d}v.$

In the interference-limited regime where the background thermal noise is ignored ($\sigma^2 \to 0$), the coverage probability can be simplified as $p_{c\_MU} = \frac{1}{1+\rho(\Gamma_{MU},\alpha)}$.

Proof The details of the derivation can be found in [9], and we just provide here the key points. From Equation 3, the definition of SINR and the assumption that $h_0 \sim \exp(1)$, the coverage probability can be rewritten as

$$\begin{aligned} p_{c\_MU} &= \mathbb{P}(h_0 \geq \Gamma_{MU} r^{\alpha} \left(I_{MU} + \sigma^2\right) / P_{MU}) \\ &= \int_{r>0} \mathbb{P}(h_0 \geq \Gamma_{MU} r^{\alpha} \left(I_{MU} + \sigma^2\right) / P_{MU} | r) f_r(r) \mathrm{d}r \\ &= \int_{r>0} \exp\left(-\Gamma_{MU} r^{-\alpha} \sigma^2 / P_{MU}\right) L_{I_{MU}} \\ &\quad \times \left(\Gamma_{MU} r^{\alpha} / P_{MU}\right) f_r(r) \mathrm{d}r \end{aligned} \quad (5)$$

where $L_{IMU}(s)$ is the Laplace transform of I_{MU}.

Considering the definition of Laplace transform and the assumption that $h_i \sim$ exp(1), $L_{IMU}(s)$

is given by

$$\begin{aligned} L_{I_{MU}}(s) &= \mathbb{E}_{I_{MU}}\left[e^{-sI_{MU}}\right] \\ &= \exp\left(-2\pi\lambda_M \int_r^{\infty} \left(1 - \frac{1}{1+sP_{MU}u^{-\alpha}}\right) u\mathrm{d}u\right) \end{aligned} \quad (6)$$

The result of Equation 5 is deduced from the property of the probability generating functional (PGFL) for PPP, which for some function $g(x)$, it satisfies $E[\prod_{x\in\Theta} g(x)]=\exp(-\lambda\int_{R^2}(1-g(x))\mathrm{d}x)$ [13], and λ is the density of the interference nodes.

Combining the results of Equations 5 and 6, Equation 4 is obtained.

When $\sigma^2 \to 0$, it can be easily derived for Equation 4 that

$$p_{c\_MU} = \frac{1}{1+\rho(\Gamma_{MU},\alpha)}.$$

Using the Theorem 1, the following corollaries can be easily obtained.

Corollary 1 In the interference-limited regime, the coverage probability p_{c_MU} increases with increasing P_{MU}, and when P_{MU} goes to infinity, $\lim_{P_{MU}\to\infty} p_{c\_MU} = \frac{1}{1+\rho(\Gamma_{MU},\alpha)}$.

Corollary 2. With $\sigma^2>0$, the coverage probability p_{c_MU} decreases with increasing Γ_{MU}.

3.1.2 Coverage probability for the MBS-RS link

Suppose the relay can successfully decode the signals from the MBS if the received SINR is larger than or equal to Γ_{MR}, then the coverage probability for the MBS-RS link can be written as

$$p_{c\_MR} = \mathbb{P}\left(\mathrm{SINR_{MR}} \geq \Gamma_{MR}\right) \tag{7}$$

where $\mathrm{SINR_{MR}} = \frac{P_{MR} g_0 r^{-\alpha}}{I_{MR}+\sigma^2}$, $I_{MR} = \sum_i P_{MR} g_i r_i^{-\alpha}$, for i in all other active MRSs in the MBS-RS frequency band except the target MBS m_0.

Theorem 2. The coverage probability for the MBS-RS link p_{c_MR}

is given by

$$\begin{aligned} p_{c\_MR} = 2\pi\lambda_M \int_{r>0} & \exp\left(-\pi r^2 \left(\lambda_M + \lambda\rho\left(\Gamma_{MR},\alpha\right)\right)\right) \\ & \times \exp\left(-\Gamma_{MR} r^{\alpha}\sigma^2/P_{MR}\right) r\mathrm{d}r \end{aligned} \tag{8}$$

where $\lambda = \min\{\lambda_M, \lambda_R\}$ and $\rho(\Gamma_{MR},\alpha) = \Gamma_{MR}^{2/\alpha} \int_{\Gamma_{MR}^{-2/\alpha}}^{\infty} \frac{1}{1+v^{\alpha/2}} \mathrm{d}v$.

In the interference-limited regime, i.e., $\sigma^2\to 0$, the coverage probability can be simplified as $\frac{\lambda_M}{\lambda_M+\lambda\rho(\Gamma_{MR},\alpha)}$.

The derivation of the coverage probability p_{c_MR} is almost the same as that of p_{c_MU}

in the previous subsection except that the density used for the derivation of the Laplace transform of I_{MR} is λ instead of λ_M. As we assume that each MBS serves at most one RS at any time in each cell, the density of interference nodes is the smaller of λ_M and λ_R. Thus, the density used for the derivation of the Laplace transform of I_{MR} is $\lambda = \min\{\lambda_M, \lambda_R\}$.

Similar corollaries can be obtained as follows.

Corollary 3. With $\sigma^2>0$, the coverage probability p_{c_MR} increases with increasing P_{MR}, and when P_{MR} goes to infinity, $\lim P_{MR}\to\infty p_{c\_MR}=\lambda M\lambda M+\lambda\rho(\Gamma_{MR},\alpha)$.

Corollary 4. With $\sigma^2>0$, the coverage probability p_{c_MR} decreases with increasing Γ_{MR}.

3.1.3 Coverage probability for the RS-UE link

We assume that R-UEs receiving messages by the aid of RSs are located according to some stationary point process in a circular area around the RS and of radius R. We denote P_R as the transmit power of the RS, Γ_{RU} as the receive threshold, and I_{RU} as the sum of interference of all other active RSs except the target RS r_0. The coverage probability for a typical RS-UE link is defined as

$$p_{c\_RU} = \mathbb{P}(\mathrm{SINR}_{RU} \geq \Gamma_{RU}) \tag{9}$$

where $= \frac{P_R l_0 r^{-\alpha}}{I_{RU}+\sigma^2}$ and $= \sum_i P_R l_i r_i^{-\alpha}$, for i denoting all other active RSs except r_0.

As the distance between RS and UE follows a different distribution, the form of p_{c_RU} is not the same as for p_{c_MR} and p_{c_MU}.

Theorem 3. *The coverage probability for the MBS-RS link p_{c_RU}* is given by

$$p_{c\_RU} = \frac{2}{R^2}\int_0^R \exp\left(-\pi\lambda r^2 p_{c\_MR}\mu\left(\Gamma_{RU},\alpha\right)\right) \times \exp\left(-\frac{\Gamma_{RU} r^{\alpha}\sigma^2}{\left(1+\Gamma_{RU}\right)P_R}\right) r dr \tag{10}$$

where $\lambda=\min\{\lambda_M,\lambda_R\}$ and $= \left(\frac{\Gamma_{RU}}{1+\Gamma_{RU}}\right)^{2/\alpha}\int_0^\infty \frac{1}{1+v^{\alpha/2}}dv.$

In the interference-limited regime, the coverage probability can be expressed as $p_{c\_} = \frac{1+\rho(\Gamma_{MR},\alpha)}{\lambda\pi R^2\mu(\Gamma_{RU},\alpha)}\left(1-\exp\left(-\frac{\lambda\pi R^2\mu(\Gamma_{RU},\alpha)}{1+\rho(\Gamma_{MR},\alpha)}\right)\right).$

Proof. From the definition of p_{c_RU}, SINR_{RU}, and I_{RU}, we can rewrite p_{c_RU} as

$$p_{c\_\mathrm{RU}} = \mathbb{P}(\frac{P_{\mathrm{R}} l_0 r^{-\alpha}}{I_{\mathrm{RU}} + \sigma^2} \geq \Gamma_{\mathrm{RU}})$$
$$= \int_0^R \mathbb{P}\left(l_0 \geq \Gamma_{\mathrm{RU}} r^{\alpha} \left(I_{\mathrm{RU}} + \sigma^2\right) / P_{\mathrm{R}} | r\right) f_R(r)\, \mathrm{d}r. \tag{11}$$

For analytical convenience, we assume that $\tilde{I}_{\mathrm{RU}} = I_{\mathrm{RU}} + P_R l_0 r^{-\alpha}$, which encompasses the signals from all the transmitting RSs including r_0. We substitute I_{RU} with $(\tilde{I}_{\mathrm{RU}} - P_R l_0 r^{-\alpha})$ and the coverage probability p_{c_RU} is given by

$$p_{c\_\mathrm{RU}} = \int_0^R \mathbb{P}\left(l_0 \geq \frac{\Gamma_{\mathrm{RU}} r^{\alpha} \left(\tilde{I}_{\mathrm{RU}} + \sigma^2\right)}{\left(1 + \Gamma_{\mathrm{RU}}\right) P_R} | r\right) f_R(r)\, \mathrm{d}r$$
$$\overset{(a)}{=} \frac{2}{R^2} \int_0^R \mathbb{E}_{\tilde{I}_{\mathrm{RU}}} \left[\exp\left(-\frac{\Gamma_{\mathrm{RU}} r^{\alpha} \left(\tilde{I}_{\mathrm{RU}} + \sigma^2\right)}{\left(1 + \Gamma_{\mathrm{RU}}\right) P_R}\right)\right] r\mathrm{d}r$$
$$= \frac{2}{R^2} \int_0^R \exp\left(-\frac{\Gamma_{\mathrm{RU}} r^{\alpha} \sigma^2}{\left(1 + \Gamma_{\mathrm{RU}}\right) P_R}\right) L_{\tilde{I}_{\mathrm{MR}}}$$
$$\times \left(\frac{\Gamma_{\mathrm{RU}} r^{\alpha}}{\left(1 + \Gamma_{\mathrm{RU}}\right) P_R}\right) r\mathrm{d}r \tag{12}$$

where step (a) follows from the fact that $f_R(r) = 2r/R^2$ and $l_0 \sim \exp(1)$.

The derivation of $L_{\tilde{I}\mathrm{MR}}$ is similar to that of $L_{I\mathrm{MU}}$ except some differences as shown below.

$$L_{I_{\mathrm{RU}}}\left(\Gamma_{\mathrm{RU}} r^{\alpha} / P_R\right) = \exp\left(-\pi r^2 \lambda p_{c\_\mathrm{MR}}\right) \mu\left(\Gamma_{\mathrm{RU}}, \alpha\right)\right) \tag{13}$$

where $\lambda = \min\{\lambda_M, \lambda_R\}$ and $\mu\left(\Gamma_{\mathrm{RU}}, \alpha\right) = \Gamma_{\mathrm{RU}}^{2/\alpha} \int_0^{\infty} . \frac{1}{1 + v^{\alpha/2}} \mathrm{d}v.$

One difference is the density used in the expression. At any time, there is at most only one RS communicating with the target MBS, and only the RSs with received SINR larger than the target SINR Γ_{MR} forward messages to the R-UEs, so the density used in the above formula is $\lambda p_{c\_\mathrm{MR}}$.

Besides, the distance between R-UE and active RSs can be any value, so the integration limits start form 0 (not r) to ∞ in $\mu(\Gamma_{RU},\alpha)$. Combining Equations 11 and 13, the expression of pc_RU is obtained.

When $\sigma^2 \to 0, p_{c\_MR} = \frac{1}{1+\rho(\Gamma_{MR},\alpha)}$

and then pc_RU can be written as

$$p_{c\_RU} = \frac{2}{R^2}\int_0^R \exp\left(-\pi r^2 \lambda \frac{\mu(\Gamma_{RU},\alpha)}{1+\rho(\Gamma_{MR},\alpha)}\right) r dr = \frac{1+\rho(\Gamma_{MR},\alpha)}{\lambda \pi R^2 \mu(\Gamma_{RU},\alpha)}\left(1-\exp\left(-\frac{\lambda \pi R^2 \mu(\Gamma_{RU},\alpha)}{1+\rho(\Gamma_{MR},\alpha)}\right)\right) \quad (14)$$

Using Theorems 2 and 3, the following corollaries can be acquired.

Corollary 5. With $\sigma^2>0$, the coverage probability p_{c_RU} increases with decreasing P_{MR} or increasing P_{RU}.

Corollary 6. With $\sigma^2>0$, the coverage probability p_{c_RU} increases with increasing Γ_{MR} or decreasing Γ_{RU}.

Corollary 7. When $\lambda_R < \lambda_M$, the coverage probability p_{c_RU} decreases with increasing λ_R. When $\lambda_R \geq \lambda_M$, the coverage probability p_{c_RU} remains the same with increasing λ_R.

Proof. When $\lambda_R < \lambda_M$, $\lambda = \lambda_R$, increasing λ_R can increase λ and decrease p_{c_RU}. However, when $\lambda_R \geq \lambda_M$, $\lambda = \lambda_M$ as λ_R rises, as a result, p_{c_RU} remains the same with increasing λ_R.

3.2 Power consumption model

In this section, we model the power consumption of MBSs and RSs in downlink relay-assisted cellular networks. The relation between total power consumption P_{tot} and transmit radiated power P_T is modeled as [14–16]$P_{tot} = \beta P_T + P_0$, where $1/\beta$ is the efficiency of the power amplifier, and P_0 is the static power consumption, which includes signal processing overhead, battery backup, cooling power consumption, etc. As a result, for MBS and RS, the total power consumption P_{M_tot}, P_{R_tot} can be given by

$$P_{M\_tot} = \beta_M P_M + P_{M0} \quad (15)$$

$$P_{R\_tot} = \beta_R P_R + P_{R0} \quad (16)$$

where $1/\beta_M$, $1/\beta_R$ denote the efficiency of the power amplifier for MBS and RS, P_M and P_R account for the total transmit power for MBS and RS, and P_{M0} and P_{R0} are the static power consumption for MBS and RS, respectively.

3.3 Performance metrics

We define the energy efficiency for the relay-assisted cellular networks as

$$\eta_{EE} = \frac{\textit{Area spectral efficiency}}{\textit{Average network area power consumption}} = \frac{\tau_M + \tau_R}{P_{M\_ave} + P_{R\_ave}} (bps/Hz/W) \tag{17}$$

where τ_M and τ_R are the area spectrum efficiency over all the MBS-UE links and MBS-RS-UE links, respectively. P_{M_ave} and P_{R_ave} denote the average network area power consumption for MBSs and RSs, respectively.

The throughput attained at the MBS-UE link is given by $p_{c\_MU} \frac{w_M}{w_M + w_R}$ log2(1+ΓMU)

where $\frac{w_M}{w_M + w_R}$ is the rate of occupancy of the frequency bands for the MBS-UE link. Thus, the area spectrum efficiency over all MBS-UE links is

$$\tau_M = \lambda_M p_{c\_MU} \frac{w_M}{w_M + w_R} \log_2 (1 + \Gamma_{MU}) . \tag{18}$$

The MBS sends messages to MBS-UE with transmit power P_{MU}, and to the RS with transmit power P_{MR} only in the one of the two phrases, so the average macrocell network area power consumption is

$$P_{M\_ave} = \beta_M (\lambda_M P_{MU} + \lambda P_{MR}/2) + \lambda_M P_{M0}. \tag{19}$$

As at most $\lambda = \min(\lambda_M, \lambda_R)$ MBSs per square meter transmit signals to the RSs, the density used for P_{MR} is λ.

As a DF strategy is adopted by the RS, then the MBS-RS-UE link operates only if both the relay and the user decode the messages received successfully, namely, the SINR attained at the RS and the R-UE must be larger than Γ_{MR} and

Γ_{RU}, respectively. Thus, the area spectrum efficiency for all the MBS-RS-UE links is given by

$$\tau_R = \frac{1}{2}\lambda p_{c\_MR} \times p_{c\_RU} \frac{w_R}{w_R + w_H} \min\left\{\log_2(1+\Gamma_{MR}),\right.$$
$$\left.\log_2(1+\Gamma_{RU})\right\}. \tag{20}$$

Note that the 1/2 factor is due to half-duplex operation of the relay nodes, and $\frac{w_R}{w_M + w_R}$ is the rate of occupancy of the frequency bands for the MBS-RS-UE link.

At any time, the number of RSs taking part in the communication is at most the same as the number of MBSs (λ relays per square meters), and only the relays decoding signals for MBSs successfully forward signals to corresponding R-UEs (λpc_MR relays per square meter), the other relays ($\lambda_R - \lambda p_{c\_MR}$ relays per square meter) do not send signals but only consume the static power part. Thus, the average network area power consumption for RSs is as follows:

$$P_{R\_ave} = \lambda p_{c\_MR}(\beta_R P_R/2 + P_{R0}) + \left(\lambda_R - \lambda p_{c\_MR}\right) P_{R0}. \tag{21}$$

Combining Equations 17 to 21, the expression of energy efficiency for the relay-assisted cellular networks is obtained.

The following theorem demonstrates the quasiconcavity of energy efficiency function.

Theorem 4. The energy efficiency function η_{EE} is strictly quasi-concave on P_{MU} or P_{RU}.

Proof. We prove η_{EE} is strictly quasi-concave on P_{MU} here, and the process of the proof that η_{EE} is strictly quasi-concave on P_{RU} is similar. As defined in [17], a function f:$\mathbf{R}^n \to \mathbf{R}$ is said to be strictly quasi-concave if its sublevel set S_α $=\{\mathbf{x}|\mathbf{x}\in \mathbf{dom}\, f, f(\mathbf{x})\geq\alpha\}$ is strictly convex for every α. When $\alpha\leq 0$, S_α is obviously convex on P_{MU}. When $\alpha>0$, S_α is equivalent to $S\alpha=\{PMU|\alpha(PM\_ave+PR\_ave)-\tau M-\tau R\leq 0\}$. It can be easily proved that p_{c_MU} is strictly concave on P_{MU}, so $-\tau_M$ is strictly concave on P_{MU}. Besides, PM_ave linearly increases with P_{MU}, PR_ave and τ_R are not relevant with P_{MU}. As a result, S_α is also strictly convex on P_{MU} and Theorem 4 follows.

Due to the strict quasi-concavity of η_{EE} on P_{MU} or P_{RU}, the optimal energy efficiency exists for various values of P_{MU} or P_{RU}.

We now study the following optimization problem:

$$\min_{P_{\mathrm{MU}}} \eta_{\mathrm{EE}}$$
$$s.t.\ p_{c\_\mathrm{MU}} \geq \theta \tag{22}$$

The above problem aims at finding the optimal energy efficiency under the constraint that the coverage probability for the MBS-UE link is larger than or equal to a certain threshold ϑ. As η_{EE} is strictly quasi-concave on P_{MU}, the bisection method can be used to find the optimal η_{EE} and the corresponding P_{MU} (denoted $P_{\mathrm{MU}\_o\ p\ t}$). If $p_{c\ \_\mathrm{MU}}(P_{\mathrm{MU}\_o\ p\ t})\geq\vartheta$, then $\eta_{\mathrm{EE}}(P_{\mathrm{MU}\_o\ p\ t})$ is the optimal value. Else, as $p_{c\ \_\mathrm{MU}}$ strictly increases monotonically on P_{MU}, $P_{\mathrm{MU}\_o\,p\,t}$ is less than the power P MU_*opt* which makes *pc*_MU(P MU_*opt*)=ϑ and ηEE(P MU_*opt*) is the optimal value according to the quasi-concavity of η_{EE}. The detailed algorithm to find the optimal η_{EE} based on bisection method is described in Algorithm 1.

Algorithm 1 The algorithm to find optimal η_{EE} based on the bisection method

---

Input: the upper bound m and the lower bound n ($m \geq n \geq 0$), the tolerance ε.
Output: the optimal η_{EE} and the corresponding P_{MU}
While: $(m - n \geq \varepsilon)$
$t = \frac{m-n}{2}$
Solve the feasibility problem $\eta_{\mathrm{EE}} \geq t$
if the problem is feasible
then $n = t$
else
$m = t$

if $p_{c\_\mathrm{MU}}\left(P_{\mathrm{MU}\_opt}\right) \geq \theta$ (P_{MU_opt} is the value that makes $\eta_{\mathrm{EE}}\left(P_{\mathrm{MU}\_opt}\right) = t$)
return t and P_{MU_opt}
else solve the equation $p_{c\_\mathrm{MU}}\left(\bar{P}_{\mathrm{MU}\_opt}\right) = \theta$
return $\eta_{\mathrm{EE}}\left(\bar{P}_{\mathrm{MU}\_opt}\right)$ and $\bar{P}_{\mathrm{MU}\_opt}$

---

4 SIMULATION RESULTS

In this section, we evaluate the derived analytical results through simulation. We use the default values of the system model in Table 1 unless otherwise stated. The simulations are divided into two parts: coverage probability and energy efficiency.

Table 1 Parameter values used in the simulations

Parameter	Value
λ_M, λ_R	10^{-5} m^{-2}, 10^{-4} m^{-2},
P_{MU}, P_{RU}, P_{MR}	43 dBm, 30 dBm, 33 dBm
R	40 m
α	4
W_M, W_R	40 MHz, 4 MHz
$\Gamma_{MR}, \Gamma_{RU}, \Gamma_{MR}$	-10 dB, -10 dB, -10 dB
σ^2	−60 dBm
β_M, β_R	5.32, 4.8
P_{MO}, P_{RO}	118.7 W, 7.5 W

4.1 Simulation results for coverage probability

In this part, we evaluate the relationship between coverage probability and transmit power of MBS or MBS density. Figure 3 shows the coverage probability for the MBS-UE link p_{c_MU} versus the transmit power of MBS P_{MU} for different target SINRs Γ_{MU}. It can be seen from the figure that the coverage probability increases with the transmit power, and the growth rate is high when the transmit power is relatively small, whereas it approaches zero when the transmit power is relative large. Besides, as the target SINR increases, the coverage probability decreases, and higher target SINR results in larger changes of coverage probability with the variation of transmit power. Note that for Γ_{MU} = -20 dB, the curve is almost flat, while for Γ_{MU} up to 0 dB, the curve becomes the most curved.

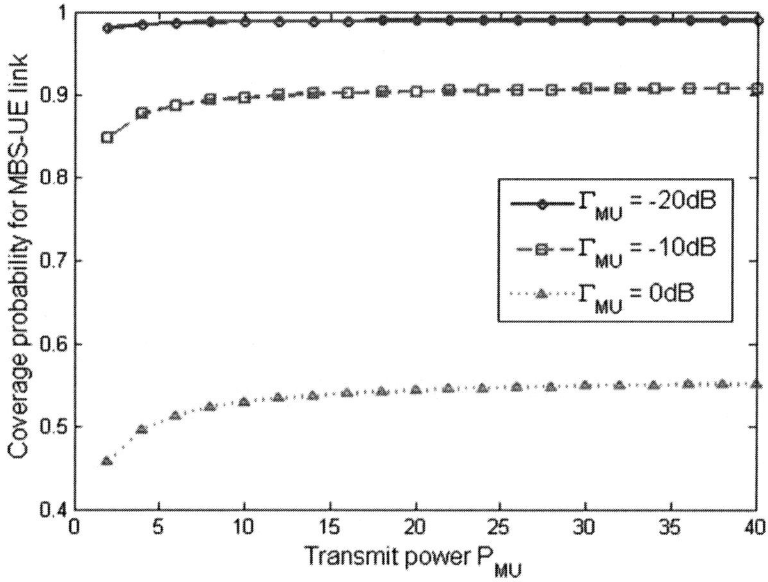

Figure 3 Coverage probability for the MBS-UE link vs. transmit power of MBS for different target SINR.

Figure 4 plots the coverage probability for the MBS-RS link p_{c_MR} versus the MBS density λ_M for different values of pathloss exponent α. The curves can be divided into two parts due to the value of λ. When λ_M is smaller than that of RS λ_R, λ equals to λ_M, and for pathloss exponent increasing, the curves become more and more curved. When the pathloss exponent is 3, the coverage probability remains almost the same as the density changes in this range. However, when the pathloss exponent becomes 4 or 5, the coverage probability increases as the density grows. Moreover, when $\lambda_M \geq \lambda_R$, λ equals to the constant value λ_R, then all the curves increase slowly with λ_M.

The relationship between coverage probability for RS-UE link p_{c_RU} and density of MBS λ_M for different pathloss exponents α is shown in Figure 5. Compared to p_{c_MR} or p_{c_RU}, the expression of p_{c_RU}, which is relevant to p_{c_MR} and the density of both MBSs and RSs, is quite different and causes different curve shapes. When λ_M is smaller than λ_R, the value of λ is λ_M, and the coverage probabilities almost linearly decrease as λ_M increases. However, when λ_M becomes larger than λ_R, the value of λ is the constant value λ_R, and then the coverage probabilities just slightly decrease as the density of MBS increases. This is because when λ_M becomes larger than λ_R, p_{c_RU} only varies with p_{c_MR}. From Figure 4, we can see that p_{c_MR} slightly increases when λ_M is larger than λ_R, which results in the shape of p_{c_RU}.

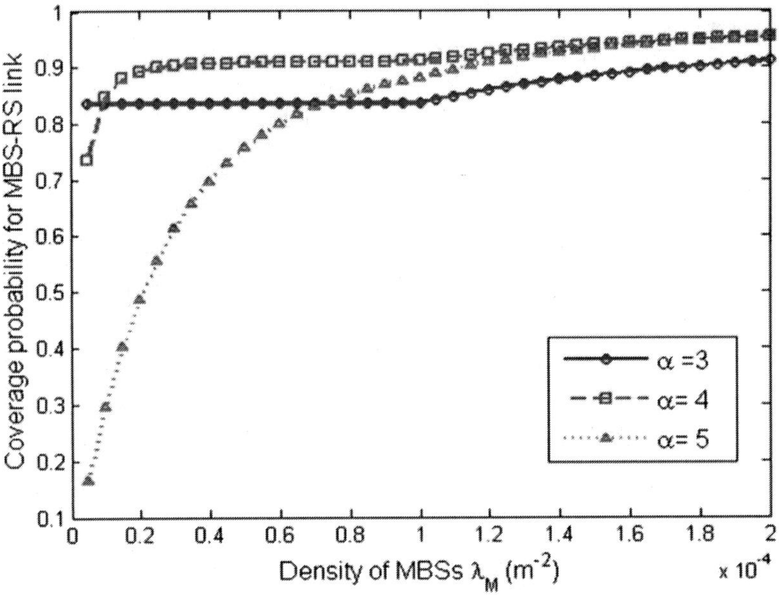

Figure 4 Coverage probability for the MBS-RS link vs. density of MBSs for different pathloss exponents.

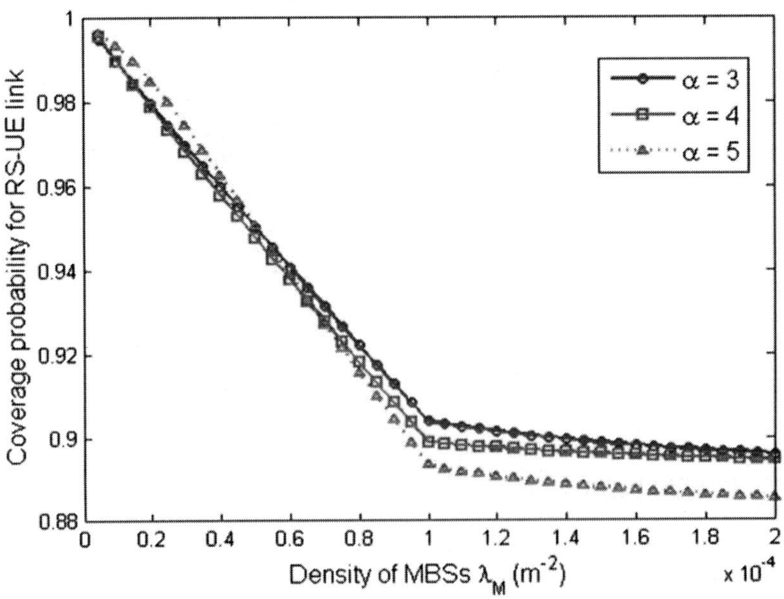

Figure 5 Coverage probability for the RS-UE link vs. density of MBSs for different pathloss exponents.

4.2 Simulation results for energy efficiency

The energy efficiency for downlink relay-assisted cellular networks is simulated in this part. Figure 6 shows how the energy efficiency varies with the density of MBSs for different densities of RSs. It can be seen that the energy efficiency increases with the density of MBSs and saturates when the density goes to infinity. That is to say, the growth of area spectral efficiency is faster than the growth of average network power consumption as the density of MBSs increases. But when the density of MBSs is relatively high, continuously increasing MBSs cannot further increase the energy efficiency. In addition, higher density of RSs brings about lower energy efficiency in the case of all other parameters constant. This is due to the access assumption in our model. We assume that each MBS serves at most one M-UE and RS at any time, and also, each RS serves one R-UE at any time. As a result, if the density of RSs is larger than that of MBSs, on average only λ_M RSs per square meter can be served at a time, the area spectral efficiency remains the same but the average area power consumption, mainly the average area static power consumption of RSs increases with the growth of density of RSs.

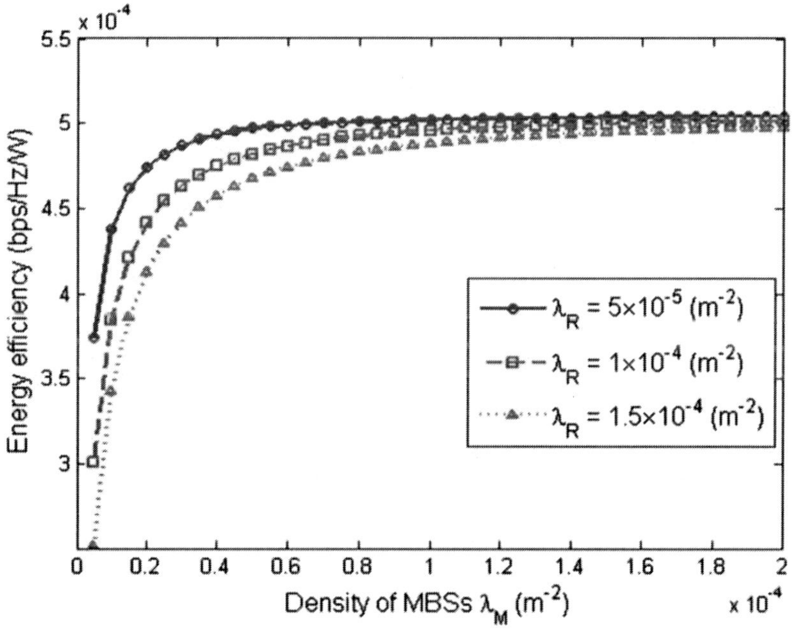

Figure 6 Energy efficiency vs. density of MBSs for different density of RSs.

Figure 7 displays the energy efficiency with respect to transmit power of MBS P_{MU} for different values of RS densities. As predicted in Theorem 4, the energy efficiency first increases and then decreases, showing a quasi-convex trend as the transmit power varies. The optimal energy efficiency can be found using the algorithm shown in Table 1. If the threshold ϑ is set to zero and the tolerance ε is 10^{-5}, the optimal energy efficiency is 6.40×10^{-4}bps/Hz/W, 5.26×10^{-4}bps/Hz/W and 4.48×10^{-4}bps/Hz/W corresponding to the three curves when $\lambda_R = 5\times10^{-5}m^{-2}, 1\times10^{-4}m^{-2}, 1.5\times10^{-4}m^{-2}$, respectively. Besides, the figure also shows that while the density of RSs grows, the energy efficiency drops, which means that the growth of area spectrum efficiency by increasing the density of RSs cannot compensate for the growth of average network power consumption.

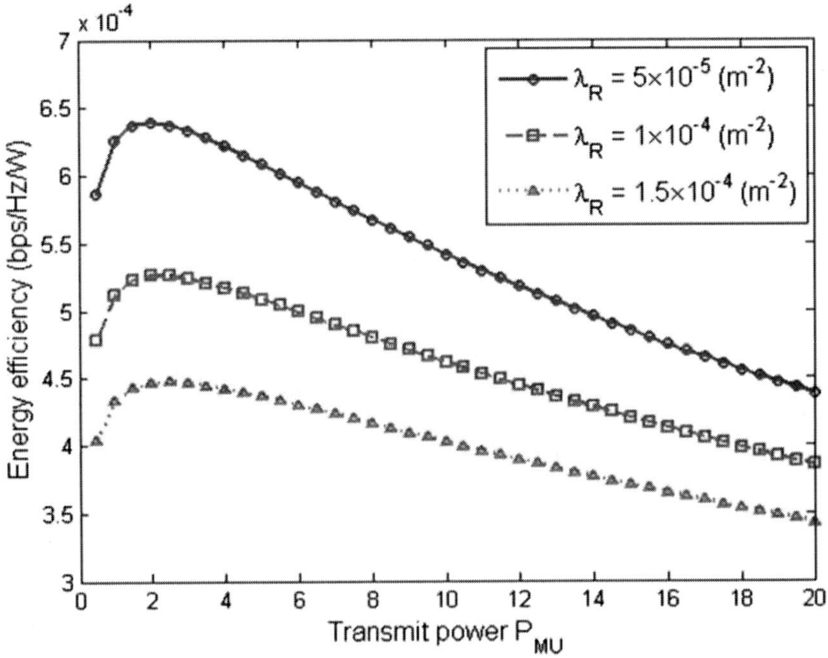

Figure 7 Energy efficiency vs. transmit power of MBS for different density of RSs.

5 CONCLUSIONS

In this paper, we study the energy efficiency of downlink relay-assisted cellular networks where MBSs and RSs are distributed according to independent HPPPs. First, we build on a tractable stochastic geometry-based model to derive the coverage probabilities for MBS-UE, MBS-RS, and RS-UE links. Second, using power consumption models for both macrocell and relay networks, we obtain expressions for the network energy efficiency. Our results reveal that there is a fundamental tradeoff between energy efficiency and transmit MBS power, and the main takeaway is that deploying more MBSs generally triggers higher energy efficiency; however this gain hits a ceiling as the density of MBSs increases.

DECLARATIONS

Acknowledgements

This work is supported by the National Basic Research Program of China (No. 2012CB316000), the National Natural Science Foundation of China (No. 61201192), the National S&T Major Project (No. 2013ZX03004007), International Science and Technology Cooperation Program (No. 2012DFG12010), China's 863 Project (No. 2012AA011402), Keygrant Project of Chinese Ministry of Education (No. 313005), the Open Research Fund of National Mobile Communications Research Laboratory, Southeast University (No. 2012D02), and Tsinghua-Qualcomm Joint Research Program.

Competing interests

The authors declare that they have no competing interests.

REFERENCES

1. Laneman JN, Tse DNC, Wornell GW: Cooperative diversity in wireless networks: efficient protocols and outage behavior. IEEE Trans. Inf. Theory 2004, 50(12):3062-3080. 10.1109/TIT.2004.838089
2. Li GY, Xu Z, Xiong C, Yang C, Zhang S, Chen Y, Xu S: Energy-efficient wireless communications: tutorial, survey, and open issues. IEEE Wireless Commun. Mag 2011, 18(6):28-35.
3. Yu H, Qin H, Li Y, Zhao Y, Xu X, Wang J: Energy-efficient power allocation for non-regenerative OFDM relay links. Sci. China Information Sci 2013, 56(2):185-192.
4. Sun C, Yang C: Is two-way relay more energy efficient? in Proceedings of the Global Telecommunications Conference (GLOBECOM 2011), 2011 IEEE . Houston, Texas, USA: IEEE; 2011:1-6.

5. Sun C, Yang C: Energy efficient hybrid one-way and two-way relay transmission strategy. In Proceedings of the Acoustics, Speech and Signal Processing (ICASSP), 2012 IEEE International Conference on. Kyoto, Japan: IEEE; 2012:2497-2500.

6. Yu H, Li Y, Zhong X, Wang L, Wang J: The analysis of the energy efficiency for the decode-and-forward two-way relay networks. 2013.

7. Madan R, Mehta N, Molisch A, Zhang J: Energy-efficient cooperative relaying over fading channels with simple relay selection. IEEE Trans. Wireless Commun 2008, 7(8):3013-3025.

8. Huang R, Feng C, Zhang T, Wang W: Energy-efficient relay selection and power allocation scheme in AF relay networks with bidirectional asymmetric traffic. In Proceedings of the Wireless Personal Multimedia Communications (WPMC), 2011 14th International Symposium on. Brest, France: IEEE; 2011:1-5.

9. Andrews JG, Baccelli F, Ganti RK: A tractable approach to coverage and rate in cellular networks. IEEE Trans. Commun 2011, 59(11):3122-3134.

10. Soh YS, Quek TQS, Kountouris M, Shin H: Energy efficient heterogeneous cellular networks. IEEE J. Selec. Areas Commun 2013, 31(5):840-850.

11. Deng N, Zhang S, Zhou W, Zhu J: A stochastic geometry approach to energy efficiency in relay-assisted cellular networks. In Proceedings of the Global Telecommunications Conference (GLOBECOM 2012), 2012 IEEE. Anaheim, California, USA: IEEE; 2012:3484-3489.

12. Wei H, Deng N, Zhao M, Zhou W, Dong P: Station density effect on energy efficiency of relay-assisted cellular networks. In Proceedings of the 1st IEEE International Conference on Communications in China (ICCC). Beijing, China: IEEE; 2012:411-415.

13. Stoyan D, Kendall WS, Mecke J, Ruschendorf L: Stochastic Geometry and Its Applications, Volume 2, Chichester, UK. Chichester: Wiley; 1995.

14. Auer G, Giannini V, Desset C, Godor I, Skillermark P, Olsson M, Imran MA, Sabella D, Gonzalez MJ, Blume O, Fehske A: How much energy is needed to run a wireless network? IEEE Wireless Commun. Mag 2011, 18(5):40-49.

15. Fehske AJ, Richter F, Fettweis GP: Energy efficiency improvements through micro sites in cellular mobile radio networks. In GLOBECOM Workshops, 2009 IEEE. Hawaii, USA: IEEE; 2009:1-5.

16. Arnold O, Richter F, Fettweis G, Blume O: Power consumption modeling of different base station types in heterogeneous cellular networks. In Future Network and Mobile Summit, 2010. Florence, Italy: IEEE; 2010:1-8.

17. Boyd SP, Vandenberghe L: Convex Optimization. Cambridge, UK: Cambridge University Press; 2004.

Chapter 9

Impact of Service Rates and Base Station Switching Granularity on Energy Consumption of Cellular Networks

Josip Lorincz[1*], Antonio Capone[2] and Dinko Begusic[1]

[1]*FESB-Faculty of Electrical Engineering, Mechanical Engineering and Naval Architecture, University of Split, Croatia, R. Boskovica b.b, Split 21000, Croatia.*

[2]*DEI, Politecnico di Milano, Via Ponzio 34/5, 20133, Milano, Italy.*

ABSTRACT

Over the last two decades we have witnessed significant growth of the cellular network energy consumption caused by a rapid increase in the number of mobile users and data volumes. This is contributed to by a tenfold increase of data rates every 5 years, and such a trend in up growth of energy consumption will continue with the introduction of a new throughput demanding services. Hence, development of new energy-saving techniques and estimation of the influence on energy savings when ensuring different service rates are the focus of this article. For the purpose of reducing cellular network energy consumption, this article proposes a new approach to energy-efficient management of network resources. The energy-efficient management is based on adaptive changes of on/off states of a complete base station (BS) site in accordance with the traffic pattern variations. Besides adaptations to the temporal traffic variations, the BS can adapt capacity to the spatial traffic variations through dynamic scaling of transmitted power according to capacity demand. We formulate the problem of energy-efficient management as a binary integer programming problem dedicated to energy consumption minimization of a complete network. Proposed methodology is simulated on a set of real size Universal Mobile Telecommunications System network instances consisting of different radio propagation environments. In addition, this article analyzes the influence on the energy savings potential of BSs switching granularity. Obtained results show that the proposed optimization approach offers significant reductions in the network energy consumption while preserving the most important QoS constraints like full area coverage and guaranteed service rates.

Keywords: Green Energy Wireless, Mobile Cellular Management Switching Base station Power Network Optimization Rate UMTS

1. INTRODUCTION

Fast growth of wireless cellular networks in the past decades raises a critical problem of energy consumed by network equipment. The number of mobile devices in cellular networks explosively increases, and by 2013 mobile devices will surpass fixed computers (PCs) as the most common web access device [1]. This imposes deployment of additional network infrastructures, primarily more base stations (BSs), which significantly increases energy consumption of the whole network [2]. On the other hand, it is expected that data volume of Information and Communication Technologies (ICT) increases approximately by a factor of ten every 5 years. This triggers energy consumption upgrowth on yearly bases for 16–20% [3]. Considering only the sector of mobile communications, it will contribute with 15–20% to the entire ICT sector energy consumption, and 0.3–0.4% of the annual world carbon dioxide (CO_2) footprint [4]. All mentioned results in both academia and industry reach a consensus on the growing need to develop more energy-efficient wireless access networks. In addition, new services which are more throughput intensive, like multimedia and data services, begin dominating in practical usage. For cellular networks, satisfying the tenfold increase of data rates every 5 years comes with the cost of high-energy consumption [3]. Because of this, investigating the influence of the transition to higher data rates on network energy consumption becomes imperative.

From the perspective of cellular network operators, increasing network energy efficiency is not only a matter of being green and responsible, but also an important economical issue since reducing energy consumption translates to lower operational expenditures. It has been estimated that the radio access part of a cellular network, more specifically the BS, is a major energy consumer [5]. As pointed out in [6], energy consumption of the BS contributes to more than 55% of total cellular network energy consumption. In this article, therefore, we focus on energy-efficient operation of the cellular BS. Increasing energy efficiency of the BS can be accomplished through a holistic approach considering improvements on: component, link, and network levels [7]. On the component level, special attention is dedicated to improving power amplifier efficiency [8], minimizing feeder losses [9], and introducing novel air cooling approaches [10]. Energy saving potential on the link level can be found in link adaptation methods [11] and discontinuous transmission realized through BS sleep modes [12]. Although component and link levels will be important contributors to "greener" wireless cellular networks, we believe that the network level approach can offer the highest energy savings. This is because the component level approach limits improvements to individual components of the BS only. On the other hand, the link level approach, based on turning off some of the components like radio transceivers on the BS during low traffic periods, provides some relief, which is still not sufficient. For significant energy savings, the network level approach based on the dynamic management of BS resources seems to be the most promising [7]. This approach allows the system to shut down an entire BS and transfer the corresponding traffic to a neighboring BS during the periods of low user activity.

In this article, we propose such an energy-efficient management scheme and as an important element we emphasize possible implementation in already deployed cellular networks. Such a possibility lies in a conservative approach to the coverage and capacity planning of analyzed cellular networks. In practical scenarios, cellular coverage is characterized with multiple overlapping layers of cells while capacity has been estimated for accommodating the highest traffic load during peak hours. However, the traffic load in cellular networks can have spatial and temporal fluctuations due to user mobility and daily variations of user activity. To save energy, during the periods of low user activity, it is possible to switch off unutilized or underutilized BSs without compromising coverage and service quality. Therefore, we propose a model that takes into account slow temporal variations of traffic through dynamic switching on and off of complete BS sites (BSSs) according to the daily traffic variations.

In addition, to consider highly dynamic spatial traffic fluctuations, we assume an energy-efficient network management system which can adaptively scale BS transmitted (Tx) power according to the capacity demands. The proposed *site capacitating* concept adaptively increases or decreases capacity and coverage of the BS in the case of transmission at higher or lower Tx power levels. Such a concept can be implemented since analyses have been performed on a set of real size Universal Mobile Telecommunications System (UMTS) networks based on Wideband Code Division Multiple Access (WCDMA) technology. By knowing that the capacity of a WCDMA BS directly depends on Tx power, and the Tx power of BSs influence on total network energy consumption, it is possible to estimate how the introduction of higher service data rates will influence on network energy consumption. In addition, the prevalence of dynamic BSs switching between on and off states is another issue that has influence on the energy-saving potential of a complete cellular network. Accordingly, in this article we perform analyses for three distinct switching approaches, differing in time discretization of a real traffic pattern. In these analyses, we also take into account differences between the traffic patterns of working and weekend days.

Hence, effectiveness of the proposed energy-efficient network management has been estimated through development of an optimization model, using principles of integer linear programming (ILP). The model minimizes monthly energy consumption of complete UMTS network, also preserving full network coverage and a fixed service rate. The main contribution of this article is the presentation of the influence of the BSs switching granularities and the guaranteed service rates on network energy consumption. Moreover, the obtained optimization results allow us to understand quantitatively the scope for potential energy savings.

The outline of the article is organized as follows: in Section 2, we give an overview of the latest research efforts dedicated to the energy-efficient resource management of wireless access networks. Description of a path loss model used for coverage planning of different radio propagation environments has been presented in Section 3. Section 4 explains the structure of the UMTS network instances used for simulating the proposed resource management approach. Also in Section 4 approximation of the traffic pattern is explained. Since coverage and capacity planning have directly been related in UMTS networks, Section 5 deals with site capacitating in order to estimate the correlation between BS capacity and the level of Tx power. Mathematical formulation of the optimization problem tending to minimize network energy consumption has been explained in Section 6. In addition, description of a software solution developed for the generation of the analyzed network instances is given also in Section 6. In Section 7, obtained optimization results in terms of energy savings have been discussed. Section 8 eventually concludes the article.

2. RELATED WORK

Cognition related to high energy consumption of mobile cellular networks contributed a non-negligible amount of greenhouse gas emissions, encouraging broad research efforts dedicated to increasing energy efficiency of the radio part of wireless access networks. In this section, we give a brief overview of these efforts, focusing primarily on energy-saving techniques realized through dynamic radio resource management based on the traffic statistics. To the best of the authors' knowledge, we have proposed for the first time, in [13, 14], a binary integer programming (BIP) optimization approach focused on energy consumption minimization of a complete large-scale wireless local area network (WLAN). Results showing significant energy savings of WLANs have been extended in [15, 16], through development of novel, practically implementable heuristic algorithms based on greedy and local search methods.

On the other hand, novel solutions contributing to the energy savings of wide area cellular access networks have been reported in [17–23]. Chiaraviglio et al. [17] show the possibility of switching off some cells and Node B's in the UMTS network during low-traffic periods, while still guaranteeing quality of service constraints in terms of blocking probability and electromagnetic exposure limits. Obtained results have been extended in [18], with the investigation of possible energy reductions of a cellular network by reducing the number of active cells of different topologies during the periods in which they are underutilized. Although authors claim energy savings potential of up to 30%, results have been obtained for small and isolated example cases. Analyses performed in [19], on a somewhat larger network example, exploits the idea of switching off cells while neighboring active cells adopt an antenna tilt at times of low network load. Although obtained for cells arranged regularly in the two concentric circles, results clearly indicate that network coverage can be kept intact when the individual cell of one or three-sector site is switched off. This article additionally gives an overview of short-term energy saving mechanisms based on micro-sleep and dynamic spectrum reduction. Even broader overviews of relevant challenges in the area of green cellular networks and solutions considering: power control, operator's cooperation, heterogeneous cells, relay techniques, granularity control, etc., have been discussed in [20]. In addition, using real data traces from the part of the real urban area network, authors emphasize promising potential in terms of power saving that one can expect by turning off BSs during low traffic periods. The studies of [21, 22] are of particular interest since both provide ILP formulations of energy optimization problems, confirming that mathematical optimization methods have extremely been valuable tools for modeling possible energy-efficient solutions in cellular networks. The study [21] concludes that practical real-time user-cell reallocation may be faceable and may bring significant energy savings, while the study [22] points out that BS density and the mean and variance of the traffic profiles have been dominant factors in determining the amount of energy savings. The study [23] proposes the concept of cell zooming, which enables adaptive adjustment of the cell size according to the traffic load. Authors emphasize that cell zooming, besides solving the problem of traffic imbalance, can also reduce energy consumption of the cellular network. Due to the need for more throughputs in wireless cellular networks, Videv et al. [24] develop a scheduler which aims to solve the problem of energy-efficient resource allocation in orthogonal frequency division multiple access cellular systems. Huang et al. [25] propose energy-efficient semi-static power control and a link adaptation scheme for UMTS (HSDPA) networks. The proposed scheme can determine the most energy-efficient modulation and coding schemes levels and transmit power at the Node B. Finally, a comprehensive overview and classification of the research approaches based on component level, cell layout adaptation with coverage extension and resource management are presented as in [26].

According to the research results presented in the mentioned studies, it is obvious that an energy-efficient approach based on a dynamic adaptation of network resources to traffic pattern can bring significant energy savings. Therefore, it is necessary to investigate such potential more deeply, especially on real size networks having the macro BSSs allocated inside distinct propagation areas, such as urban-dense, urban, or suburban. In the light of the permanent increase of guaranteed service rates, absence of valuable research results addressing the influence of such an increase on the network energy consumption represents reasonable grounds for deeper investigation. In addition, temporal granularity of control is an important issue in dynamic BS switching and this article, as far as we are aware, gives, for the first time, insights into this issue.

3. COVERAGE PLANNING

One of the most basic considerations in radio channel modeling is determining the morphology that the analyzed network models will be applied to. This leads to general assumptions being made for an analyzed service area (SA) that corresponds to a broader region of a medium sized city. In order to consider such a morphology case, we assume three different morphology categories classified as urban-dense, urban, and suburban (rural) coverage areas (CAs). Figure 1a,b presents anticipated virtual borders of each CA for the lowest and highest Tx power levels of installed equipment at each coverage site (CS), respectively. Each type of CA is categorized according to density of obstacles assumed on the radio propagation path. It may seem obvious that classifying CAs according to these morphologies is rather ambiguous because the geographic size will never have regular CA borders in reality. Nevertheless, we assume such classification and bordering in order to more realistically model real radio propagation environment of a medium city, which can typically be characterized with decreasing obstacle density in direction from a center to a border city region.

In order to correctly and efficiently model the propagation environment of a UMTS network, we utilize the path loss model commonly used for modeling propagation environments in macro cells of the 2G and 3G networks known as the Hata (Okumura-Hata) model [27, 28]. Actually, for estimating radio coverage of BSSs in analyzed UMTS networks, we exploit extended version of the Okumura-Hata model known as COST-231, which extends Hata's model to frequencies of 2 GHz.

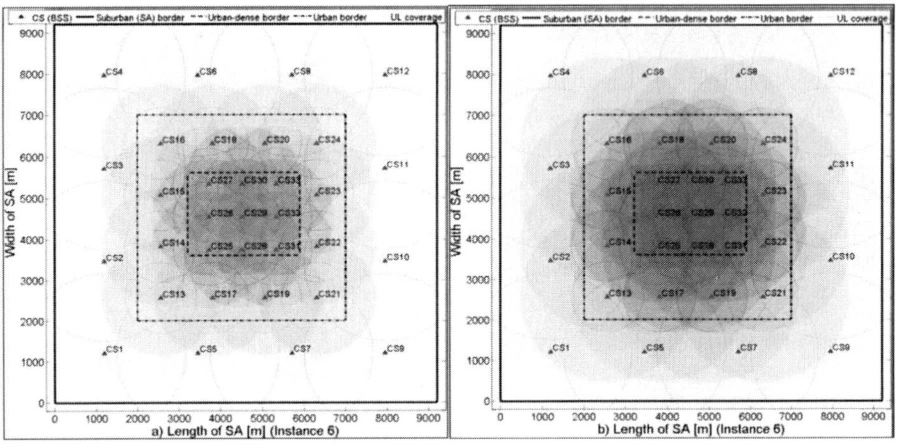

Figure 1 Guide borders among different propagation areas and DL CAs of each BSS transmitting at (a) 10 W, (b) 40 W.

Maximal coverage distances in uplink (UL) or downlink (DL) directions of BSs allocated inside urban-dense and urban CAs are performed using relation

$$d_{Murb/d} = 10^{\frac{P_{Tk} - P_{Rth} - \left[A + B\log_{10} f - 13.82\log_{10} h_{BS} - a(h_{UT})\right] - SMF}{(44.9 - 6.55\log_{10} h_{BS})}} [\mathrm{km}], \quad (1)$$

while the next relation has been used for calculating BSs coverage inside suburban CA

$$d_{Msurb} = 10^{\frac{P_{Tk} - P_{Rth} - \left[A + B\log_{10} f - 13.82\log_{10} h_{BS} - a(h_{UT})\right] + \left\{2\left[\log_{10}(f/28)\right]^2 + 5.4\right\} - SMF}{(44.9 - 6.55\log_{10} h_{BS})}} [\mathrm{km}]. \quad (2)$$

Descriptions and numerical values of each parameter used in relations (1) and (2) can be found in Table 1. In relations (1) and (2), the P_{Tk} stands for total Tx power radiated from BSS antenna. Selected parameter values are characteristic for UMTS networks and the COST 231 model [28–30]. Table 2 presents calculated values of maximal coverage distances in UL and DL directions from BS, for three different propagation environments. Selected values of the receiver sensitivity thresholds P_{Rt} equal to −117 dBm for user (mobile) terminal (UT) and −121 dBm for BS are in compliance with practical implementations and 3GPP standards [31, 32]. Obtained BS and UT coverage distances are, from a practical point of view, typical for corresponding propagation environments of the real UMTS networks.

Table 1 Path loss model (Okumura-Hata, COST-231) parameters

Parameter	Index	Value
Constant (1500–2000 MHz) (Okumura-Hata/COST-231 model)	A	46.3 dB
	B	33.9 dB
Operating frequency	f	2100 MHz
UT antenna height	h_{UT}	1.5 m
BS antenna height	h_{BS}	30 m
Correction factor for MS antenna height	$a(h_{UT})$	-9.2×0^{-3} dB
Area type correction factor	C_m	0 dB
Slow fading margin (UL/DL calculation)	SFM	13.16 dB
Standard deviation/point location probability	σ	8 dB/95%
UT (TC) sensitivity threshold	P_{Rth}	−117 dBm
BS (CS) sensitivity threshold	P_{Rth}	−121 dBm

Table 2 Maximal coverage distances for different Tx powers

Tx power P_{Tk}	Urban/urban dense area ($d_{Murb/d}$)	Sub-urban area (d_{Msurb})
Downlink BS coverage (COST-231 model)		
10 W (40 dBm)	1.416 km	3.193 km
20 W (43 dBm)	1.723 km	3.885 km
30 W (44.77 dBm)	1.935 km	4.361 km
40 W (46 dBm)	2.097 km	4.727 km
Uplink UT coverage (COST-231 model)		
Tx Power	Urban/urban dense area ($d_{Murb/d}$)	Sub-urban area (d_{Msurb})
0.7 W (28.4 dBm)	0.864 km	1.949 km

It is necessary to emphasize that coverage planning of analyzed UMTS networks has been made according to maximal coverage distances in the UL direction. This is because lower Tx power levels of UTs (up to 1 W) results in lower maximal coverage distances if compared with the coverage distances of BSs which are much higher in DL due to higher Tx powers (Table 2). In the analyses, for simplicity a fixed Tx power level of UT equal to 0.7 W is assumed and such a Tx power is characteristic for most mobile phones nowadays. As presented in Table 2, we performed analyses for four BS Tx power levels and even for the lowest Tx power level (10 W), DL coverage of the BS sector is higher than the maximal coverage range of UT in the UL direction. This motivates us to select maximal coverage in the UL direction as asset for network coverage planning. Therefore, possible wireless links between the UT and BS can be accomplished only if the UT can reach the BS in the UL direction, which is also an attribute of real UMTS networks.

4. NETWORK MODEL

Research presented in this article is primarily focused on improving energy-efficiency of the UMTS terrestrial radio access network, and more specifically, the base transceiver stations (or BS) also known as Node B's. Hence, in this article, the BS will be a term describing the hardware and software of one UMTS Node B installed inside one cabinet. In addition, the BSS will refer to a physical location hosting installed cellular network equipment consisting of one or more BS cabinets, antenna lines, power supply and if it is necessary, site cooling system.

In order to analyze possible energy savings in the UMTS networks, we have developed network instances that are similar to those of real UMTS networks. Analyses have been performed on 12 different network instances and Figure 2a,b shows the structure of the two arbitrarily selected UMTS network instances. We assume two different types of network: the first one with all BSSs having omni-directional antennas—also called one-sector (cell) BSSs. In addition, we analyze the second network type consisting of only the three-sector (cell) BSSs. Regarding the one-sector BSSs, we assume omni-directional antenna with 360° horizontal radiation pattern around the BSS, while in the case of the three-sector BSSs, the assumption is that each sector covers 120° of horizontal area.

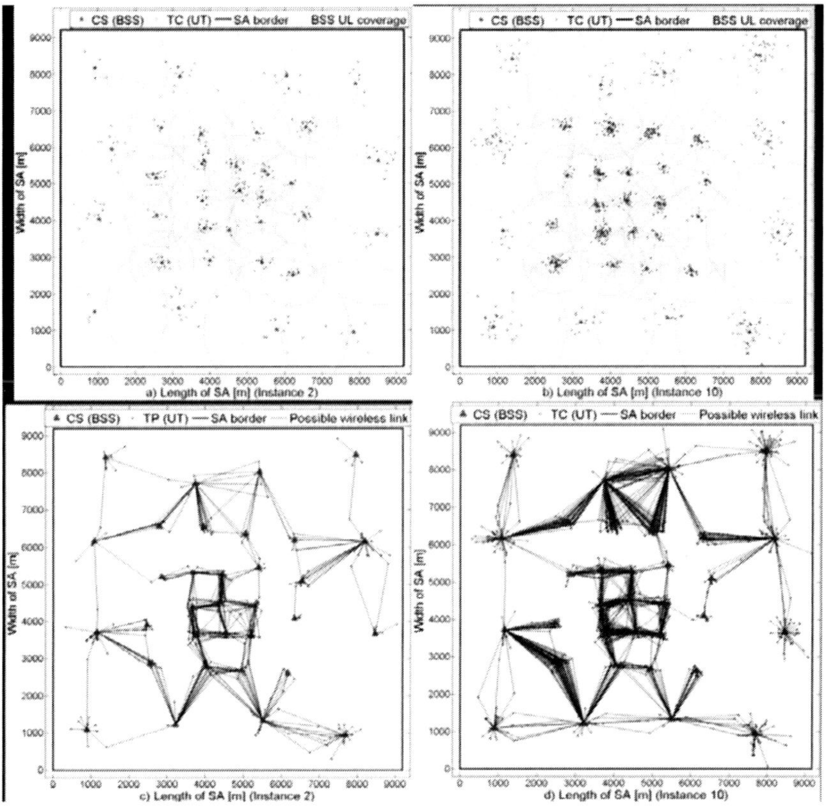

Figure 2 Allocation of CSs (BSSs) and TPs (UTs) inside SA for (a) one-sector BSSs (Instance 2), (b) three-sector BSSs (Instance 10), Visualization of potential wireless links among CSs (BSSs) and TPs (UTs) for (c) one-sector BSSs (Instance 2), (d) three-sector BSSs (Instance 10).

Hence, the first six analyzed network instances (indicated as Instances 1–6) have consisted of one-sector BSSs (e.g., Instance 2 in Figure 2a). The other six instances indicated as Instances 7–12 contain three-sector BSSs (e.g., Instance 10 in Figure 2b). Although most UMTS network architecture nowadays combine single and multi-sector BSSs, we have decided to investigate these two different network types separately, due to expected differences in energy savings potential. For each of the analyzed instances, we performed a random allocation of 33 macro BSSs inside a 9.2 × 9.2 km^2 SA (Figure 2a,b). Such a square SA size equal to approximately 85 km^2 corresponds to a broader area of a medium sized city. From a practical point of view, it is reasonable to believe that the allocated number of BSSs inside an SA of that size fairly corresponds to a number of BSSs that will be allocated by a single operator in the real coverage planning case.

Furthermore, inside the SA of each network instance we have allocated a different number of user terminals (UTs) representing users demanding a specific type of network service (Figure 2a,b). Modeling of each user as an individual UT in the UMTS network can significantly increase the computational complexity of the optimization model. In order to keep computational complexity on an acceptable level whilst ensuring precise information about number and user positions in the network, representations of UTs have been performed through the introduction of traffic clusters (TCs). The TC is envisioned as the cluster (group) comprised of one or more UTs allocated at a close proximity to each other (Figure 2). In addition, we have decided to use two types of TCs denoted as

- TC1—representing a group of five UTs (users) always demanding voice (speech) service, and
- TC2—representing one data UT (user) always demanding a corresponding type of service.

For simplicity, we use in the analyses only two different types of service (voice and data) simultaneously offered to users allocated inside SA of the analyzed UMTS instances. Although in the real-life situations a user with an advance mobile device like a smart phone can in one moment demand voice and in another moment data service, we assume for simplicity that voice users always demand voice service while data users request only data service. Table 3 gives precise information about the allocated TCs and corresponding number of voice and data users for each of the analyzed UMTS network instances. Allocation of the TCs performed randomly inside CAs of BSSs have been illustrated in Figure 2a for the network with one-sector BSSs (Instance 2) and in Figure 2b for the network having three-sector BSSs (Instance 10). It can be noticed that the allocation densities of TCs have been higher around locations of the BSSs. This is related to the selection of BSS locations, which, in the real networks, are performed such that they are always placed on traffic hot spots. As such they offer the best power budget for the mobile users served by those BSSs. Therefore, with this allocation approach, we want the analyzed instances to be as close as possible to the real UMTS networks.

For the same network instances visualized in Figure 2a,b, potential wireless links between BSSs and TCs can be seen for instances with one-sector BSSs in Figure 2c and for instances with three-sector BSSs in Figure 2d. Potential wireless links can be obtained only if the coverage criteria explained in previous section has been satisfied. Therefore, potential wireless links indicate on which BS(s) each of TCs (UTs) can be connected. In Figure 3a,b, it can be perceived that TCs located inside urban-dense CAs have higher possibilities in terms of potential connections with BSs if compared with TCs located within sub-urban

CAs. This is fully in compliance with practical coverage situations of real UMTS networks.

Table 3 Statistics of allocated voice and data TCs inside SA for each of analyzed instances

Types of traffic centroids (TCs)	Number of voice users in TC	Number of data users in TC	Total number of users in each TC	Guaranteed bit rates for data/voice users in TC1/TC2 (kb/s)	Instance number					
					1/7 (1 s BSS/ 3 s BSS)	2/8 (1 s BSS/ 3 s BSS)	3/9 (1 s BSS/ 3 s BSS)	4/10 (1 s BSS/ 3 s BSS)	5/11 (1 s BSS/ 3 s BSS)	6/12 (1 s BSS/ 3 s BSS)
TC1 (type 1)	0	1	1	384/128/64						
TC2 (type 2)	5	0	5	12.2						
Total number of TCs inside SA (one sector BSS)					389	345	359	287	391	379
Total number of voice users inside SA (one-sector BSS)					1660	1365	1470	1005	1680	1600
Total number of data users inside SA (one-sector BSS)					57	70	65	86	55	59
Total number of TCs inside SA (three-sector BSS)					1092	907	849	796	945	1109
Total number of voice users inside SA (three sector BSS)					4725	3570	3210	2880	3810	4830
Total number of data users inside SA 3 (three-sector BSS)					147	193	207	220	163	143
Total number of CSs (BSSs) representing 1 or 3 sector BSS					33	33	33	33	33	33
SA size (km^2)/SA dimensions							84.64 km^2/9.2 × 9.2 km			
Number of MPs/MP density (m × m)							17,161/70 × 70 m			

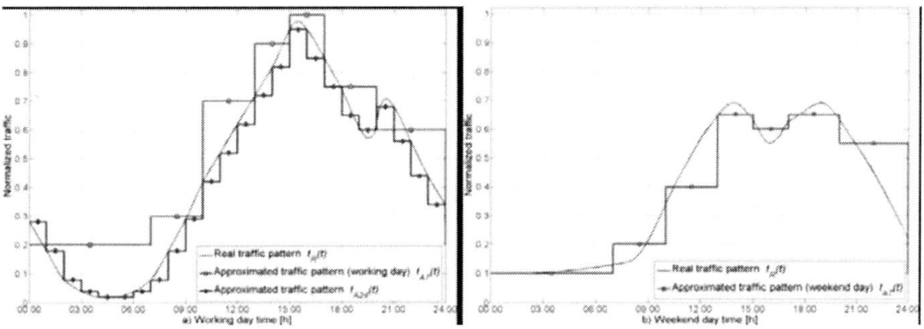

Figure 3 Approximation of the real traffic pattern for (a) working day using seven ($\mathbf{f}_{A7}$) and 24 ($\mathbf{f}_{A24}$) time periods, (b) weekend day using seven ($\mathbf{f}_{A7}$) time periods.

According to Table 4, we assume that an active BS dynamically changes Tx power among four discrete power levels (10, 20, 30, and 40 W) denoted with index *k* in Table 4. Based on 3rd Generation Partnership Project (3GPP) standards [32], we analyzed macro BSs belong to the category of wide area BSs. In general, instantaneous BSS power consumption can be divided into two components: fixed (P_j) and variable (P_k). Fixed power consumption P_j (W) is constant in time and is dedicated to signal processing, analog/digital (A/D) conversion, battery supply, etc. Power amplifier and site cooling are the two most significant contributors to the variable power consumption P_k (W) which increases with enlargement of the Tx power. Therefore, the total instantaneous power consumption of a BSS in any moment can be expressed as a summation of these two components

$$P_{tot} = P_j + P_k \quad [\mathrm{W}]. \tag{3}$$

Table 4 Instantaneous BSS power consumption (one-carrier/sector)

Tx powers		1 s BSS fixed (P_j)/variable (P_k) power consumption (W)	3 s BSS P_{tot}—total instantaneous power consumption (W)	1 s BSS fixed (P_j)/variable (P_k) power consumption (W)	3 s BSS P_{tot}—total instantaneous power consumption (W)
Tx power level *k*	Tx power P_{Tk} (W/dBm)				
4	10/40	330/66.66	396.66	819/259.97	1,087.97
3	20/43	330/133.33	463.33	819/519.9	1,338.9
2	30/44.77	330/200	530	819/780	1,599
1	40/46	330/266.66	596.66	819/1039	1,858

Total power consumptions for one- and three-sector BSSs transmitting at different Tx power levels are presented in Table 4. Assuming *one frequency carrier* per BSS sector, instantaneous BSS power consumptions for each of the four Tx power levels have been presented in Table 4.

4.1. Traffic approximation

In order to model the influence on network energy consumption of the BS switching granularity, we performed an approximation of the real traffic pattern obtained for the cellular access networks through on-site measurements [33]. Approximation is based on discretization of the real traffic pattern $f_R(t)$ using three distinct approximation approaches. The first approach assumes equal traffic patterns for working and weekend days. According to Figure 3a, the discretization function $f_{A\,7}(t)$ is based on seven discrete time periods during 1 day. The second approximation approach is also based on the same number of discrete time periods (seven), but it takes into account differences among traffic profiles of the working and weekend days (Figure 3a,b). The third approximation, also presented in Figure 3a, assumes finer-grained discretization with 24 time periods ($f_{A\,24}$) in the case of a working day only. Table 5 lists for each approximation the beginning and ending times of each time period with the time duration of each time period and percentages of active users. Using such approximation approaches, besides the impact of switching granularity, we can also perceive the influence of different working-day and weekend-day traffic profiles on network power consumption.

Table 5 Different approaches in approximations of the real traffic pattern

Traffic approximation with 7 time periods						Traffic approximation with 24 time periods					
Time period t for $f_{A7}(t)$	T_t (h)	T_{t+1} (h)	$\Delta T_t = T_{t+1} - T_t$ (h)	% of active users (TCs) working day	% of active users (TCs) weekend day	Time period t (*hh–hh*) for $f_{A24}(t)$	% of active users (TCs) $f_{A24}(t)$	Time period t (*hh–hh*) for $f_{A24}(t)$	% of active users (TCs)	Time period t (*hh–hh*) for $f_{A24}(t)$	% of active users (TCs)
1 (8)	00	07	7	20	10	1 (00–01)	28	9 (08–09)	18	17 (16–17)	85
2 (9)	07	10	3	30	20	2 (01–02)	18	10 (09–10)	29	18 (17–18)	75
3 (10)	10	13	3	70	40	3 (02–03)	8	11 (10–11)	42	19 (18–19)	65
4 (11)	13	15	2	90	65	4 (03–04)	4	12 (11–12)	52	20 (19–20)	60
5 (12)	15	17	2	100	60	5 (04–05)	2	13 (12–13)	62	21 (20–21)	68
6 (13)	17	20	3	75	65	6 (05–06)	2	14 (13–14)	72	22 (21–22)	56
7 (14)	20	24	4	60	55	7 (06–07)	4	15 (14–15)	82	23 (22–23)	44
/	/	/	/	/	/	8 (07–08)	8	16 (15–16)	95	24 (23–00)	34

5. CAPACITY PLANNING

In order to make a realistic estimation of the number of users that can potentially be served by the UMTS network, complete network capacity planning including capacity planning of each BSS needs to be performed. To do this, distribution of traffic to the different voice and data calls inside each BSS CA should be known as accurately as possible. Hence, in the dimensioning phase, a fixed load for all BSSs within the SA is assumed. The value for the load can be the *maximum acceptable load* (MAL) per sector (cell) or it can be the *predicted load* (PL) per sector during the busy hour. Both of these loads will be used in this article for estimation purposes.

The MAL is maximal load that can be accepted by BSS sector(s) according to the technological hardware and software capabilities of BSS. Values of MAL define the maximal number of TCs (voice and data users) that can simultaneously be served by each BSS while transmitting at some Tx power level. Actually, the idea is to use the complete BSS capacity contained within the MAL when a BSS needs to take over the load of those BSSs that go to off state due to energy-efficient resource management. On the other hand, we use the term PL to express a realistic number of TCs that will be allocated for simulation purposes inside the CA of a BSS during busy hours. Hence, the PL is the highest load served by BSS which periodically occurs each day during busy hour. Values of the PL define maximal number of TCs which will be allocated inside each BSS CA of analyzed network instances. The PL is always lower than the MAL and in practice ranges between 70 and 90% of the MAL. Therefore, in the performed capacity planning we consider a value of average PL $\bar{\eta}$ equal to 80%.WCDMA *capacity planning* is directly related to the power budget (Tx power) and thus to the BSS CA. For that reason, coverage and capacity planning have been made together in this article. Two approaches in the capacity planning known as *uplink* (UL) or *downlink* (DL) capacity planning can be used for estimating the number of users simultaneously served by UMTS BSS. Since the DL load is generally higher than the UL load, our capacity planning for analyzed networks has been done based on DL capacity estimation. In the DL, there are two aspects that should be considered. First, the interference from other cells $\bar{\iota}$, and, the Tx power P_{Tk} of the BS. Computation of the theoretical WCDMA load is straightforward in the DL and is based on the so-called capacity load (pole) equations. For the calculation of an average BS load $\bar{\eta}$ in the DL direction we use the following equation [30, 34]

$$\bar{\eta} = \sum_{z=1}^{q} \sum_{i=1}^{n} \frac{(E_b/N_o)_{z,i} \nu_{z,i}}{W/R_{z,i}} [(1-\bar{\alpha}) + \bar{\iota}] \tag{4}$$

where index q defines maximal number of different services offered to the users, and n corresponds to the maximal number of simultaneously served users of z th service. In addition, meanings of each parameter used in relation (4) and corresponding numerical values can be found in Table 6. The selected values are typical for UMTS networks and references for the selected values can be found in [30, 34]. Recoiling previous assumptions, we envisioned two different types of service (q = 2); voice (TC1) and data (TC2). Although UMTS systems use *adaptive multirate* speech coder [27], which is actually several coders in one providing dynamic changes of different speech coding rates during a call, for simplicity we performed analyses assuming that UMTS network always ensures fixed rates of 12.2 kb/s for voice service. In addition, we performed analyses assuming that UMTS network guarantees inside a complete SA one of three fixed data rates equal to: 384, 128, and 64 kb/s. These voice/data rates are ensured in each part of the BSS CA and they are typical for modern UMTS networks deployed worldwide. Although a network adopts the data rates of each individual user according to the quality of the radio-propagation link, we neglect this fact since one of our goals is to determine the influence of a guaranteed service rate on network energy consumption. Therefore, we perform analyses assuming that in any moment, the network simultaneously offers to the voice users a rate of 12.2 kb/s for ensuring voice service, and only one of the three data rates for guaranteeing data service to the data users. From relation (4), interdependence between number of voice (n_V) and data (n_D) users that can simultaneously be served by a single BSS can be expressed as

$$n_D = \frac{\bar{\eta}W - [(1-\bar{\alpha})+\bar{\iota}]\ n_V (E_b/N_o)_V R_V \nu_V}{[(1-\bar{\alpha})+\bar{\iota}]\ (E_b/N_o)_D R_D \nu_D}. \tag{5}$$

Because the output Tx power (P_{Tk}) of the BS transmitter is detected among all users inside a sector, it is possible that the power of the transmitter(s) limits the capacity in the DL direction. The required BS Tx power (P_{Tk}) for the case of a single service type is given in the following equation [27, 31]

$$P_{Tk} = \frac{N_{rf} W \bar{L} \sum_{i=1}^{n} \frac{(E_b/N_o)_i \nu_i}{W/R_i}}{1 - \sum_{i=1}^{n} \frac{(E_b/N_o)_i \nu_i}{W/R_i} [(1-\bar{\alpha})+\bar{\iota}]} \quad [\text{dBm}]. \tag{6}$$

Table 6 Parameters used for capacity estimation of BSS

Parameter description	Index	Value
Mean other-to-own cell interference factor	$\bar{\iota}$	0.55 (1 s BSS)/0.65 (3 s BSS)
Maximal number of different services in the cell	q	2
System chip rate (bandwidth)	W	3,840,000.00 chips/s
Bit energy per noise spectral density (interference) ratio	E_b/N_o	6.5 dB (12.2 kb/s)
		5.5 dB (64 kb/s)
		5 dB (128 kb/s)
		4.5 dB (386 kb/s)
Bit rate of user i demanding zth type of service	$R_{z,i}$	12.2/64/128/384 kb/s
Activity factor in DL for ith TC demanding zth type of service	$V_{z,i}$	0.58 (voice)/0.7 (data)
Mean orthogonality factor	$\bar{\alpha}$	0.5
Noise spectral density of the receiver front end	N_{rf}	−167 dB
Mean path loss margin		10 dB
Path loss at cell border		149.4 dB
Mean path loss between BS and UT	$\bar{L}$	139.4 dB
DL mean cell (sector) load (PL)	$\bar{\eta}$	80%

Values of newly introduced parameters such as noise spectral density N_{rf} of the receiver front end and the average path loss between BS and user (mobile) terminal L^{---}

can be found in Table 6. Without loss of generality, we assume that the feeder and combiner losses have been compensated with the antenna gain, and therefore an effective isotropic radiated power of BS, denoted as Tx power P_{Tk}, is equal to the power at the output of a power amplifier.

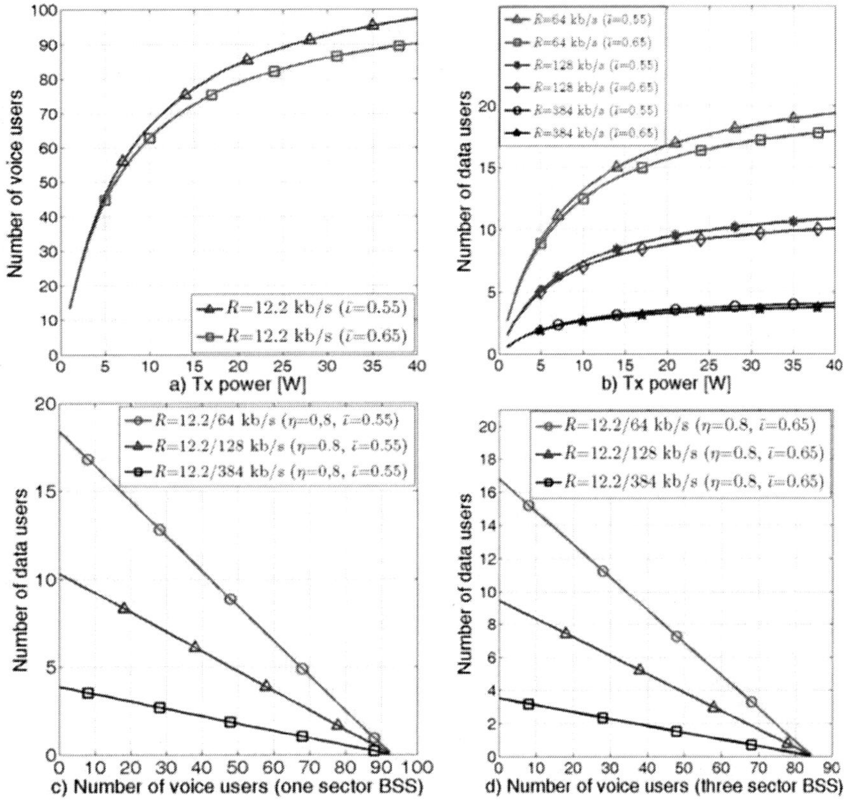

Figure 4 Estimated capacity of 3G BS for different Tx powers in terms of (a) maximal number of voice users, (b) maximal number of data users, estimated interdependence of simultaneously connected voice and data users for case of (c) one-sector 3G BSS, (d) three-sector 3G BSS.

Table 7 Estimated maximal number of voice and data TCs that can independently be served by single BSS for different service rates

Level of sector Tx power	Sector Tx power	One-sector BSS capacity c_{kz} service rates: voice/data 12.2/384 kb/s		Three-sector BSS capacity c_{kz} service rates: voice/data 12.2/384 kb/s		One-sector BSS capacity c_{kz} service rates: voice/data 12.2/128 kb/s		Three-sector BSS capacity c_{kz} service rates: voice/data 12.2/128 kb/s		One-sector BSS capacity c_{kz} service rates: voice/data 12.2/64 kb/s		Three-sector BSS capacity c_{kz} service rates: voice/data 12.2/64 kb/s	
k	P_k (W)	TC2 number	TC1 number	TC2 number	TC1 number	TC2 number	TC1 number	TC2 number	TC1 number	TC2 number	TC1 number	TC2 number	TC1 number
1	40	4	20	12	54	11	20	30	54	19	20	54	54
2	30	3	19	9	51	10	19	27	51	18	19	51	51
3	20	3	17	9	48	9	17	24	48	17	17	48	48
4	10	2	13	6	36	8	13	21	36	14	13	39	36

According to relation (5), it is possible to develop an equation which defines the maximal number of users (MAL) that can be served in the cell for different Tx powers of BS as

$$n_{D(V)} = \frac{0.85 \times P_{Tk}}{(E_b/N_o)_i v_i R_i \left\{ \frac{0.85 \times P_{Tk}}{W} [(1-\bar{\alpha}) + \bar{\imath}] + N_{rf}\bar{L} \right\}} \cdot \quad (7)$$

In relation (7), the total BS Tx power P_{Tk} is scaled with a value equal to 0.85, since 15% of the total Tx power of BS goes to the WCDMA control channels. Using relations (5) and (7), we manage to perform capacity planning of the analyzed UMTS networks. The capacity estimations in terms of maximal number of voice and data users (MAL) which can independently be served by single BS sector at different Tx powers have been presented in Figure 4a,b, respectively. Estimations have been performed according to the relation (7) for the case of different other-to-own cell interference factors that are characteristic for one (omni-directional) sector ($\bar{\imath}$ = 0.55) and three-sector ($\bar{\imath}$ = 0.65) sites [30, 34]. Obviously, for the site capacitating concept it can be said that the higher the Tx power, the higher the number of voice or data users that can be (separately) served by BSS.

Based on the estimations presented in Figures 4a,b, we can calculate the maximal number of voice and data TCs which can independently be served by a BSS while transmitting at one of the Tx power levels k. Capacities have been presented for one-sector and three-sector BSSs in Table 7. These maximal BSS capacities c_{kz} are calculated for the cases of guaranteed voice/data rate combinations equal to 12.2/384, 12.2/128, and 12.2/64 kb/s. Table 7 indicates that guaranteeing service of lower data rates results in higher number of data TCs (TC2) which can simultaneously be served by a single BSS for the same Tx power levels. Although BSS capacity significantly depends on many factors such as interference (which is the most dominant), obtained capacity values are typical for most one (omni-directional) and three-sector BSSs nowadays.

In addition, assuming 80% of the PL for a single BSS sector, Figure 4c,d presents interdependence between the number of voice and data users that can simultaneously be served by single sector ($\bar{\imath}$ = 0.55) and three-sector ($\bar{\imath}$ = 0.65) BSSs, respectively. Figure 4 indicates that with an increase in the number of served data users, the number of simultaneously served voice users decreases, and vice versa. Also, extreme points on the abscissas and ordinates of the graphs in Figure 4c,d present two adversary cases indicating the maximal number of voice and data users that can be served independently by BS for a PL equal to 80% ($\bar{\eta}$=0.8).

6. OPTIMIZATION MODEL

In order to mathematically formulate the problem of energy-consumption optimization of a whole UMTS network, we use ILP methods. More specifically, we develop an optimization model that is based on BIP, assuming that the network scales energy consumption in accordance with quantity of guaranteed rates for data (voice) service and daily variations in number of active users. Scaling of energy consumption is based on the possibility to completely turn off some of the BSSs in situations when the number of active users inside the CA of those BSSs becomes low. In addition, to accommodate guaranteed service rates for a diverse number of voice and data TCs, each BSS can dynamically adopt a Tx power by choosing from the set of Tx power levels denoted as $k \in K$ (Table 4). Since dynamic adaptation of Tx power has a direct influence on instantaneous BSS power consumption, such an approach contributes to reductions of network energy consumption.

Mathematical formulation of the optimization problem is based on a straightforward approach typical for radio coverage problems. Using such an approach, we define with $j \in J$ a set of CSs hosting BSS equipment and with $i \in I$ a set of TCs where each UT is located. We model the affinity between TCs and CSs (Figure 2c,d) by means of a bipartite and undirected graph $G(I,J)$. Let $G(I,J)$ be a graph with vertex set $V = \{I \cup J\}$ and such that there is an edge between $i \in I$ and $j \in J$ if and only if TC i can reach BSS j in the UL direction. Since each CS j can ensure the transport of voice and data traffic, we define with $z \in Z$ a set of network services simultaneously offered to TCs. Each of the network services is delivered to the TCs during a set of time periods $t \in H$ (Table 5), and the time difference between consecutive time periods is denoted as $T_{t+1} - T_t$. Annotation of all sets, subsets, and indices used in the mathematical model has been listed in Table 8.

Table 8 Annotation of sets and indexes used in mathematical model

Set description	Set index	Set structure	Maximum number of set elements
Set of all CSs: J	j	$j \in J = \{1, \ldots, m\}$	m
Set of all test points (TCs): I	i	$i \in I = \{1, \ldots, n\}$	n
Set of all Tx power levels: K	k	$k \in K = \{1, \ldots, l\}$	l
Set of all time periods: H	t	$t \in H = \{1, \ldots, p\}$	p
Set of all service types (voice/data) offered to users	z	$z \in Z = \{1, \ldots, q\}$	q
Set of all measurement points (MPs): S	s	$s \in S = \{1, \ldots, u\}$	u
Subsets of Tx power levels detected for (i,j) combination: $K(ij)$	k	$k \in K = \{1, \ldots, l_{ij}\}$	l_{ij}
Subsets of TCs covered with (j,k) combination: $I(jk)$	i	$i \in I(jk) = \{1, \ldots, n_{jk}\}$	n_{jk}
Subsets of TCs covered with (j,k,r) combination: $I(jkr)$	i	$i \in I(jkr) = \{1, \ldots, n_{jkr}\}$	n_{jkr}
Subset of CSs allocated at points of basic grid (BG): J^*	j	$j \in J^* = \{1, \ldots, m_{BG}\}$	m_{BG}
Sets of CS configuration pairs covering TC i: $JK(i)$ Indexing of $JK(i)$ set members (pairs): $b(i)$	j_b, k_b, h	$j_b k_b \in JK(i) = \{(j_1,k_1), \ldots, (j_{B(i)},k_{B(i)})\}$ $h \in b(i) = \{1, \ldots, B(i)\}$	$j_{B(i)}, k_{B(i)}, B(i)$
Subsets of TCs covered with (j,k) combination in time period t: $I(jkt)$	i	$i \in I(jkt) = \{1, \ldots, n_{jkt}\}$	n_{jkt}
Subsets of TCs covered with (j,k,r) combination in time period t: $I(jkrt)$	i	$i \in I(jkrt) = \{1, \ldots, n_{jkrt}\}$	n_{jkrt}

To model the capacity of a BSS for a specific type of service z when the BSS transmits at k th Tx power, we introduce the parameter c_{kz} (Table 7). According to Table 7, the capacity parameter c_{kz} defines the maximal number of voice and data TCs that can simultaneously be served by a single three-sector or one-sector BSS, when transmitting at different Tx power levels k. As explained in the previous section, numerical values of the parameter c_{kz} for different combinations of guaranteed voice/data service rates have been calculated according to relations (5) and (7). In addition to the capacity parameter c_{kz}, we use four binary parameters which are actually 0–1 incidence matrices defined as

$$a_{ijkt} = \begin{cases} 1 & \text{if TC } i \text{ is covered during time period } t \\ & \text{by CS } j \text{ transmitting at } k - th \text{ Tx power} \\ 0 & \text{otherwise} \end{cases}$$

$$v_{iz} = \begin{cases} 1 & \text{if TC } i \text{ belongs to } z - th \text{ type of network} \\ & \text{service} \\ 0 & \text{otherwise} \end{cases}$$

$$d_{it} = \begin{cases} 1 & \text{if TC i is in active (on) state during} \\ & \text{time period t} \\ 0 & \text{otherwise} \end{cases}$$

$$b_{sjk} = \begin{cases} 1 & \text{if s} - \text{th measurement point is covered} \\ & \text{by j} - \text{th CS transmitting at k} - \text{th Tx} \\ & \text{power level} \\ 0 & \text{otherwise} \end{cases}$$

Binary parameter a α_{ijkt} defines affiliation of TC i inside the CA of the j th CS transmitting at the k th Tx power during time period t. v_{iz} is binary parameter which defines belonging of the i th TC to the specific service type z, while d_{it} denotes activity state of the i th TC during each time period t. If the i th TC is active during time period t, we assume an active state of all voice (TC1) or data (TC2) users belonging to that TC. More precisely, it can be said that all users in an active TC demand corresponding capacity resources of the UMTS network.

Coverage parameter b_{sjk} indicates the determination of the s th measurement point (MP) inside the CA of the j th CS, in the case when the CS transmits at the k th Tx power level. For ensuring full coverage of a complete SA during all time periods, we introduce in our work [14] the concept of MPs which is based on a set $s \in S$ (Table 8) of MPs virtually allocated inside the SA. In this article, we also exploit this concept assuming regular dispersion of MPs inside the SA in a grid manner with 70 × 70 m^2 distances between grid points (Table 3). In each of the 17,161 MPs virtually allocated inside the SA, the signal strength in the UL direction during each time period must be in compliance with the sensitivity threshold stated in Table 1.

The problem is to find a minimum cardinality subset of powered on CSs (BSSs) with minimal energy consumption while enabling coverage and capacity demand of all TCs (UTs) which are active in each time period. To solve the stated optimization problem, we use three different binary decision variables defined as

$$w_{ijkt} = \begin{cases} 1 & \textit{if TP i is assigned to jth CS transmitting} \\ & \textit{at kth Tx power level during time period t} \\ 0 & \textit{otherwise.} \end{cases}$$

$$x_{\text{jkt}} = \begin{cases} 1 & \text{if an j} - \text{th CS during time period t} \\ & \text{transmits at k} - \text{th Tx power level} \\ 0 & \text{otherwise} \end{cases}$$

$$w_{ijkt} = \begin{cases} & \text{if TP i is assigned to j} - \text{th CS} \\ 1 & \text{transmitting at k} - \text{th Tx power level} \\ & \text{during time period t} \\ 0 & \textit{otherwise.} \end{cases}$$

The first *decision variables* y_{jt} are typical for covering problems, denoting, as part of a final solution, which CSs are active in each time period. *Decision variables* x_{jkt} must be introduced in order to define at which Tx power level each powered on the *j* th CS transmits during time period *t*. Variables w_{ijkt} are *assignment variables* defining which TP *i* is in a particular time period *t* connected to which CS. A mathematical expression of the optimization problem is given with the next relations

$$\begin{aligned} \text{Min} & \left[\sum_t \sum_j P_j y_{jt} (T_{t+1} - T_t) \right. \\ & \left. + \sum_t \sum_j \sum_k P_k x_{jkt} (T_{t+1} - T_t) \right] \times C_1 \\ & + \left[\sum_t \sum_j P_j y_{jt} (T_{t+1} - T_t) \right. \\ & \left. + \sum_t \sum_j \sum_k P_k x_{jkt} (T_{t+1} - T_t) \right] \times C_2 \end{aligned} \quad (8)$$

s.t.

$$\sum_k x_{jkt} \leq y_{jt} \qquad \forall(j,t) : j \in J = \{1,\ldots,m\}, \quad \forall t \in H = \{1,\ldots,p\} \tag{9}$$

$$\sum_j \sum_k a_{ijkt} x_{jkt} \geq 1 \quad \forall(i,t) : i \in \{1,\ldots,n\}, \quad \forall t \in H = \{1,\ldots,p\} \wedge \forall \mathrm{d}_{it} \neq 0 : \mathrm{a}_{ijkt} \neq 0 \tag{10}$$

$$\sum_{i \in I(jkt)} \sum_z \frac{d_{it} v_{iz} w_{ijkt}}{c_{kz}} \leq 1 \qquad \forall(j,k,t) : j \in J = \{1,\ldots,m\}, \quad \forall k \in \{1,\ldots,l\}, \quad \forall t \in H = \{1,\ldots,p\} \tag{11}$$

$$x_{j_b k_b t} + \sum_{h=b+1}^{B} w_{ij_h k_h t} \leq 1 \quad \forall(i,t) i \in \{1,\ldots,n\}, \quad t \in \{1,\ldots,p\}, d_{it} \neq 0, \quad s.t \forall : \quad 1 \leq b \leq B-1 \tag{12}$$

$$w_{ij_h k_h t} \leq x_{j_b k_b t} \quad \forall(i,t) : i \in \{1,\ldots,n\}, \quad t \in \{1,\ldots,p\}, d_{it} \neq 0, \quad s.t \forall : \quad 1 \leq b \leq B-1 \quad h = b \tag{13}$$

$$\sum_{h=1}^{B} w_{ij_h k_h t} = 1 \quad \forall(i,t) : i \in \{1,\ldots,n\}, t \in \{1,\ldots,p\}, \quad d_{it} \neq 0, \qquad h = b \tag{14}$$

$$\sum_j \sum_k b_{sjk} x_{jkt} \geq 1 \qquad \forall(s,t) : s \in S = \{1,\ldots,u\}, \quad \forall t \in H = \{1,\ldots,p\} \tag{15}$$

$$y_{jt} \in \{0,1\} \quad \forall j \in \{1,\ldots,m\}, \forall t \in \{1,\ldots,p\} \tag{16}$$

$$x_{jkt} \in \{0,1\} \quad \forall j \in \{1,\ldots,m\}, \forall k \in \{1,\ldots,l\}, \forall t \in \{1,\ldots,p\} \tag{17}$$

$$w_{ijkt} \in \{0,1\} \quad \forall i \in \{1,\ldots,n\}, \forall j \in \{1,\ldots,m\}, \forall k \in \{1,\ldots,l\}, \forall t \in \{1,\ldots,p\} \tag{18}$$

Objective function (8) minimizes the monthly energy consumption (kWh/monthly) of a complete UMTS network. In the case of traffic approximations neglecting differences among traffic patterns of working and weekend days realized with seven and 24 time periods, the expression in brackets is multiplied by C_1 and C_2 transformation factors (C_1 = 0.03 1/month C_2 = 0 1/month). When approximation of the weekend and working day traffic patterns using seven time periods took place, transformation factors C_1 and C_2 have next values: C_1 = 0.022 1/month and C_2 = 0.008 1/month. They are used for transforming energy consumption from daily into monthly. The first term in brackets of the objective function (8) refers to the fixed component (P_j), while the second term concerns the variable component (P_k) of the daily network power consumption.

Ten remaining constraints (9–18) ensure that the characteristics of the proposed optimization model are as close as possible to those of the real word UMTS network. Constraints:

- (9) are coherence constraints forcing each CS to transmit using only one Tx power level,
- (10) are coverage constraints mandating that in any moment each TC must be inside the CA of at least one activated CS,
- (11) are capacity constraints which limit the total capacity of each CS according to Tx power,
- (12) are best power selection constraints forcing connection of each TC to CS from which it detects the highest signal strength,
- (13) are configuration constraints mandating that an active TC can be connected to a CS only if this CS transmits at some Tx power,
- (14) are connection constraints ensuring connection of each active TC to maximally one CS,

- (15) are coverage constraints ensuring permanent signal coverage of each MP allocated inside a complete SA,
- (16–18) are integrality constraints.

Such a set of constraints ensures that the proposed model takes into account some of the most important parameters of the UMTS networks, like capacity and coverage. Capacity constraints (11) define the maximal number of simultaneously connected voice and data users on one CSs (BSSs) transmitting at some specific Tx power. For different Tx power levels, capacity constraints perform the site capacitating through concurrent adaptation of BSS capacities regarding the maximal number of voice and data users served in parallel. By satisfying coverage constraints (15) at the position of each densely allocated MP, it is reasonable to believe that 95% coverage of the SA is guaranteed during each time period. The presented optimization model is programmed using a prominent programming language known as A Modeling Language for Mathematical Programming.

6.1. Instance generator

In order to generate the network instances described in Section 4, we have developed an instance generator (IG). The IG is programmed using C++, and data obtained at the end of the generation process are used as input data for the optimization solver. Based on a predefined set of input parameters, the IG generates analyzed UMTS network instances.

Decisions about the number of voice (TC1) and data users (TC2) that will be allocated inside the CA of a single CS (BSS) have been performed based on Table 9 developed with respect to the graphs shown in Figure 4c,d. According to these graphs, we assume capacity planning for the case of 80% network load (PL) with guaranteed 12.2/384 kb/s voice/data rates. On graphs presented in Figure 4c,d, we select five graph points indicating the relationship between the maximal number of simultaneously served voice and data users. Table 9 presents chosen combinations used for allocation. The IG randomly picks one of five possible voice/data (TC1/TC2) combinations, and performs a random allocation of TCs inside the CA of each BSS. Such an approach offers randomness in dispersion of voice and data users, ensuring similarity with user dispersion characteristics for real networks. In addition, with this approach to TC allocation, we ensure consistency in network capacity planning which is close to the real ones.

Table 9 Combinations for generation of TCs inside CA of a single BSS

TC type	One-sector BSS					TC type	Three-sector BSS				
TC1	0	1	2	3	4	TC1	0	3	6	9	10
TC2	18	13	9	4	0	TC2	51	36	21	6	0

7. NUMERICAL RESULTS

Optimization results have been obtained at the output of a state of the art CPLEX optimization solver. For each analyzed network instance, we obtain three optimization results; one per envisioned service type defined as combination of voice (12.2 kb/s) and one (64/128/384 kb/s) data service. Since we separately analyzed six different one-sector and three-sector instances for three different combinations of traffic pattern approximations, we obtained 108 optimization results in total. We believe that such a number of obtained results and analyzed instances is enough to prove consistency of the proposed optimization concept. In order to reduce the complexity of the proposed optimization model, in our analyses we neglected the influence of transients from the off to the on state of BSSs. Hence, optimization problem of every analyzed network instance has been solved to the optimum, within a reasonable computational time.

7.1. Visualization of obtained results

Visualization of the optimization results obtained for traffic approximation with 24 time periods has been presented in Figure 5. This example presents visualization during the second time period ($t = 2$) having only 18% of active TCs (Table 5). For the case of Instance 4, which consisted of one-sector BSSs, Figure 5a,c,e presents visualization of results obtained while ensuring voice/data service rates equal to 12.2/64, 12.2/128, and 12.2/384 kb/s, respectively. Similarly, Figure 5b,d,f visualizes optimization results of the Instance 10 with three-sector BSSs offering the same service rate combinations. Two reasons motivate us to visualize optimization results for the time period ($t = 2$) with such low percentage of active users (TCs).

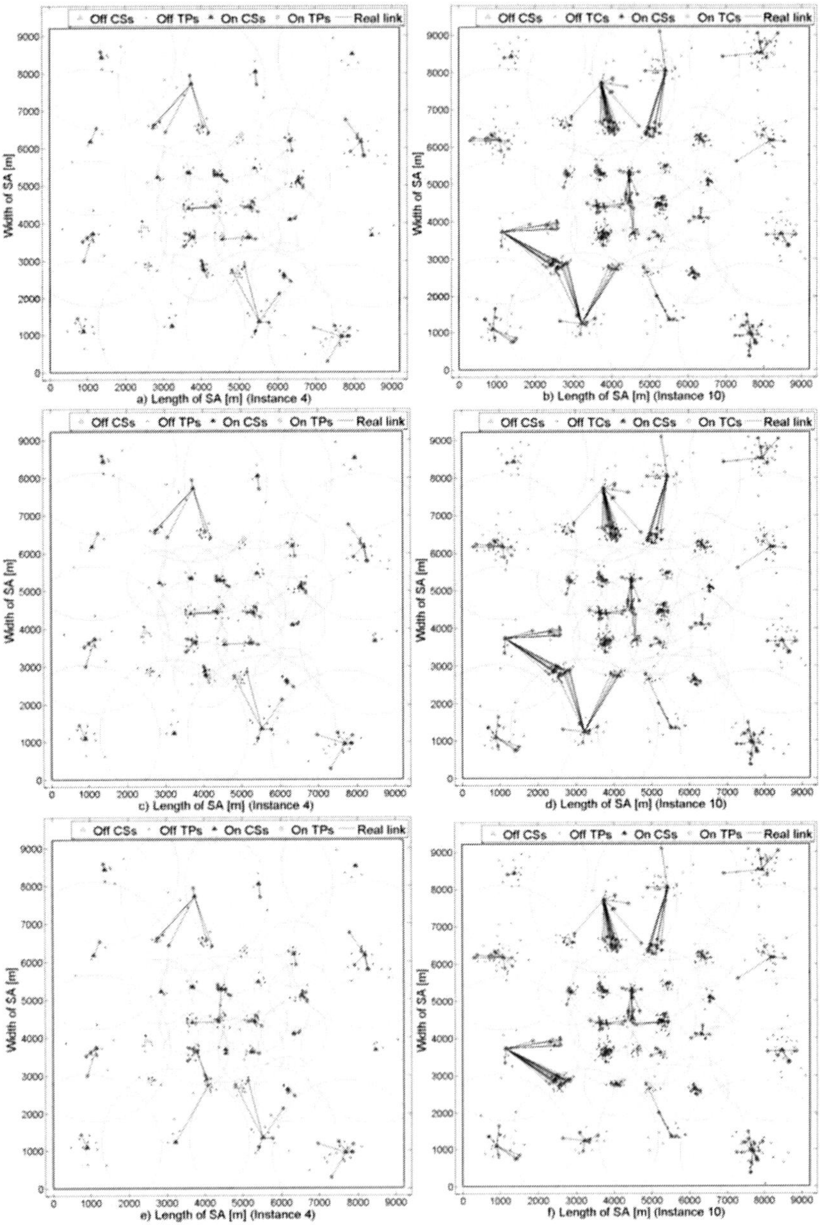

Figure 5 Visualization of optimization results for Instance 4 (one-sector BSSs) with guaranteed voice/data rates: (a) 12.2/64 kb/s, (c) 12.2/128 kb/s, (e) 12.2/384 kb/s; visualization of optimization results for Instance 10 (three-sector BSS) with guaranteed voice/data rates: (b) 12.2/64 kb/s, (d) 12.2/128 kb/s, (f) 12.2/384 kb/s.

Table 10 Statistics of powered off CSs (BSSs) during period of 1 day for instance 10

Time period *t (hh–hh)*	Number/% of powered off BSSs		
	12.2/64 kb/s	**12.2/128 kb/s**	**12.2/384 kb/s**
1(00–01)	8/24.2	8/24.2	8/24.2
2(01–02)	8/24.2	8/24.2	8/24.2
3(02–03)	8/24.2	8/24.2	8/24.2
4(03–04)	8/24.2	8/24.2	8/24.2
5(04–05)	8/24.2	8/24.2	8/24.2
6(05–06)	8/24.2	8/24.2	8/24.2
7(06–07)	8/24.2	8/24.2	8/24.2
8(07–08)	8/24.2	8/24.2	8/24.2
9(08–09)	8/24.2	8/24.2	8/24.2
10(09–10)	8/24.2	8/24.2	8/24.2
11(10–11)	7/21.2	7/21.2	6/18.1
12(11–12)	7/21.2	5/15.1	4/12.1
13(12–13)	5/15.1	5/15.1	3/9.1
14(13–14)	5/15.1	5/15.1	1/3
15(14–15)	4/12.1	3/9.1	0/0
16(15–16)	3/9.1	3/9.1	0/0
17(16–17)	3/9.1	3/9.1	0/0
18(17–18)	5/15.1	4/12.1	2/6.1
19(18–19)	5/15.1	5/15.1	3/9.1
20(19–20)	7/21.2	6/18.1	2/6.1
21(20–21)	5/15.1	5/15.1	1/3
22(21–22)	618.1	5/15.1	4/12.1
23(22–23)	8/24.2	8/24.2	6/18.1
24(23–00)	8/24.2	8/24.2	7/21.2

The first reason is to show results for the time period during which the highest number of CSs (BSSs) will be in the powered off state. This is presented in Table 10 for traffic approximation performed using 24 discrete time periods in 1 day.

According to Figure 5, a considerable number of macro CSs (8 out of 33) are in the powered off state. Active users (TCs) inside CAs of those powered off CSs have been connected to the conceivable powered on CSs. However, slight differences among instances offering different service rates can be noticed regarding the locations of the powered off CSs. This consequently influences on the topology of established wireless connections between TPs and CSs. For example, in the case of results presented for one-sector and three-sector BSSs in Figure 5e,f, somewhat different establishments of wireless connections can be perceived if compared with those on Figure 5c,d, respectively. This is due to the differences in guaranteed service rates, where Figure 5e,f shows optimization results for network instances that must ensure the highest data rates (384 kb/s).

Regarding the second reason, it must be emphasized that with the presented visualization we confirm that even during the time period having the highest percentage (24.5%) of powered off BSSs, each of the analyzed networks preserved full SA coverage. Table 10 gives detailed information about the number and percentage of the powered off CSs for Instance 10 during 1 day. Even if we take into consideration time periods with the lowest number of active users (e.g. t = 5, 6 in Table 5), the percentage of powered off CSs according to Table 10 remains unchanged (24.2%). Therefore, full coverage of the SA with radio signal will not be lost in any moment during a day. This is one of the most important criteria when implementing energy-efficient network management.

To ensure full SA coverage, the network management system is forced to keep powered on CSs in those areas lacking any active TCs (Figure 5). This is the reason why during time periods with lowest user activity, the number of powered off CSs presented in Table 10 cannot be higher than eight. Hence, it is not possible, for the purpose of increasing energy efficiency, to simply shut down those BSSs temporarily missing active users, since this can lead to coverage holes lacking any service accessibility. Conclusions obtained for the two presented network instances are equal to those obtained for other analyzed instances.

7.2. Impact of service rates on power consumption

In order to evaluate the quality of the obtained optimization results, we have introduced a reference model (RM) as a term describing modern cellular networks. The RM is a network model with working properties typical of the radio side of already deployed UMTS networks. This means that for the RM we assume permanent transmission of BSs at the maximal Tx power equal to 40 W. In addition, BSs cannot dynamically be shut down and powered on in accordance with the traffic pattern changes and there are no adaptations of Tx power in order to ensure targeted service rates. Consequently, it is assumed that each one-sector BSS in the RM continuously consumes 596 W, while each three-sector BSS consumes 1858 W (Table 4). The selected values are taken from Table 4 and they correspond to the instantaneous power consumptions of the one-sector and three-sector BSSs in the case of transmitting at the maximal Tx power (40 W). With such an approach, it is possible to perform a comparison of the obtained optimization results with the power consumption of modern day networks represented with the RM.

As an example case, Figure 6 shows changes in instantaneous power consumption of a complete UMTS network for Instances 6 and 12. Changes are presented on hourly bases during 1 day and similar patterns have been obtained for other analyzed instances. On both Figure 6a,b, maximal instantaneous power consumptions during the peak hours are lower than the constant instantaneous power consumption of the RM. This is because the RM represents a traditional UMTS network planning approach, primarily dedicated to accommodating the MAL. In practice, the PL during peak hours is generally lower than the MAL. This is the reason for the optimization model for Instances 6 and 10 resulting in lower instantaneous power consumptions when compared with the RM.

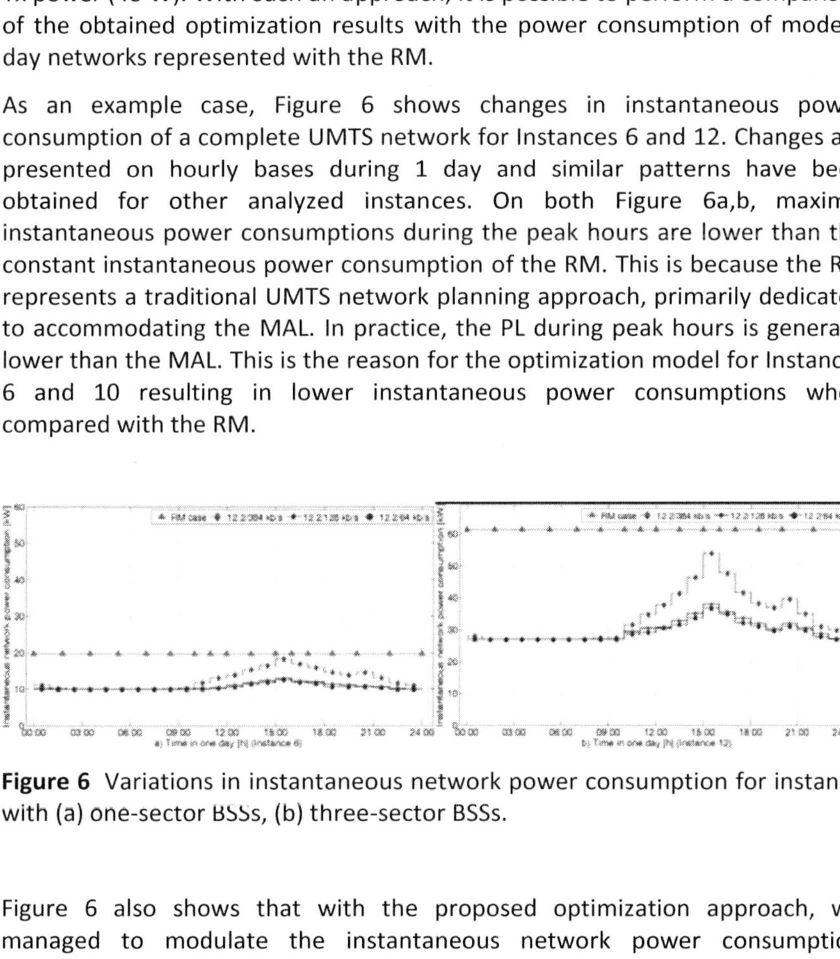

Figure 6 Variations in instantaneous network power consumption for instance with (a) one-sector BSSs, (b) three-sector BSSs.

Figure 6 also shows that with the proposed optimization approach, we managed to modulate the instantaneous network power consumption according to the traffic pattern changes. This results in significantly lower average instantaneous power consumption of the network during 1 day when compared with the RM power consumption. general, network power consumption becomes higher during periods with a higher percentage of active users and decreases when user activity diminishes (Figure 6). This decrease in the network power consumption is primarily followed with an increase in the number of powered off CSs, as presented in Table 10. Since the one-sector BSS consumes nearly three times less power than the three-sector BSS (Table 4), instantaneous network power consumptions expressed in kWh on Figure 6 are almost three times higher for Instance 10, having three-sector BSSs.

Modulation of the instantaneous network power consumption in accordance with the traffic pattern variations is directly reflected in a coefficient of network energy savings, expressed on samples with 1-h duration. Changes in the coefficients of energy savings on an hourly basis during a day have been presented in Figure 7a,b for the one-sector and three-sector BSSs, respectively. Such coefficients are calculated as the ratio of hourly network energy consumptions obtained for analyzed optimization model and RM during each time period. According to Figure 7, lower energy savings have been obtained for the time periods characterized with higher percentages of active users (TCs) and vice versa. During hours of very high user activity (from 13:00 to 18:00 h), the coefficient of energy savings becomes low. For the case of guaranteed 12.2/384 kb/s service rates, Figure 7b shows a fall in the energy savings coefficient to approximately 10%. This is because the network must accommodate increased user traffic, which results in reductions in the number of powered off CSs and consequently in a lower energy savings coefficient. Such a decrease in the number of powered off CSs during peak hours can even reach the value of zero, as presented in Table 10 for the case 12.2/384 kb/s rates. Therefore, dynamic adaptation of BS Tx power should not be neglected, since such a factor contributes to reductions of the monthly energy consumption.

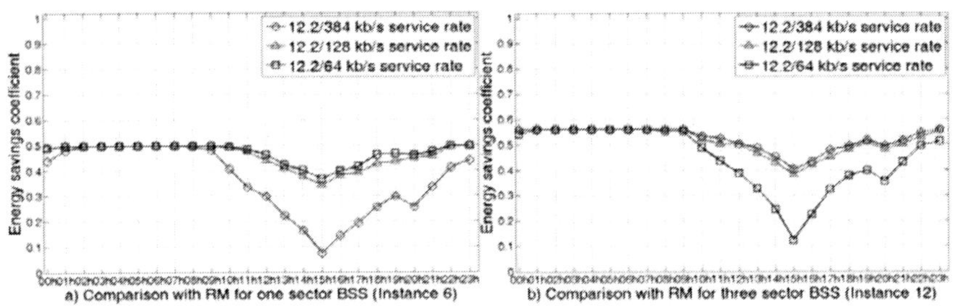

Figure 7 Changes of energy savings coefficient for different service rates in case of instance with (a) one-sector BSSs, (b) three-sector BSSs.

On the other hand, the highest hourly energy savings, reaching 50%, have been obtained during morning hours from 02:00 to 07:00 h. If we consider only traffic pattern variations, the coefficient of energy savings could hypothetically reach even higher values during low activity hours. But, for the reason of ensuring full SA coverage explained previously, the coefficient in Figure 7a,b remains unchanged for all time periods with less than 20% of active TCs.

7.3. Impact of service rates on energy savings

Influence of guaranteed service rates on instantaneous network power consumption is significant, which can be perceived from Figure 6. For the same network instance in terms of allocated TPs and BSSs, ensuring higher service rates along a complete SA results in a higher network power consumption. This is directly reflected in the values of monthly energy consumption presented for each of the analyzed instances in Figure 8. Reason why 12.2/384 kb/s voice/data service causes the highest monthly energy bill can be found in the necessity of activating the highest volume of network resources during one month to accommodate more demanding service rates. In the case of ensuring lower service rates, the volume of network resources regarding the average number of powered on CSs and the level of Tx power engaged during a month can be lower. This results in a lower monthly energy consumption, presented in Figure 8 for 12.2/64 and 12.2/128 kb/s services.

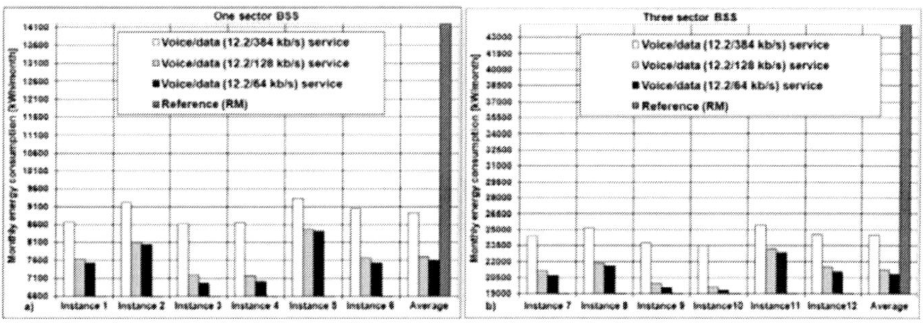

Figure 8 Statistic of monthly energy consumption for instances with (a) one-sector BSSs and (b) three-sector BSSs.

Reasons for the relatively small differences among monthly energy consumptions of the same instances offering 12.2/64 and 12.2/128 kb/s rates (Figure 8) can be found in the network capacity planning approach. In Sections 5 and 7, we emphasize that the allocation of TCs inside CAs of each BSS has been performed assuming guaranteed 12.2/384 kb/s service rates for an 80% network load (PL). Since we use the same instance in terms of number and allocations of TCs when examining influence on network consumption of the other two rate combinations, the network will be over-capacitated in the case of ensuring a 12.2/64 kb/s service rate. Due to excessive network capacity, guaranteeing 12.2/64 kb/s rates will have a slightly lower influence on monthly energy consumptions when compared with the influence caused by 12.2/128 kb/s rates. To have higher influence on monthly energy consumption, a higher number of users should be allocated inside the SA when 12.2/64 kb/s rates will be guaranteed. But, an increase in the number of allocated TCs results in a completely new instance, which eliminates the possibility of making a comparison among guaranteed service rates for the same instance.

Nevertheless, it can clearly be noticed from Figure 8 that ensuring higher service rates imposes higher network energy consumption, even when energy-efficient resource management is used. This is an extremely important cognition when developing future energy-efficient cellular network systems which must satisfy the constant pressure of ensuring higher service rates to mobile users. Statistics presented in Figure 8a,b show variations in monthly energy consumption among analyzed one-sector and three-sector instances, respectively. These differences are primarily due to the fact that we analyze different one-sector and three-sector instances in terms of allocation of BSSs and TCs (users). In addition, the differences among instances regarding the overall number of randomly allocated TCs (Table 3) also contribute to the variations in the total monthly energy consumption. The rightmost vertical bars in Figure 8 indicate the monthly energy consumption of the RM and average consumptions of all analyzed instances while guaranteeing corresponding service rates. We performed averaging of the monthly energy consumption since each of the one-sector and three-sector instances have the same number of macro BSSs with the possibility of transmitting using the same set of Tx powers.

When compared with monthly energy consumption of the RM, each one-sector or three-sector instance has significantly lower monthly consumptions (Figure 8). Obviously, higher differences in monthly consumption can be noticed in Figure 8b for instances with three-sector BSSs, simply due to a higher initial power consumption of the three-sector BSS (Table 4). For example, differences between the monthly energy consumption of the RM and averaged one-sector and three-sector instances when ensuring 12.2/128 kb/s service rates are equal to 6,468.85 and 22,929.35 kWh, respectively. Translated into ecological benefits and financial savings, this equates to an annual reduction in emitted CO_2 of approximately 46 and 162 metric tons and with yearly savings on the energy bill equal to 124,201.92 € and 440,243.52 € (1.6 €/kWh), respectively. Since most modern day networks have been deployed as a combination of a smaller number of one-sector and a higher number of multi (three) sector macro sites, operator savings in terms of energy consumption are much closer to the values obtained for the averaged case having three-sector BSSs. Figure 9 shows percentages of monthly energy savings for each of the analyzed instances, obtained through comparison with the monthly energy consumption of the RM. According to Figure 9, the proposed energy-efficient network management scheme, based on dynamic scaling of Tx power and on/off state of BSSs, gives remarkable savings in monthly network energy consumption. For the case of one-sector instances, reductions in monthly energy consumption range from 35 to 50%, while even higher percentages ranging from 43 to 57% have been obtained for instances with three-sector BSSs.

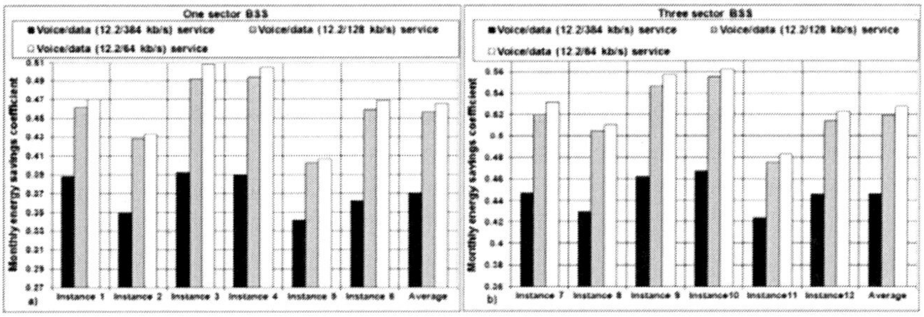

Figure 9 Coefficients of monthly energy savings for instances with (a) one-sector BSSs and (b) three-sector BSSs.

7.4. Impact of switching granularity on power consumption

Figure 10 presents four instances with one-sector and three-sector BSSs changes in the daily power consumption for different traffic pattern approximations. It can be seen that switching granularity influences on instantaneous network power consumption. When compared with traffic approximation based on 7 time periods, approximation with 24 time periods offers lower energy consumption during periods of higher user activity (08:00–24:00). This is because finer-grained temporal control of network resources enables better adaptation to time and space traffic pattern changes. As a consequence, this results in lower instantaneous power consumption of the network. In the period of low user activity (24:00–08:00), the network can keep QoS in terms of coverage and capacity with minimum activated resources. As a result, higher switching granularity in such periods does not have significant influence on the instantaneous power consumption.

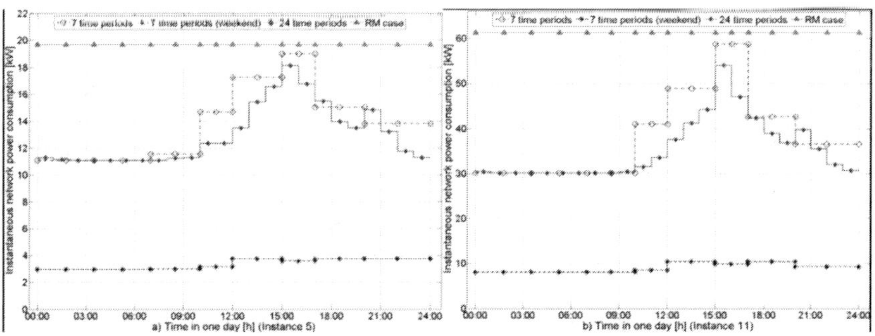

Figure 10 Variations in instantaneous network power consumption for guaranteed voice/data service rates equal to 12.2/384 kb/s in case of (a) one-sector BSS (Instance 5), (b) three-sector BSS (Instance 11).

According to the results presented in Figure 10, the optimal solution in terms of switching frequency during a day could be higher granularity in the periods of higher user activity and very low granularity during periods characterized with low user activity. In addition, from the results presented in Figure 10, it can be perceived that one-sector or three-sector network instances have significantly lower instantaneous power consumption during a weekend day, when compared with consumption during a working day. Therefore, switching granularity in periods of higher user activity during a weekend day can be lower than switching granularity during a working day. Changes in the coefficient of energy savings during a day for three traffic pattern approximations presented in Figure 11 confirm these statements. Figure 11 indicates higher energy savings during a weekend day. This is due to the lower amount of network resources that need to be activated to accommodate lower user activity during a weekend day. However, higher energy-saving coefficients in the period of higher user activity during a working day have been obtained when higher switching granularity is used. This is further confirmation that higher energy savings can be obtained if finer-grained temporal control is used in these periods.

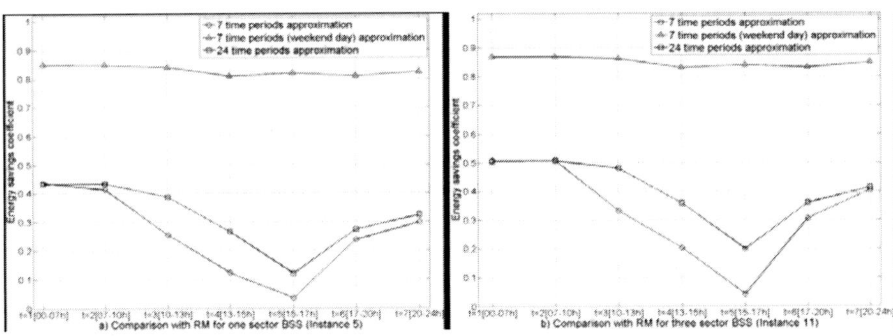

Figure 11 Trends in changes of energy savings coefficient for guaranteed voice/data service rates equal to 12.2/384 kb/s in the case of (a) one sector BSS (Instance 5), (b) three sector BSS (Instance 11).

7.5. Impact of switching granularity on energy savings

Figure 12 indicates the influence of switching granularity on monthly energy savings for all the combinations of voice/data service rates. In general, it can be said that higher switching granularity leads to larger monthly energy savings. It is worth emphasizing that vertical bars for 7/7 time periods indicate coefficients of monthly energy savings taking into account different traffic patterns of weekend and working days. In comparison with traffic approximation neglecting weekend days (7 time periods), higher energy savings between 0.5 and 2.5% per month can be perceived when traffic patterns of weekend days are considered. Even larger differences, ranging between 3 and 5.5%, can be seen among coefficients of energy savings obtained for traffic approximations with lower (7 time periods) and higher (24 time periods) switching granularity. As an example, we emphasize three-sector Instance 10 for 12.2/384 kb/s voice/data service rates with differences in energy consumption on a yearly level equal to 42,590.16 kWh. Translated into financial savings, this leads to a 68,144.25 € annual reduction at the network level. Considering the obtained energy savings, switching granularity will play an important role in the development of energy-efficient network management algorithms. In order to achieve the highest monthly energy savings, switching frequencies of BSs must be closely integrated with the intensity of traffic changes.

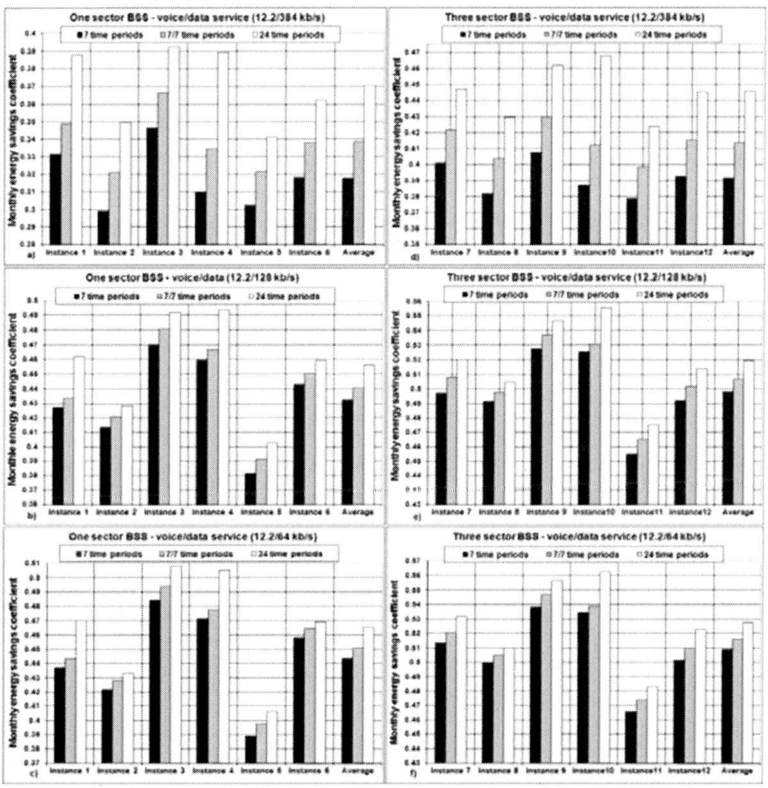

Figure 12 Coefficient of monthly energy savings for each of analyzed one and three sector BSS instance in case of voice/data service with guaranteed rates equal to (a, d) 12.2/384 kb/s, (b, e) 12.2/128 kb/s, (c, f) 12.2/64 kb/s.

8. CONCLUSION

In this article, we have presented a novel optimization model which can be used for energy-saving purposes at the level of a complete cellular access network. The proposed model was tested on a set of real size UMTS network instances, generated using a specially developed IG. This article shows that the proposed model ensures full SA coverage and the targeted service rate requirement through dynamic radio resource management. According to obtained numerical results, remarkable monthly energy savings between 35 and 57% can be obtained, if dynamic BSS switching accompanied with Tx power scaling is implemented in UMTS networks.

The main contribution of this article is the presentation of the influence of BSs switching granularity on the possible reductions of network energy consumption in the time domain. We show that appropriate selection of BSs switching granularity during different periods of working and weekend days can result in significant monthly energy savings. In addition, in this article, we analyze the influence of introducing higher service rates on energy savings potential. Providing higher service rates to the users results in a higher network energy consumption, even when the same energy-efficient network management is in use. Such a cognition additionally emphasizes the need for development of the holistic approaches focused on maximizing energy savings, since future cellular networks will certainly be even more throughput demanding.

Our further research activity will be focused on investigating energy-saving potential when implementing energy-efficient management in the heterogeneous cellular networks consisted of macro, micro, and pico/femto cells. Also, the influence on energy savings of switching the BSs transients from the off to the on states will be investigated more deeply.

DECLARATIONS

Acknowledgments

The authors would like to thank Massimo Bogarelli from Politecnico di Milano for assisting us during research process.

Competing interests

The authors declare that they have no competing interests.

REFERENCES

1. Gammage B: Gartner's Top Predictions for IT Organizations and Users, 2010 and Beyond: A New Balance, (Gartner 2009). 2011. . Accessed 20 September http://www.gartner.com
2. Etoh M, Ohya T, Nakayama Y: Energy consumption issues on mobile network systems. Proceedings of the International Symposium on Applications and the Internet SAINT 2008 2008.
3. Fettweis G, Zimmermann E: ICT energy consumption-trends and challenges. In Proceedings of the 11th International Symposium on Wireless Personal Multimedia Communications – WPMC. Lapland, Finland; 2008.
4. Tafazolli R, Mattila V: eMobility Mobile and Wireless Communications Technology Platform: Strategic Research Agenda, Version 7 (2008). , Accessed 1 May 2011 http://www.emobility.eu.org
5. Chen T, Zhang H, Zhao Z, Chen X: Towards green wireless access networks. In Proceedings of the 7th International ICST Conference on Communications and Networking in China (CHINACOM). Beijing, China; 2010.
6. Ha C, Harrold T, Armour S, Krikidis I, Videv S, Grant PM, Haas H, Thompson JS, Ku I, Wang C-X, Tuan AL, Nakhai MR, Jiayi Z, Hanzo L: Green radio: radio techniques to enable energy-efficient wireless networks. IEEE Commun Mag 2011, 49(6):46-54. 10.1109/MCOM.2011.5783984
7. Correia LM, Zeller D, Blume O, Ferling D, Jading Y, Gódor I, Auer G, Van Der Perre L: Challenges and enabling technologies for energy aware mobile radio networks. IEE Commun Mag 2010, 48(11):66-72. 10.1109/MCOM.2010.5621969
8. Ferling D, Bitzer T, Bohn T, Wiegner D, Pascht A: Power efficient transceivers to enable energy-efficient mobile radio systems. Bell Labs Tech J 2010, 15(2):59-76. 10.1002/bltj.20441
9. Roy S: Energy logic for telecommunications, A White Paper Emerson Network Power Energy Systems. 2008. , Accessed 28 June 2011 http://www.eqlfiles.com/20090615/pdf/Emerson-Q2-Energy-Logic-fo.pdf
10. Hernon D: Enhanced energy efficiency and reliability of telecommunication equipment with the introduction of novel air cooled thermal architectures. Bell Labs Tech J 2010, 15(2):31-51. 10.1002/bltj.20439

11. Bogucka H, Conti A: Degrees of freedom for energy savings in practical adaptive wireless systems. IEEE Commun Mag 2011, 49(6):38-45. 10.1109/MCOM.2011.5783983

12. Saker L, Elayoubi S-E, Chahed T: Minimizing energy consumption via sleep mode in green base station. In IEEE Wireless Communications and Networking Conference – WCNC. Sydney, Australia; 2010. 18–21 April 2010

13. Lorincz J, Capone A, Bogarelli M: Energy savings in wireless access networks. In Proceedings of 5th International Symposium on Wireless Pervasive Computing – ISWPC 2010. Modena, Italy; 2010.

14. Lorincz J, Capone A, Begusic D: Optimized network management for energy savings of wireless access networks. Comput Netw J 2011, 55(3):514-540. 10.1016/j.comnet.2010.09.013

15. Lorincz J, Capone A, Begusic D: Heuristic algorithms for optimization of energy consumption in wireless access networks. KSII Trans Internet Inf Syst J 2011, 5(4):626-648. 10.3837/tiis.2011.04.001

16. Lorincz J, Bogarelli M, Capone A, Begusic D: Heuristic approach for optimized energy savings in wireless access networks. In Proceedings of the 18th International Conference on Software, Telecommunications and Computer Networks – SoftCOM 2010. Split-Bol, Croatia; 2010.

17. Chiaraviglio L, Ciullo D, Meo M, Marsan MA: Energy-aware UMTS access networks. In Proceedings of the 11th International Symposium on Wireless Personal Multimedia Communications – WPMC 2008. Lapland, Finland; 2008.

18. Marsan MA, Chiaraviglio L, Ciullo D, Meo M: Optimal energy savings in cellular access networks. In First International Workshop on Green Communications - GreenComm'09. Dresden, Germany; 2009.

19. Blume O, Eckhardt H, Klein S, Kuehn E, Wajda WM: Energy savings in mobile networks based on adaptation to traffic statistics. Bell Labs Tech J 2010, 15(2):77-94. 10.1002/bltj.20442

20. Oh E, Krishnamachari B, Liu X, Niu Z: Toward dynamic energy-efficient operation of cellular network infrastructure. IEEE Commun Mag 2011, 49(6):56-61. 10.1109/MCOM.2011.5783985

21. Dufková K, Bjelica M, Byongkwon M, Kencl L, Le Boudec J-Y: Energy savings for cellular network with evaluation of impact on data traffic performance. In Proceedings of the 16th European Wireless Conference – EW. Lucca; 2010. 12–15 April 2010

22. Oh E, Krishnamachari B: Energy Savings through Dynamic Base Station Switching in Cellular Wireless Access Networks. In Proceedings of the Global Communication Conference (GLOBECOM 2010). Miami, FL, USA; 2010.

23. Niu Z, Wu Y, Gong J, Yang Z: Cell zooming for cost-efficient green cellular networks. IEEE Commun Mag 2010, 48(11):74-79. 10.1109/MCOM.2010.5621970

24. Videv S, Thompson JS, Haas H, Grant PM: Resource allocation for energy efficient cellular systems. EURASIP J Wirel Commun Netw 2012, 181: 1-15. 10.1186/1687-1499-2012-181

25. Huang Y, Xu J, Qiu L: An energy efficient semi-static power control and link adaptation scheme in UMTS HSDPA. EURASIP J Wirel Commun Netw 2012, 87: 1-13. 10.1186/1687-1499-2012-87

26. Suarez L, Nuaymi L, Bonnin JM: An overview and classification of research approaches in green wireless networks. EURASIP J Wirel Commun Netw 2012, 142: 1-18. 10.1186/1687-1499-2012-142

27. Smith C, Collins D: 3G Wireless Networks. McGrow-Hill Companies; 2002.

28. Seybold JS: Introduction to RF Propagation. Wiley & Sons, New York; 2005.

29. Rappaport TS: Wireless Communications: Principles and Practice. 2nd edition. Prentice Hall PTR, Englewood Cliffs, NJ; 2001.

30. Lempiäinen J, Manninen M: Radio Interface System Planning for GSM/GPRS/UMTS. Kluwer Academic Publishes, Amsterdam; 2001.

31. 3GPP TS 25.101, User Equipment (UE) radio transmission and reception (FDD), 3rd Generation Partnership Project. , Accessed 1 May 2012 http://www.3gpp.org/ftp/Specs/html-info/25-series.htm

32. 3GPP TS 25.104, Base Station (BS) radio transmission and reception (FDD), 3rd Generation Partnership Project. 2012. , Accessed May 1 http://www.3gpp.org/ftp/Specs/html-info/25-series.htm

33. Lorincz J, Garnma T, Petrovic G: Measurements and modelling of base station power consumption under real traffic loads. Sensors J 2012, 12(4):4281-4310. 10.3390/s120404281

34. Laiho J, Wacher A, Novosad T: Radio Network Planning and Optimization for UMTS. 1 and 2 edition. Wiley & Sons Ltd., New York; 2002.

Chapter 10

Cross-Tier Interference Management with a Distributed Antenna System for Multi-Tier Cellular Networks

Massa Ndong* and Takeo Fujii

ABSTRACT

In a multi-tier cellular communication system, the interference from one tier to another, denoted as cross-tier interference, is a limiting factor for the system performance. In spectrum-sharing usage, we consider the uplink cross-tier interference management of heterogeneous networks using femtocells overlaid onto the macrocells. We propose a variation of the cellular architecture and introduce a novel femtocell clustering based on interference cancellation to enhance the sum rate capacity. Our proposal is to use a distributed antenna system (DAS) as an interface to mitigate the cross-tier interference between the macrocell and femtocell tiers. By placing a DAS remote antenna unit (RAU) near a set of femtocells that experience interference from a macrocell user, the DAS can retrieve the interference symbols and feed them back to the femtocells, where each cell can perform interference cancellation when necessary. In addition, the DAS can forward the recovered data to the macrocell base station (MBS); thus, the macrocell user can reduce its transmit power to reach a RAU located closer than the MBS. By distributing the sensor nodes within the macrocell coverage, the proposed scheme can mitigate the cross-tier interference at different locations for several femtocell clusters. Our simulation results show substantial improvement in the network sum rate capacity.

Keywords: Macrocell Femtocell DAS Cross-tier interference mitigation Heterogeneous networks Spectrum sharing Small cell networks

1 INTRODUCTION

Next-generation wireless communication demands enhancement of the cooperation at the multi-tier level to improve the end-user data rate. Cooperation among wireless systems has been demonstrated to be a better alternative for coexisting networks[1]. Contemporary novel wireless network concepts such as heterogeneous networks (HetNets) and multi-tier networks such as femtocells overlaid onto the macrocells require advanced management of the cell load, cross-tier interference, and user access to the spectrum. Spectrum sharing increases to the complexity of meeting the demand in wireless broadband access, thus prompting the emergence of these networks. This process requires the design of spectrum-sharing self-enforcing rules compatible with each individual system[2]. Considering the differences in the transmit power and the near-far problem in two-tier networks[3], our proposal contributes to the design of a coordinator between the two systems, which manages the cross-tier interference and enhances the ergodic sum rate capacity scaling.

In this study, we focus on the cross-tier interference management between a primary system (PS) and a secondary system (SS) in a spectrum-shared scenario. The PS or macrocell tier is composed of a macrocell base station (MBS) and the mobile users that communicate directly with the MBS by default, denoted as macrocell user equipment (MUE). The SS or femtocell tier consists of femtocells. A femtocell is a low-powered user-deployed base station (FBS) that operates in co-channel with the macrocell to deliver high spectral efficiency in closed- or open-access regime[3] to one or two indoor mobile users, denoted as femtocell user equipment (FUE). The femtocell tier constitutes the core technology of small cell networks[4–6].

Femtocells constitute an underlay cell network with the macrocells. Because each FBS is exclusively managed by its end-user, mitigating the cross-tier interference coordination similar to that in traditional cellular networks (individually handled by the operators) becomes difficult. Furthermore, the backhaul exchange information between the MBS and the FBSs becomes tedious for a large number of femtocells because the MBS communicates with the femtocells through a gateway. Our proposal aims to address these challenges through the advanced HetNets scheme and innovative femtocell clustering consequent to the proposed introduction of the distributed antenna system (DAS) interface.

We consider the DAS as referred in our references such as[7]. A DAS consists of signal processing modules and remote radio frequency (RF) antenna element modules denoted as remote antenna units (RAUs). Multiple RAUs can share the same signal processing unit denoted as common processing unit (CPU). Each RAU is connected to a CPU by optical fiber. The RAUs are located nearby premises which shelter FBSs. Therefore, different RAUs can cover several small areas in a heterogeneous coverage while sharing the same processing unit. The wired part of the system, made of optical fiber, presents a negligible error rate and delay. The DAS has a common platform role because it can accommodate different wireless service operators and different protocols. Thus, the DAS is an asset for heterogeneous small cell deployment and can be functionally compared to a radio access network (RAN) aggregation system[8]. As RAN aggregation is considered between operators in[8] to improve throughput and spectrum efficiency; this study considers femtocell integration by the DAS interface.

In our system model, the FBSs are small base stations located inside the premises of the customers. In a close access operating regime, such FBSs allow communication with only their registered mobile users. The RAUs of the DAS are antenna elements located outside the premises sheltering the FBSs. We consider a path loss exponent difference between the wireless communications of each FBS and its registered FUE and each RAU and an unregistered (to the FBS) MUE.

Reference[9] proposed a cellular architecture based on the DAS to address the coverage scarcity in HetNets. The sum rate of the network improved because of the DAS, which added more degrees of freedom to the network. Recently, HetNets have been considered[10] to implement efficient architecture for broadband access in relation to LTE[11] where DAS is used as HetNet technology enabling system. Interference mitigation in HetNets was considered in[12] through orthogonal resource allocation and relaying. The ergodic capacity analyses can be found in[7], where downlink DAS was considered in multicell environment. Analysis that features cooperation between DAS antenna element modules and the femtocells located in tall building was presented in[13].

The study in[14] consisted of the cooperation of the base stations equipped with multiple antennas within a cellular system. The proposed interference cancelation was based on multiple antenna signal processing and the exchange of strongly interfering terminal data with the base stations where the received signal decoding remained unsuccessful. The latter base stations performed successive interference cancellation, which used additional multiple antenna processing. Reference[14] assumed that some base stations in the cell can decode multiple user data using their multiple antennas and then exchange the estimated data with the other base stations. Then, the forwarded data were used for interference cancellation when the signal-to-interference-plus-noise ratio (SINR) at the femtocell due to the MUE interference was negligible.

The concept in[14] reflected more of the version of the traditional successive interference cancellation combined with the network MIMO. The traditional successive interference cancellation involved the extraction of the estimated interfering signal from the combined signals and the estimation of the desired signal from the difference. The system model in[14] lacked reliability because one of the base stations could be unable to estimate the signal of a strongly interfering terminal from the received combined signals. In contrast to[14], the study in this paper uses signaling without resorting to multiple antennas. Because femtocells are indoor systems managed by their end users, it is difficult to apply the schemes in[14] or network MIMO to the different femtocell base stations. In addition, given the recent ubiquity of small cell networks[6], our proposal in this paper effectively addresses the cross-tier interference issue in a novel way to the best of the authors' knowledge. Because DAS has already been introduced into cellular systems, we propose to use the DAS as an interface between the macrocell and femtocell tiers, considering uplink transmission to mitigate the cross-tier interference as follows:

- The mitigation at the femtocell side considers a cluster of femtocells. Near the cluster, an MUE that transmits to its MBS drastically subjects the cluster into a high outage probability. When the DAS is connected to the MBS and to each femtocell in the cluster through the optical line, information exchange can be performed for interference cancellation. By placing a RAU in the cluster, the DAS can retrieve the symbols transmitted by the MUE and feed them back to each FBS, thus cross-tier interference can be cancelled at the FBS. The DAS can multicast the feedback to the FBSs in the cluster affected by the same MUE on the basis of each FBS request for feedback. The feedback request can be based on the evaluation of the SINR at each FBS in the cluster of femtocells.
- At the macrocell side, the interference from the MUE to the femtocells in the cluster can be mitigated by reducing the MUE transmit power because the DAS retrieves the symbols transmitted by the MUE through the nearby RAU located closer to the MUE than the MBS. Such reduction is constrained by the MUE outage probability at the RAU. The introduction of the DAS shortens the radio transmission distance of the MUE. This implies a reduction of its transmit power[15].

In addition to the mitigation of the cross-tier interference made possible by the DAS interface, we can derive consequent benefits from our proposal. Instead of exchanging the related information of the strongly interfering terminal as performed in[14] or resorting to network MIMO as done in[7], our proposed interface directly forwards the MUE transmitted symbols where needed in the femtocell cluster. The traditional MBS-FBS direct connection can be used for interference cancellation[16]; however, as the number of femtocell clusters increases, the macrocell off-loading role of the femtocells[6] is compromised by the induced traffic delay in the backhaul. Furthermore, the signal decoding complexity individually involves each femtocell that requests interference cancellation, in contrast to the schemes in[7, 14]. As stated in the outage analysis, to ensure that the feedback system retrieves the symbols of the MUE(s) causing interference at the FBSs, we propose a cross-tier interference avoidance which consists of prohibiting any FUE transmission susceptible of creating significant interference at the RAU involved in the feedback system. The overall novelty and contribution of our proposal is the use of the distributed direct (wired) links between the DAS (through CPUs and RAUs) and the femtocell tier (FBSs). Consequent to our proposal, the sum rate of the small cell networks improves substantially. Such result is confirmed in this study by computer simulations.

The rest of this paper is structured as follows. The system model and the proposed architecture are described in Section 2. The femtocell clustering concept based on the feedback request is presented in Section 3. Section 4 is devoted to the proposed interference management. The computer simulations are discussed in Section 5, and this paper ends with the conclusion in Section 6.

2 SYSTEM MODEL AND PROPOSED ARCHITECTURE

Figures1 and2 illustrate our conventional and proposed systems respectively. $Op_x, x = n,m,u$ are telecommunications operators operating in the same region, each having his own MBS. For each operator, frequency sharing between macrocell and femtocells is considered. Considering the work in[8], we assume that operators can share the DAS through wired optical links connected to the CPUs. We further assume that the RF part of the DAS sensor supports the operator frequency bands. We do not consider the situation where different MBSs share simultaneously the same RAU. Therefore, this study focuses on a macrocell coexistence with femtocells with regard to the cross-tier interference mitigation in presence of the DAS which is an independent interface to the operators. We assume TDD-OFDMA in this paper. The cases presented in this paper are feasible situations for our proposal performance evaluation. As the MUEs are managed by the MBS, if several MUEs are located in the same area, the MBS allocates different frequency bands or time slots for them to communicate. Consequently, we considered one or two MUEs for the cases studied in this paper since we consider frequency sharing in our system model.

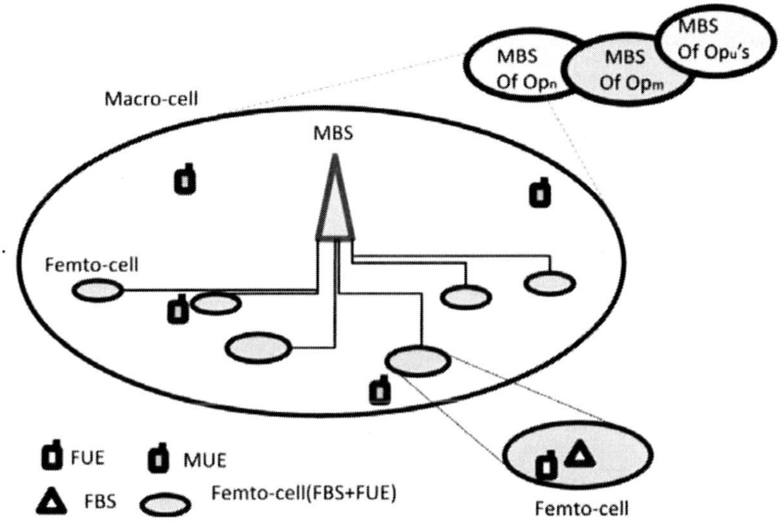

Figure 1 Femtocells overlaid onto legacy macrocells.

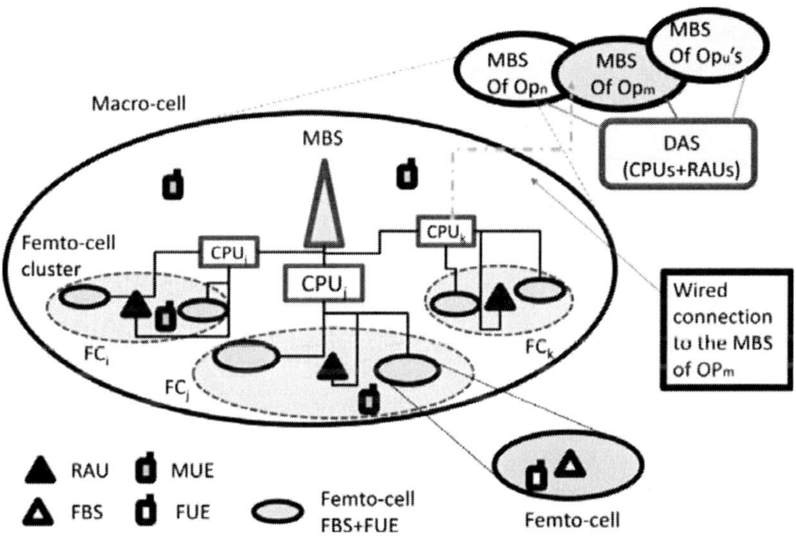

Figure 2 Femtocells overlaid onto legacy macrocells in uplink interference with the added DAS.

2.1 Conventional system: traditional coexistence of the femtocell and macrocell tiers

Each femtocell constitutes an FBS and an FUE (FUE + FBS) as shown in Figure1. The FBS carries the radio access control of its FUE. Because we assumed a closed regime, the MUE is unable to communicate directly on the wireless link with any FBS. The femtocell radius is on the order of 10 to 50 m. The macrocell tier is composed of the MBS and each MUE randomly distributed in the tier. Each MUE attempts to transmit directly to the MBS.

2.2 Cellular architecture modification

Our proposed modification of the conventional system involves the insertion of the DAS in the benchmark. The DAS is represented in Figure2 by the CPUs and the RAUs. Each RAU is connected directly to a CPU, which is linked to the MBS. We assume that all RAU-CPU and CPU-MBS links are fiber optic links with negligible delay. Signal processing such as minimum mean square estimation (MMSE) through recursive least square algorithm and symbol demodulation can be performed at each FBS, and each CPU. A logical implementation of a DAS that cooperates with a network operator can be found in[11].

3 PROPOSED CONCEPT OF AN FEMTOCELL CLUSTERING BY DAS

In[16], the interference cancellation considered information exchange directly between MBS and FBS. Because of the number of femtocells in a macrocell, we consider DAS cooperation as a solution to address the drawbacks such as signal decoding complexity and the traffic constraint on a single MBS.

3.1 Femtocell clustering concept

Our proposal considers a virtual cluster concept. Our concept of 'cluster' is defined as an aggregation of several femtocells (FBSs + FUEs) around a DAS sensor node which assists the FBS at the femtocells in the cross-tier interference cancellation. Due to the small radius of a femtocell, we can obtain several femtocells that form a cluster in a random location in the macrocell. Figure2 shows such clusters, and we denote three of them as FC_i,FC_j, and FC_k associated each with at least a RAU. In FC_j, F_j is a femtocell and R_j is a RAU. Considering such cluster, we derive the following femtocell clustering concept relative to the DAS elements and the interfering MUE near the cluster. The MUE transmission generates interference at the cluster. Each FBS evaluates its SINR to assess the need to mitigate the interference. We define a femtocell in outage as a femtocell whose FBS has a SINR below a threshold above which symbol decodability is possible. The femtocells in outage can request the cooperation of the DAS for interference mitigation. The DAS selects the femtocells that request interference mitigation. This selection consists of the femtocell clustering by the DAS. The interference mitigation requires that the

DAS retrieves the symbols transmitted by the MUE through its RAU(s) and feed them back to the set of selected femtocells. Thus, each femtocell in the femtocells clustered by the DAS can retrieve its desired signal by interference cancellation which uses the feedback symbols of the interfering terminal(s). Given the distribution of the DAS elements within the macrocell-tier, such cross-tier interference management can be performed in different clusters. This concept generalizes the interference cancellation scheme in[16].

3.2 Interference cancellation patterns

The introduction of the DAS allows cross-tier interference management under different situations. The channel gains are modeled as independently and identically distributed variables with zero mean and a unit variance. The interference cancellation for the femtocells selected by the DAS is illustrated in the following cases:

- Figure3 shows MUE1 and MUE2 interference at the FBS. On the wireless link, the FBS receives its desired signal S_0 weighted by the channel gain h_{ff} in addition to the signals S_1 and S_2 which are weighted by the channel coefficients $h_{m\ 1f}$ and $h_{m\ 2f}$, respectively. Simultaneously, the feedback system constituted by CPU1 and CPU2 retrieves the symbols S_1 and S_2 through signal processing and forward them to the FBS. CPU1 receives MUE1 transmitted signal S_1 without interference from MUE2 assumed to be located far enough from CPU1. Then, it demodulates the received signal in order to forward S_1 to both the FBS and CPU2. After receiving S_1, CPU2 performs interference cancellation as detailed in Figure4 in order to retrieve S_2 from the wireless signal $h_{m\ 1R}S_1 + h_{m\ 2R}S_2$, where $h_{m\ 1R}$ ($h_{m\ 2R}$) is the channel coefficient between MUE1 (MUE2) and the RAU of CPU1. The FBS can perform interference cancellation using the forwarded symbols S_1 and S_2 to recover S_0.

- Figure4 shows the feedback system composed of the MBS and a RAU, where $h_{m\ 1R}$ ($h_{m\ 2R}$) is the channel coefficient between MUE1 (MUE2) and the RAU of the CPU, and $h_{m\ 1M}$ is the channel coefficient between MUE1 and the MBS. MUE1 is close to the MBS such that interference from MUE2 to MBS is negligible, and MUE2 is close to the RAU linked to the CPU. When both MUEs interfere at the RAU, the MBS can recover S_1 and forward it to the CPU, which can recover the symbol S_2 of MUE2. This process involves a single feedback scheme at the CPU. Then, the two MUE symbols can be forwarded at the femtocells (FBSs connected to the CPU) that experience the interference from the two MUEs.

- Figure5 shows a similar situation to Figure4 with the difference that the MBS in Figure4 is replaced by a RAU. Owing to the presence of the two MUEs and two RAUs sharing a CPU, we propose a power control of the

MUEs through the following scheme: MUE2 reduces its transmit power to reach its closest RAU, RAU2. MUE1 transmits to RAU1. The CPU can demodulate the symbol received by RAU1 independently of the signal received at RAU2. The CPU connecting the two RAUs retrieves the MUE symbols to feed them back to the FBS for interference cancellation.

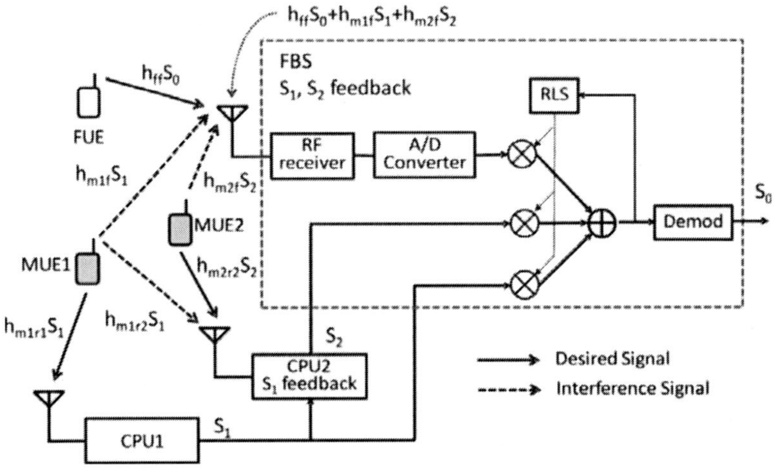

Figure 3 Interference pattern and interference cancellation principle at the FBS.

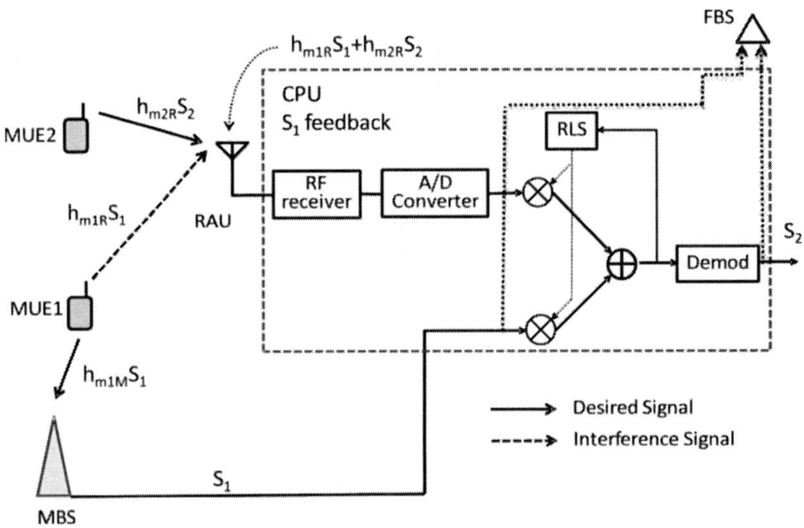

Figure 4 Feedback system with CPU and MBS.

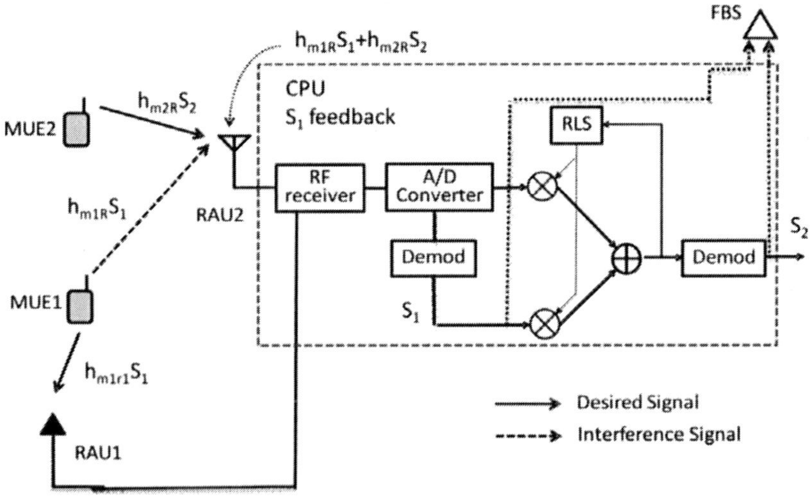

Figure 5 Feedback system with two RAUs.

In Figure5, RAU1 receives the symbol S_1 without interference. Thus, the RLS module can receive the symbol S_1 after demodulation. The equation represents the signals received at the RLS input for interference cancellation. As represented in Figure5, $h\ m\ 1r\ 1$ is not required at the RLS input, therefore is not necessary in the following matrix representation of the signals for modeling the interference cancellation process. Such interference cancellation scheme has been proposed in our previous work in[16]. In Figure5, the signals received by the CPU can be represented as follows:

$$\begin{pmatrix} S_2 h_{m2R} + S_1 h_{m1R} + n_R \\ S_1 \end{pmatrix} = \begin{pmatrix} h_{m2R} & h_{m1R} \\ 0 & 1 \end{pmatrix} \begin{pmatrix} S_2 \\ S_1 \end{pmatrix} + \begin{pmatrix} n_R \\ 0 \end{pmatrix},$$

where $h_{m\ 2R}$ and $h_{m\ 1R}$ are the channel coefficients from MUE2 and MUE1, respectively, to RAU2; $h_{m\ 1r\ 1}$ is the channel coefficient from MUE1 to RAU1, and n_R is an AWGN. MUE1 signal at RAU1 is received without interference. S_1 and S_2 can be independently decoded by the signal processing at the CPU because the channel matrix

$$\begin{pmatrix} h_{m2R} & h_{m1M} \\ 0 & 1 \end{pmatrix}$$

is invertible.

Considering that the feedback system can be one of the situations represented in either Figures4 or5, we use the notation in Figure3 to effectively describe the interference management schemes for the rest of this section. FUE transmits the symbol S_0 under channel gain h_{ff} to its related FBS. S_1 and S_2 are transmitted from the two MUE1 and MUE2, respectively. h_{m1f} and h_{m2f} represent the channel gains from MUE1 and MUE2 to the FBS. We consider MUE1 and MUE2 in Figure3 for the analysis of the S_0 recovery under the interference of two MUEs. We assume that the estimation process of S_0 can be performed as described in[16] by considering the signal feedback from the feedback system to the FBS that receives the combined signal $S_0h_{ff} + S_1h_{m1f} + S_2h_{m2f}$. The interference case that must be considered before the recovery of S_0 by the FBS is described as follows: $(h_{m1r2}S_1 + h_{m2r2}S_2)$ represents the signal received at the RAU resulting from the transmission of the two MUEs. The feedback of S_1 from the MBS to CPU2 allows a single-symbol detection as in[16]. Assuming correct recovery of S_1 and S_2 at the CPU, each femtocell in the DAS femtocell cluster can perform interference cancellation using the following relationship:

$$\begin{pmatrix} S_0h_{ff} + S_1h_{m1f} + S_2h_{m2f} + n_f \\ S_1 \\ S_2 \end{pmatrix} = \begin{pmatrix} h_{ff} & h_{m1f} & h_{m2f} \\ 0 & 1 & 0 \\ 0 & 0 & 1 \end{pmatrix} \times \begin{pmatrix} S_0 \\ S_1 \\ S_2 \end{pmatrix} + \begin{pmatrix} n_f \\ 0 \\ 0 \end{pmatrix},$$

where n_f is an AWGN. The symbols S_1 and S_2 are provided to the FBS by the CPU. The FBS can retrieve S_0 by channel matrix inversion. Such channel matrix inversion is approximated by the RLS algorithm with minimum mean square error equalization. Therefore, the FBS approximates h_{m1f} and h_{m2f} by the RLS algorithm. The detection of S_0 using the above matrices notation was explicitly derived in[16].

3.3 Reception at the RAUs and FBSs

On one hand, the proposed interference management based on the DAS relies on the correct reception of the interfering MUE symbols at the RAU located in the vicinity of both the femtocell cluster and the MUE. On the other hand, the inter-femtocell interference may remain at the FBS although the cross-tier interference is cancelled with our proposed feedback. The restriction on the femtocells transmission due to their closeness to the DAS sensor node is made to allow the DAS sensor node to receive the interfering MUE signal without interference from the close femtocell users.

Thus, we evaluate the probability of successful reception of the transmission of an MUE (FUE) at a RAU (FBS). This evaluation is subject to the MUE (FUE) transmit power control and the combined interference from the set of femtocells that transmit within the cluster. We denote the cardinality of this set by $|\Omega|$. We assume that the femtocells involved in the outage derivation are each located at a distance r from the RAU. Such combined interference is subject to the following analysis:

We denote the set of transmission occurrences in the vicinity of the RAU as $E = \{x_1, x_2 ...\}$. Therefore, we obtain $\Sigma = \{\emptyset, \{x_1\}, \{x_1, x_2\} ...\}$ as the set of the subsets that can be constructed from E, where $\emptyset$ is an empty set. We define the following function over Σ:

$$
\begin{aligned}
f : \Sigma &\rightarrow R^{+} \\
\Omega &\rightarrow P(X \in \Omega),
\end{aligned}
\tag{1}
$$

where Ω is the set of FUE transmission occurrences from the cluster, $P(X \in \Omega)$ is the total probability of the events constituting Ω and R^{+} is the set of positive real numbers. To express $P(X \in \Omega)$, we must rely on the DAS femtocell clustering which defines the set of femtocells that request interference mitigation. The cardinality of Ω is given by:

$$
\begin{aligned}
|\Omega| &= |\{X_s \in \Omega_M\}| + |\{X_n \in \Omega_c\}| \\
&= |\Omega_M| + |\Omega_c| - |\Omega_M \cap \Omega_c|,
\end{aligned}
\tag{2}
$$

where $s,n \in \{1,2,..|\Omega|\}$ are the indexes, Ω_M is the set of transmission occurrences interfering at the RAU, and Ω_c is the set that requires symbol feedback from a CPU for interference cancellation. The probability of successful reception at the RAU denoted as $P_s(r,|\Omega_M|)$ is derived from[17] as follows:

$$
P_s(r, |\Omega_M|) = e^{-\theta \frac{N_0}{P_{\text{tm}}} d^{\lambda_{\text{out}}}} \prod_{i=1}^{|\Omega_M|} \frac{1}{1 + \frac{P_i}{P_{\text{tm}}} \frac{d^{\lambda_{\text{out}}}}{r^{\lambda_{in}}}},
\tag{3}
$$

where P_i is the transmit power of the FUE whose transmission is received at the RAU considered in Equation 3, ϑ is the threshold SINR for successful reception, N_0 is the noise power, P_{tm} is the transmit power of the MUE, d is the distance from the MUE to the RAU, and λ_{out} and λ_{in} are the outdoor and indoor path loss exponents,respectively.

Proof. The SINR of the MUE at the RAU is expressed as

$$
\text{SINR} = \frac{P_{\text{tm}} d^{-\lambda_{\text{out}}} |h_M|^2}{N_0 + \sum_{i=1}^{|\Omega_M|} P_i r^{-\lambda_{\text{in}}} |h_i|^2},
\tag{4}
$$

where h_M is the channel coefficient between the MUE and the RAU and h_i is the channel coefficient from the i th femtocell to the RAU. The outage probability is given by

$$P(\text{SINR} < \theta) = 1 - P(\text{SINR} \geq \theta), \tag{5}$$

where $P((x) < (y))$ indicates the probability that x is lower than y. Furthermore,

$$P(\text{SINR} \geq \theta) = P(|h_M|^2 \geq \frac{N_0\theta}{P_{\text{tm}}d^{-\lambda_{\text{out}}}} + \frac{\sum_{i=1}^{|\Omega_M|} P_i r^{-\lambda_{\text{in}}} |h_i|^2}{P_{\text{tm}}d^{-\lambda_{\text{out}}}}), \tag{6}$$

We note that $|h_M|^2$ and $|h_i|^2$ are exponentially distributed with a variance of one; thus, by computing the moment generating function[18], we can obtain

$$P(\text{SINR} \geq \theta) = e^{-\theta \frac{N_0}{P_{\text{tm}}} d^{\lambda_{\text{out}}}} M_t(Y), \tag{7}$$

where t is a real number parameter of the moment generating function and Y is defined as:

$$Y = \sum_{i=1}^{|\Omega_M|} P_i r^{-\lambda_{\text{in}}} |h_i|^2 . \tag{8}$$

$M_t(Y)$ is the moment generating function of the random variable Y and parameter t is defined as

$$\begin{aligned} M_t(Y) &= E[e^{tY}] \\ &= \prod_{i=1}^{|\Omega_M|} \int_0^{\infty} e^{-st} e^{-t} dt \\ &= \prod_{i=1}^{|\Omega_M|} \frac{1}{s+1}, \end{aligned} \tag{9}$$

Where $s = \frac{\theta P_i r^{-\lambda_{\text{in}}}}{P_{\text{tm}} r^{-\lambda_{\text{out}}}}$

and e^{-t} is the density function of an exponentially distributed random variable with a variance of one. Replacing $M_t(Y)$ by its value in Equation 9 ends the proof.

4 PROPOSED INTERFERENCE MANAGEMENT IN UNDERLAY MACROCELL

As $|\Omega|$ results partially from the cardinality of the terminals that interfere at the RAU, we propose to manage the interference by considering the situation where two different sets of DAS-selected femtocells are linked to different DAS elements as denoted by FC_j and FC_k where the MUE denoted by M_j interferes in both sets. Because interference cancellation can be performed at each element of the DAS femtocell clustered in $\{X_n \in \Omega_c\}$, the resulting sum rate capacity can be written as follows:

$$C_{\Omega_c} = \sum_{n=1}^{|\Omega_c|} \log_2(1 + \mathrm{SINR}_n), \tag{10}$$

where SINR_n is the SINR of $X_n \in \Omega_c$. Because the feedback option is unfeasible with M_j interfering in FC_k, we propose the following time slot based orthogonal resource allocation for the MUE transmissions to improve the network sum rate.

Considering the transmission of an FUE to its FBS, the previous analysis can be applied by substituting the MUE with the FUE and the RAU with the FBS. The set of femtocells surrounding the considered FBS can cause interference and degrade the FBS reception even in the case of feedback. Thus, for a perfect reception at the RAU, we propose to prohibit the FUE transmission close to the RAU as illustrated in Figure6 used for our simulation model described in Section 5.

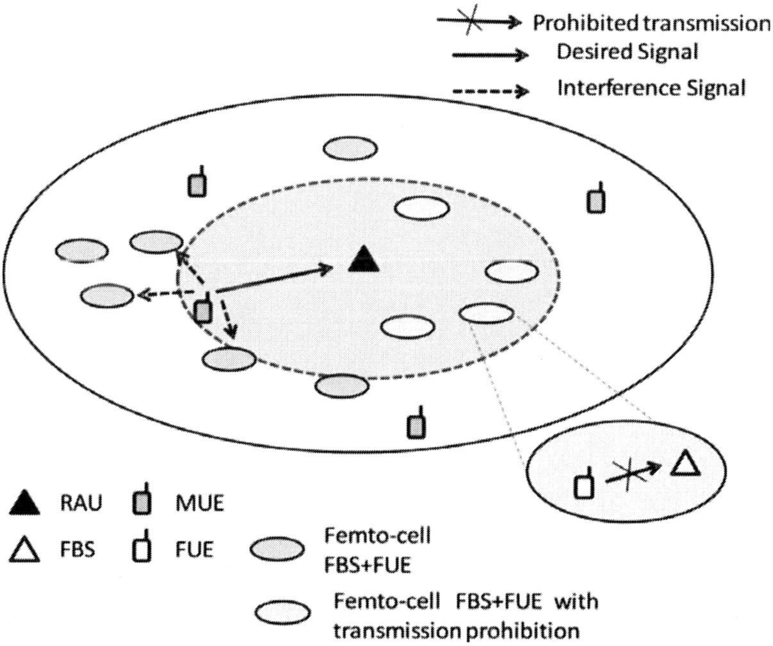

Figure 6 Interference management at the reception.

4.1 Proposed radio resource allocation for cross-tier interference management in the underlay macrocell

In Figure2, let us consider M_j and M_k are MUEs interfering at FC $_j$ and FC $_k$, respectively. In addition, we assume that M_j interferes at F_j and F_k whereas M_k interference is restricted to FC $_k$. R_j and R_k are RAUs in FC $_j$ and FC $_k$, respectively. F_j (resp. F_k) is a femtocell in FC_j (resp. FC $_k$). The interference management through feedback can be achieved as follows: R_j (resp. R_k) retrieves M_j (resp. M_k) transmitted symbols, then CPU $_j$ (resp. CPU $_k$) feeds back the retrieved symbols to F_j (resp. F_k). Such feedback process is enabled by the following radio resource allocation of the MUEs. Table1 lists the access of the femtocells and MUEs to FC $_j$ and FC $_k$ partitioned using time slots (TS). For each column, the different TSs are expressed as TS $_i$, $i = 1,2,3$. At each column TS $_i$, the simultaneously transmitting terminals are denoted as MUE and/or FUE. M_j and F_j can transmit simultaneously during TS_1 because M_j can target the MBS or the RAU. The interference of M_j at F_j can be cancelled by the feedback as illustrated by Figures4 or5. In TS_2, the interference from M_j to F_k is avoided and a transmission similar to TS_1 can occur in M_k and F_k whereas F_j is beyond the interference range. F_j and F_k can transmit simultaneously in TS_3. The DAS enables terminal access management and permits cooperation if the MUE access to any femtocell base station is restricted. In the conventional scheme, three TSs are required to avoid interference from the base stations. Because the transmit power decays with the distance, the interference of M_k with F_j is assumed to be negligible.

Table 1 Proposed orthogonal radio resource allocation for interference management

Proposed	Conventional
$TS_1 : M_j, F_j$	$TS_1 : M_j$
$TS_2 : M_k, F_j, F_k$	$TS_2 : M_k, F_j$
. . .	$TS_3 : F_j, F_k$

4.2 Simultaneous transmission without the proposed radio resource allocation

This section describes the conventional scheme of the proposal in Subsection 4.1. We consider FC $_j$ and FC $_k$ in the absence of the proposed radio resource allocation. We assume a simultaneous transmission of all femtocells and MUEs. Considering the DAS elements, the MUEs can transmit at minimum power to reach the RAU. Although F_j and F_k benefit from the interference cancellation from M_j and M_k, respectively, in the DAS femtocell clustering, F_k experiences interference from M_j. In the conventional system, F_j and F_k experience low SINR from the high transmit power of the MUEs.

5 SIMULATION RESULTS AND DISCUSSION

5.1 Simulation conditions

We use the C language for the simulations presented in this paper. The channel model is a five-path exponential Rayleigh fading. The noise is generated as AWGN. Each user transmits by QPSK modulation to generate the symbols from the binary output of the convolutional encoder. Then, OFDM is applied before the signals enter the channel. The OFDM key parameters are listed in Table2, and the simulation parameters related to the macrocell-femtocell two-tier network are presented in Table3 which are similar to those in[3]. The noise is added during signal reception whereas the signal fed back to the base station or the DAS element is free from noise. We use the RLS-based MMSE algorithm to recover the data at the receiver for interference cancellation at CPU and/or FBS. The transmit power, positions, and details of the distance path loss model of each user are listed in Table3. A fixed loss is chosen instead of considering a random lognormal shadowing. We use the SINR and capacity evaluation from[19]. The sum rate is then evaluated by Equation 10 whose cumulative distribution function (CDF) is used in the simulation results.

Table 2 Simulation conditions

Parameter	Value
Bandwidth	5 MHz
Number of subcarriers	512
Useful symbol time	6.4 μs
Guard interval	1.25 μs
Data modulation	OFDM QPSK
Small-scale channel model	Rayleigh flat fading
Weight estimation algorithm	RLS
Noise	AWGN
Convolutional code rate	1/2
Convolutional code constraint length	7

Table 3 HUE and MUE femtocell and macrocell parameters

Parameter (variable)	Value
Macrocell radius (R_c)	1,000 m
Femtocell radius (R_f)	30 m
Normalized distance (r)	0.01 to 0.95
Carrier frequency (f_c)	2,500 MHz
Wall penetration loss (P_l)	5 dB
Mobile maximum transmit power (P_{tmax})	23 dBm
Macrocell path loss exponent (λ_{out})	3.8
Femtocell path loss exponent (λ_{in})	3

5.2 Performance evaluation and discussion

Our simulation system model is illustrated by Figure6. In Figure6, MUE is located at random distances from the femtocells in its vicinity. In our simulation, we consider different values of r for the variation of the distance between the MUE and different FBSs under the MUE cross-tier interference. Therefore, we evaluate the performance at the FBS with and without feedback from the feedback system. As represented in Figure6, the reception of the MUE symbols at the RAU is performed without femtocell interference. In the case of 2 MUEs, we apply the feedback system in Figures4 or5 where the MBS or RAU(s) are protected from femtocell interference.

The performance evaluation consists of the computation of the sum rate that considers a femtocell cluster and its neighboring MUE. The sum rate is evaluated using Equation 10 for a given number of mobile terminals. Thus, we add up the rates of the femtocells and MUEs chosen in the situations that illustrate our proposal and the benchmark. For the two MUEs and two RAUs connected to the same DAS signal processing unit, the DAS can recover the MUE transmitted symbols and forward them to the FBSs that require interference cancellation. In the case of a one-sensor node and two MUEs, one MUE can target the RAU and the other transmits directly to the MBS. The DAS can demand data from the MBS for interference cancellation as depicted in Figure4. In the presence of more than two MUEs, we assume a time slot-based orthogonal resource allocation for the MUE transmissions as explained in Subsection 4.1. Therefore, the FBS can synchronize with the MUE scheduling. We present the results through the CDF of the capacity.

5.3 Spectral bit rate at RAU and FBSs with $|\Omega_M|$

The capacity of the MUE at the RAU or at any FBS is subject to the interference of the surrounding femtocells. Thus, we evaluate the capacity considering the transmitting femtocells, i.e., $\{X_s \in \Omega_M\}$. We present the performance of a receiver considering the presence of surrounding terminal interference and the proposed interference cancellation. In Figure7, *NF* stands for *No Feedback*, i.e., there is no interference cancellation of the cross-tier interference. The curves with NF represent the benchmark. *WF* indicates *With Feedback*, i.e., the receiver performs interference cancellation using the feedback of the symbols of an interfering MUE. $N = i$ indicates that i FUEs are interfering at the receiver as in the situation where the FBS retrieves its FUE data with the MUE symbols feedback and the interference of the neighboring femtocells. The curves with WF represent the simulations with interference cancellation by the CPU or the FBS using feedback symbols. The spectral bit rate performance improves with the reduction of the received interference power which decreases in these simulations with the path loss and the interfering terminals are farther located from the receiver.

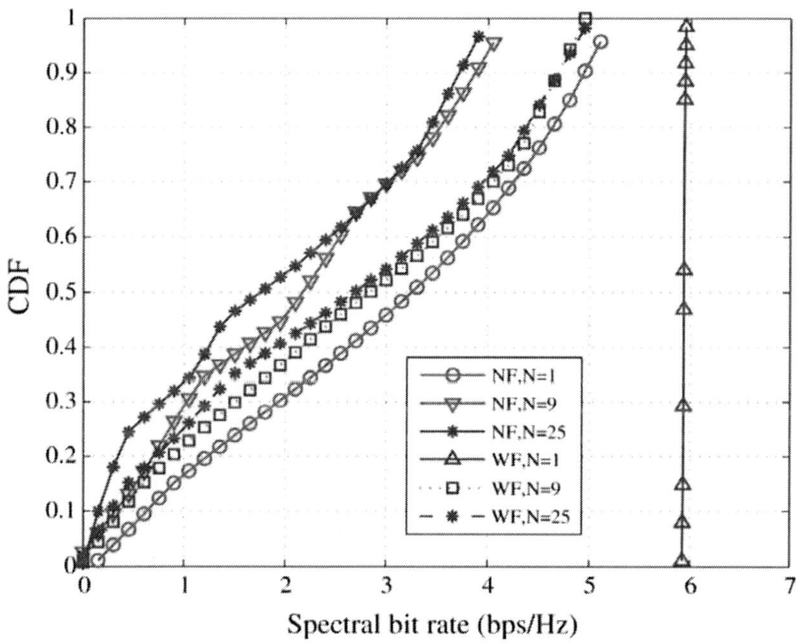

Figure 7 Spectral bit rate at sensor node and FBSs with $|\Omega_M|$.

NF, $N = 1$ evaluates the situation where the desired signal of one mobile equipment is received with interference. Such cases is represented in this study by an MUE transmission received at the RAU while another MUE interference occurs at the reception. Another illustration of NF, $N = 1$ is the interference of an MUE at an FBS receiving the transmission of its FUE without inter-femtocell interference. WF, $N = 1$ represents the interference cancellation performance where the symbols of the single interfering terminal are fed back to the receiver. In this case, the interference is completely removed. This situation illustrates the performance of the feedback system in Figures4 or5.

NF, $N = 4$ and NF, $N = 25$ evaluate each the performance at the receiver when the number of interfering mobile terminals increases and without interference mitigation. WF, $N = 4$ and WF, $N = 25$ are similar to NF, $N = 4$, and NF, $N = 25$, respectively; with the difference that for NF, $N = 4$ and NF, $N = 25$, the MUE interference at the FBS has its interference cancelled at the receiver whereas the interference from neighboring femtocells affects the performance. The interference mitigation effectively reduces the cross-tier interference to enhance the spectral bit rate.

5.4 Sum rate capacity by proposed radio resource allocation

We consider two MUEs and two femtocells to evaluate the sum rate capacity, as shown in Figure2 and described in Subsections 4.1 and 4.2. The situation of FC_j and FC_k requires our proposal in 4.1 to improve the sum rate of the two MUEs (M_k and M_j) and the two femtocells (F_j and F_k). We evaluate the CDF of the sum rate capacity resulting from the performance of the MUEs at the RAUs and the corresponding $|\Omega_c|$ femtocells for the situation presented by FC_j and FC_k, as described in Subsections 4.1 and 4.2. conv1 and proposed1 represent the performance of the conventional and proposed systems, respectively, as described in Table1 and Subsection 4.1. conv2 and proposed2 represent the transmission without the proposed orthogonal radio resource allocation, i.e., MUE simultaneous transmissions presented in Subsection 4.2. In conv2, FC_k experiences interference from the two MUEs whereas in proposed2, it receives feedback from M_k.

In Figure8, the intersections of conv1 with conv2 and proposed1 with proposed2 occur because of the power decay with the distance adopted for each interfering terminal, i.e., as the interfering terminal moves away from the receiving terminal, the interference effect decreases. The proposed resource allocation enhances the DAS femtocell clustering interference mitigation to reduce the combined interference in order to lower the outage probability at the RAU. With the increase in N due to the number of terminals that experience high interference mitigated by feedback, the joint interference cancellation and DAS femtocell clustering yields significant sum rate improvement compared with the conventional system.

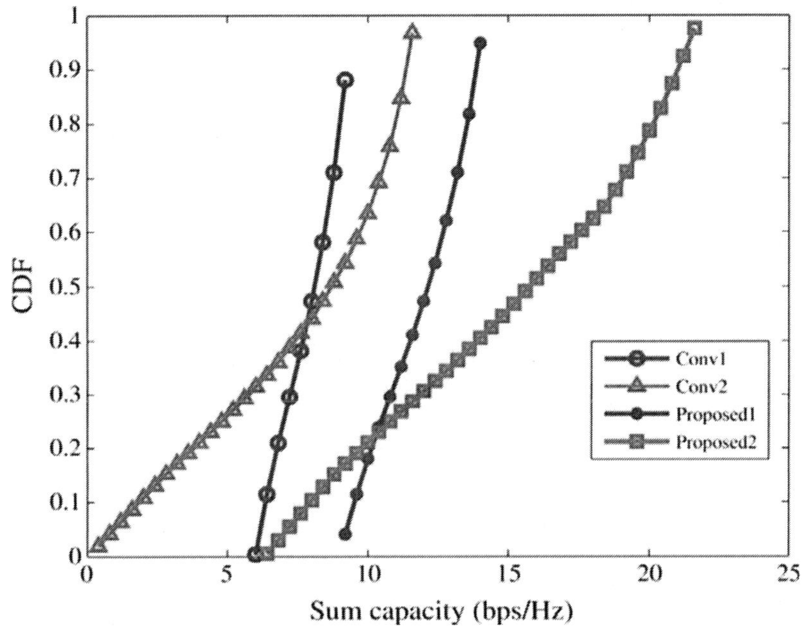

Figure 8 MUE interference management spectral bit rate at RAUs and FBSs with $|\Omega_M|$.

5.5 Average capacity across all users in two-tier network

This section shows the effect of inserting the DAS in the two-tier network. The performance metric used in the simulations is the average capacity across all users in the underlay system. The sum rate capacity of the two-tier network can be rewritten as:

$$C_f = \sum_{i=1}^{N_1} \log_2(1 + \text{SINR}_i) + \sum_{j=1}^{N_2} \log_2(1 + \text{SINR}_j), \tag{11}$$

where SINR_i is the average SINR for each femtocell in the DAS clustering, SINR_j is the SINR of an interfering MUE on the femtocells. Equation 11 is an expansion of Equation 10. The benchmark is the conventional system described in Subsection 5.3. We have $N_i \in \{4,14,25\}$. We use all the permutations of {4,14,25} to evaluate Equation 11 with the permutations of the SINR_js. In the Figure9 simulations, $N_1 = N_2 = 3$. Each element of {4,14,25} is a number of femtocells clustered by the DAS selection described in Section 3.

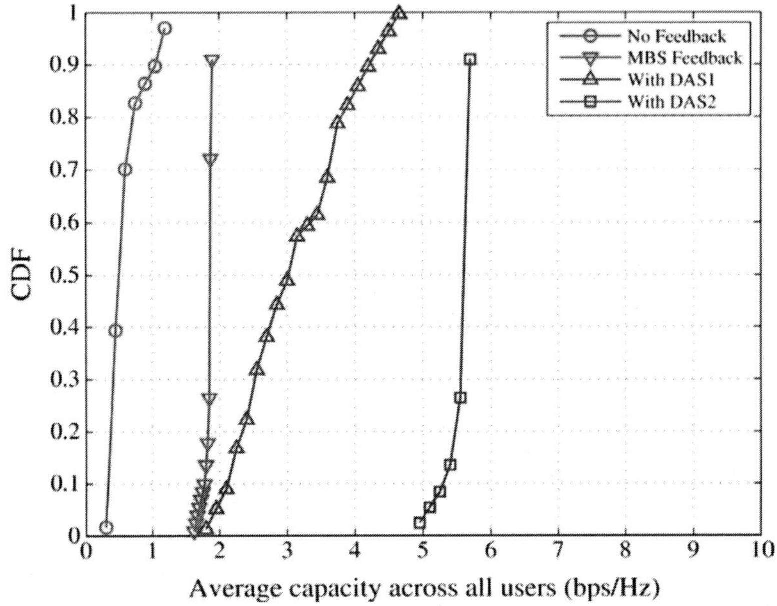

Figure 9 Average capacity across all users in the underlay system.

Figure9 shows the CDFs of the average capacity across all users in different scenarios. The average capacity across all users is defined by:

$$C_u = \frac{C_f}{N_1 + N_2}. \tag{12}$$

Without the insertion of the DAS, and $N_2 > 1$, we assume that the radio resource management (by the MBS) assigns orthogonal resource blocks for the MUESs because of their direct transmission to the MBS by default. Thus, there is MUE interference avoidance from the MUEs to the MBS. Each femtocell cluster experiences the interference from one or two MUEs. No Feedback and MBS Feedback represent each the conventional system without the DAS. With No Feedback, each femtocell experiences an MUE interference without interference mitigation. MBS Feedback considers a feedback from the MBS; each cluster which experiences an MUE interference can receive a feedback from the MBS for interference cancellation. However, the feedback is subject to the orthogonal resource allocation applied for the MUEs directly transmitting to the MBS. Because of the MUE transmission scheduling (we assume different time slot allocations in the simulations), the sum rate decreases with N_2. DAS1 and DAS2 represent our proposed modification of the two-tier network with the DAS insertion. With DAS1, the MUEs transmit to their respective RAUs. Thus, there is a simultaneous transmission of the MUEs; besides, the DAS can feedback the symbols of each interfering MUE to the femtocells which experience the interference. DAS2 is similar to DAS1 except that in DAS2, the interference cancellation removes all the cross-tier interference.

With DAS1, more than 90% of all users get each an average capacity of 2 bps/Hz, whereas with the MBS Feedback, almost all users get each less than 2 bps/Hz. The performance gap obtained by the insertion of the DAS can be interpreted as the traffic unmanageable by the MBS without the DAS.

A synchronization of the transmitting terminals is required in order to achieve our proposed interference cancellation scheme. The FBS and sensor node can adopt the macrocell synchronization by listening to the closest MUE signaling with the MBS. We propose the use of the convex combination algorithm in[20] to update the FBS timing to their nearest neighbors in the wireless interface. Additionally, we propose to feedback the synchronization signaling to the cluster of femtocells from the CPU as the CPU is directly connected to the femtocell cluster without intermediate node.

6 CONCLUSION

We have presented a novel femtocell clustering in a multi-tier network which was made possible using a proposed HetNet consisting of a DAS on the macrocell and femtocell tiers. We proposed the use of DAS as an interface between the two tiers to manage the cross-tier interference. Consequently, the network cross-tier interference was effectively mitigated, and the sum rate capacity has improved substantially because the DAS femtocell clustering linearly scaled the capacity in proportion of the proposed femtocell clustering and its cardinality. The introduction of a DAS within the macrocell overlaid with the femtocells improved the cross-tier interference management and can be used as a benchmark for future HetNet.

Competing interests

The authors declare that they have no competing interests.

REFERENCES

1. Ozgur A, Leveque O, Tse D: Hierarchical cooperation achieves linear capacity scaling in ad hoc networks. In IEEE International Conference on Computer Communications (INFOCOM),) Anchorage. New York: IEEE; 2007:382-390.
2. Etkin R, Parekh A, Tse D: Spectrum sharing for unlicensed bands. IEEE J. Selected Topics Commun 2007, 25: 517-528.
3. Chandrasekhar V, Kountouris M, Andrews JG: Coverage in multi-antenna two-tier networks. IEEE Trans. Wireless Commun 2009, 8: 4316-4328.
4. Claussen H, Ho LTW, Samuel LG: An overview of femtocell concept. Bell Labs Tech. J 2008, 13: 221-245. doi: 10.1002/bltj.20292 10.1002/bltj.20292
5. Li Y, Maeder A, Fan L, Nigam A, Chou J: Overview of femtocell support in advanced WiMAX systems. IEEE Commun. Mag 2011, 49: 122-130.
6. Andrews JG, Claussen H, Dohler M, Rangan S, Reed M: Femtocells: past, present, and future. Selected Areas Commun. IEEE J 2012, 30(3):497-508.
7. Choi W, Andrews JG: Downlink performance and capacity of distributed antenna systems in a multicell environment. IEEE Trans Wireless Commun 2007, 6: 69-73.
8. Kon Y, Ito M, Hassel N, Hasegawa M, Ishizu K, Harada H: Autonomous parameter optimization of a heterogeneous wireless network aggregation system using machine learning algorithms. In Consumer Communications and Networking Conference (CCNC), Las Vegas. New York: IEEE; 2012:894-898.
9. Castanheira D, Gameiro A: Distributed antenna system capacity scaling [coordinated and distributed MIMO]. IEEE Wireless Commun 2010, 17: 68-75.
10. Hu RQ, Qian Y, Kota S, Giambene G: Hetnets - a new paradigm for increasing cellular capacity and coverage. IEEE Trans Wireless Commun 2011, 18: 8-9.
11. Li H, Hajipour J, Attar A, Leung VCM: Efficient HetNet implementation using broadband wireless access with fiber-connected massively distributed antennas architecture. IEEE Trans. Wireless Commun 2011, 18: 72-78.
12. Peng Y, Qin F: Exploring Het-Net in, LTE-advanced system: interference mitigation and performance improvement in macro-pico scenario. In

Proceedings of the IEEE International Conference on Communications Workshops (ICC), Kyoto. New York: IEEE; 2011.

13. Alade T, Huiling Z: Joint signal processing in femtocell based distributed antenna systems in high buildings. In Proceedings of the IEEE 72nd Vehicular Technology Conference Fall, Ottawa. New York: IEEE; 2010.

14. Balachandran K, Kang JH, Karakayali K, Rege KM: NICE: a network interference cancellation engine for opportunistic uplink cooperation in wireless networks. IEEE Trans. Wireless Commun 2011, 10: 540-549.

15. Alade T, Zhu H, Osman H: Spectral efficiency analysis of distributed antenna system for in-building wireless communication. In 2011 IEEE 22nd International Symposium on Personal Indoor and Mobile Radio Communications (PIMRC), San Francisco. New York: IEEE; 2011:2075-2079.

16. Ndong M, Fujii T: Cross-tier interference mitigation for self organized-self optimized femtocells. In Wireless Telecommunications Symposium, London. New York: IEEE; 2012.

17. Vu M, Devroye N, Vahid T: On the primary exclusive region of cognitive networks. IEEE Trans. Wireless Commun 2009, 6: 3380-3385.

18. Goldsmith A: Performance of digital modulation over wireless channels. In Wireless communications, 1st edition. New York, USA, Cambridge: university press; 2005:187-197.

19. Andrews JG, Ghosh A, Muhamed R: Orthogonal frequency division multiple access. In Fundamentals of WiMAX understanding broadband wireless networking, 1st edition. Edited by Rappaport TS. Massachusetts: Prentice Hall; 2007:210-211.

20. Lien S, Lee H, Shih S, Chen P, Chen K: Network synchronization among femtocells. In the second GLOBECOM workshop on femtocell networks, Houston. New York: IEEE; 2011:248-252.

Chapter 11

Power and Rate Allocation for Video Conferencing in Cellular Networks

Chao Yang and Scott Jordan*

ABSTRACT

We consider resource allocation for semi-elastic applications such as video conferencing. We propose to model user utility as a sigmoid function of the average bit rate over multiple time slots. The goal is to maximize the total expected user utility over time through allocation of downlink power and subcarriers in each time slot. We propose resource allocation that considers both the average rate achieved so far and the future expected rate, and show how the future expected rate can be estimated by modeling the probability that a user will be allocated a subcarrier in a future time slot. The algorithm can be implemented in a distributed fashion by an exchange of price and demand amongst users, the network, and an intermediate power allocation module. To reduce the complexity, we also propose a greedy algorithm to maximize incremental utility in the current time slot. The performance is illustrated using numerical results.

Keywords: Communication system traffic control Cellular networks Video conferencing

1 INTRODUCTION

Video applications constitute a rapidly increasing proportion of the total traffic on cellular networks. It is now estimated that video comprises one third of downstream North American mobile Internet access peak period traffic[1].

Most of this video traffic is streaming encoded using either Adobe Flash or MPEG. Some of this video traffic is video conferencing, e.g., Apple's FaceTime for iPhones and Skype's video conferencing application for smartphones. For such applications, variable bit rate video encoding algorithms are typically used, e.g., MPEG. For real-time use, such as required to support video conferencing, best-effort service can result in unacceptable performance. Fourth generation (4G) cellular networks will integrate voice, video, and data applications using packet switching. As a consequence, resource allocation should take into account application characteristics when best-effort packet switching would not result in acceptable performance. There are two issues.

First, as in other study on network resource allocation, there is a potential gain from employing the concept of utility. The most common objective video quality metric is peak signal-to-noise ratio (PSNR). Many studies of subjective video quality, however, have observed that users evaluation of video quality levels off at high PSNRs. This has provided motivation for modeling utility as a sigmoid function of throughput[2, 3], i.e., convex at rates less than a threshold and concave at rates above that threshold. A sigmoid function also reflects the layered coding structure of MPEG video, which exhibits increasing returns with rate below a threshold and decreasing returns with rate above that threshold.

Second, video encoding algorithms often use a *group of pictures* as a central concept, and intentionally vary the bit rate over different frames within this group. Network resource allocation should thus be cognizant of this structure. Some researchers have proposed that resource allocation should satisfy some type of delay deadline, e.g., delay constraint[4–7]. However, none of these articles use sigmoid utility functions.

We are thus motivated to look for a resource allocation method that uses sigmoid utility functions (rather than hard performance constraints or utility proportional to PSNR) and that takes into account the group of pictures structure (rather than using long term rate constraints). We posit here that for video conferencing applications, user utility should be a sigmoid function, not of the *instantaneous* transmission rate, but of the *average* rate over *one or more groups of pictures*.

In this article, we consider allocation of downlink power and subcarriers in orthogonal frequency division multiplexing (OFDM) 4G cellular systems for semi-elastic applications. The user's satisfaction is modeled by a sigmoid function of the average rate over a time window. The goal is to create algorithms that allocate power and subcarriers in each time slot in a manner that optimizes total average user utility over time. The difficulty with maximizing utility over many time slots is that the system should consider both the current channel and the likely achievable average rates by the end of the time window. For instance, if a user's channel is poor at the beginning of a time window, should the system allocate resources for this user or should it wait to see if the user's channel improves later in the time window? If a user's accumulated rate in a time window is low, when should the system respond by increasing the user's resource allocation and when should the system give up on the user's performance in this time window and allocate resources to other users?

The article proceeds as follows. In Section 2, we pose the resource allocation problem. In Section 3, we review related work. In Section 4, we consider a non-causal system in which at the beginning of each time window the network knows the channel gains of each user in each time slot of the window. While knowing all the future information is unrealistic, this will serve as an upper bound to causal systems. We illustrate how a dual problem formulation can be used to allocate power and subcarriers in each time slot of the time window. The complexity can be reduced by distributing the resource allocation process amongst users, the network, and intermediate power allocation modules. The network allocates power and subcarriers, and charges the power allocation module a shadow price for power in each time slot. The power allocation module translates this price per unit power into a price per unit rate, and resells the system resources to users. Users choose desired rates based on the cost and the resulting utility. We pose an iterative algorithm that determines near-optimal shadow prices in each time window.

In the remainder of the article, we consider causal systems which must make resource allocation decisions in each time slot without knowledge of the channel gains of users in future time slots. The challenge is to decide how the resource allocation in the current time slot should be based on the average rate achieved so far in the time window and on the likely achievable rate during the remainder of the time window. In Section 5, we propose a resource allocation policy that considers both the average rate achieved so far and the future expected rate. We show how this future expected rate can be estimated by modeling the probability that a user will be allocated a subcarrier in a future time slot. We show that although maximization of the expected utility due to future rate is prohibitively complex, maximization of the utility of the expected average rate over the time window can be near-optimally solved in a distributed fashion using a similar exchange of price and demand. In Section 6, we design a resource allocation policy that considers the average rate achieved so far and ignores future rate, by only attempting to maximize the total *incremental utility* of the current slot. We show how this allocation can be iteratively determined in a distributed fashion amongst user, network, and power allocation modules by exchanging shadow prices and desired rates and powers.

Finally, in Section 7 we illustrate the performance of these algorithms using numerical results. We find that the non-causal algorithm is near-optimal, with some sub-optimality when resources are severely constrained and a significant number of users are unable to achieve rates in the concave portion of the utility curve. The greedy causal algorithm is similar to that of the non-causal algorithm when resources are severely constrained, but lags as resources become more plentiful. The reason is that although the greedy algorithm can react to a poor channel by allocating fewer resources, it does not consider future rates and thus may not decrease allocations enough. The algorithm with prediction of future expected rates closes a portion of this gap by allocating resources in a balanced way based on both current channel and likely future channels.

2 SYSTEM MODEL AND PROBLEM FORMULATION

2.1 System model

We consider a single cell downlink OFDM system serving K users, with N subcarriers. The bandwidth of each subcarrier is B which is assumed to be less than the coherence bandwidth of the channel so that the channel response can be considered flat. The rate of user k on subcarrier n in time slot t is:

$$r_{k,n,t}(p_{k,n,t}) = B\log_2\left(1 + p_{k,n,t}\frac{|H_{k,n,t}|^2}{\delta^2 + I}\right)$$

where $p_{k,n,t}$ is the power allocated, $|H_{k,n,t}|$ is the channel gain, I is the interference power and δ^2 is the noise power. The channel gain is assumed to be a stationary Markov process and the fading on different subcarriers is assumed to be independent to each other. The base station is assumed to know the channel gain of each user on each subcarrier in the current time slot and the conditional density$f(|Hk,n,\tau|2|||Hk,n,t|2)$

for future slots τ. The total rate of user k in time slot t is the sum of the rate over all subcarriers:

$$R_{k,t} = \sum_{n=1}^{N} r_{k,n,t}$$

2.2 Utility model

We propose to model a user's utility as a sigmoid function over the average rate over a time window. Both the choice of a sigmoid function and of an average rate over a time window deserve discussion.

There are two classes of quality assessment methods commonly proposed in the literature. Objective quality assessment typically uses PSNR as an evaluation criterion. Many articles have thus proposed that network resources should be allocated to maximize the average PSNR, see e.g.,[8, 9], sometimes subject to long term average rate constraints, see e.g.,[10, 11]. This approach essentially equates user utility with PSNR. However, a major disadvantage of PSNR is that it cannot reflect users' subjective satisfaction of performance. Many studies have shown that at high PSNRs, perception of incremental quality falls off with increasing PSNR. On this basis, many studies have turned to subjective quality assessment, usually directly measuring a user's satisfaction with video quality, and representing performance by the mean opinion score (MOS)[12].

Some articles have thus proposed that utility of multimedia applications be modeled as a sigmoid function of the throughput[2, 3]. A sigmoid utility function also reflects the layered coding structure of MPEG video. The initial convex portion reflects the rate required to transmit the base video layer. The concave portion reflects the use of incremental rate to transmit enhancement layers; every additional enhancement layer increases user satisfaction, but with decreasing returns. Sigmoid utility functions are widely used to evaluate the performance and make resource allocation of video and other semi-elastic applications[13–16]. The relationships among *PSNR, MOS* and various utility functions constitute the domain of many research projects; we refer to the reader interesting in these mappings to[3] which explores this mapping through multimedia experiments over Android-based smartphones.

In addition, video encoding algorithms often use a *group of pictures* as a central concept, and intentionally vary the bit rate over different frames within this group. Network resource allocation should thus be cognizant of this structure. Despite the common use of PSNR to rate video quality, allocating network resources subject to a long term rate constraint does not reflect the group of pictures structure of the video decoder, and thus may not maximize video quality. As a consequence, some researchers have proposed replacing long term average rate constraints with some type of delay deadline, see e.g.,[4] which minimizes wireless resource usage subject to statistical delay and loss constraints,[5] which maximizes concave utility subject to delay constraints,[6] which minimizes the expected end-to-end distortion subject to delay constraints, and[7] which minimizes the error propagation of a group of pictures subject to delay constraints. However, none of these articles use sigmoid utility functions.

We desire a model which represents user utility as a sigmoid function of rate and which recognizes the group of pictures construct. Thus, we propose to model user utility as a sigmoid function of the average rate over a time window, as pictured in Figure1. The time window consists of *T* time slots which correspond to the transmission time for one or more groups of pictures. For video, the time window is likely to be chosen to be one of more group of pictures. At time *t*, denote the contribution toward the average rate during the current time window bySk, $\frac{1}{T}\sum_{\tau=1}^{t} R_{k,\tau}$. The utility of user *k* is assumed to be a function $U_k(S_{k,T})$ which maps the average rate achieved in a time window,

$S_{k,T}$, to the level of the satisfaction perceived by the application. There exists an inflection point$S^f_{k,T}$ such that U_k is convex $S_{k,T} < S^f_{k,T}$ and concave for $S_{k,T} > S^f_{k,T}$. We denote the rate at the maximum average utility by $S'_{k,T}$, namely $S'_{k,T}$, argmax$_{S_{k,T}}$ $U_k/S_{k,T}$.

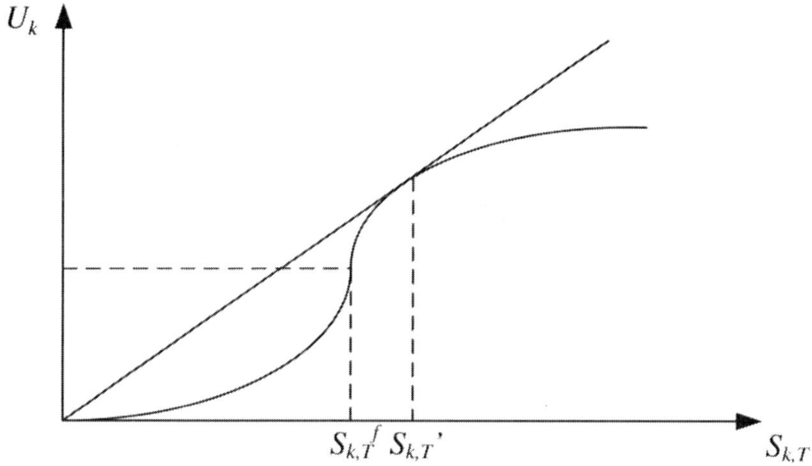

Figure 1 Sigmoid utility function.

2.3 Problem formulation

We pose a maximization problem for each time window as follows. Denote the power allocation by $\mathbf{p}_t = \{p_{k,n,t}, \forall k,n\}$. Each subcarrier can be allocated to at most one user; thus denote the feasible set of power and subcarrier allocations by $\mathbf{A}_t = \{\mathbf{p}_t$ s.t. $\forall n$, $p_{k,n,t} > 0$ for at most one user $k\}$. The goal is to maximize the total expected user utility within the time window under a constraint that the total transmitted power in the current time slot not exceed the power supply P:

$$U^*_{tot} = \max_{\mathbf{p}_t \in \mathbf{A}_t} \sum_{k=1}^{K} E[\,U_k(S_{k,T})]$$

$$\text{s.t.} \quad \sum_{k=1}^{K} \sum_{n=1}^{N} p_{k,n,t} \leq P;$$

$$p_{k,n,t} \geq 0 \ \forall k, n; \qquad (1)$$

where past allocations and rates, $\{\mathbf{p}_\tau, 1 \le \tau < t\}$ and $\{r_{k,n,\tau}, 1 \le \tau < t\}$, and past and current channels, $\{H_{k,n,\tau}, 1 \le \tau \le t, \forall k,n\}$, are known.

As discussed in the Introduction, the novel part of this maximization problem is the use of a sigmoid utility function and its dependence upon the average rate achieved over a time window. The major challenge of this problem is that power allocation in the current time slot affects not only the current rate but also the utility that may be earned at the end of the time window. The system should consider both the current channel and the likely achievable average rates by the end of the time window. For instance, suppose that it is early in the time window and a user currently faces a worse than average channel. Policies for inelastic applications would respond to a poor channel by increasing power, while policies for elastic applications would respond by allocating less power and waiting for a better channel. However, for video conferencing, utility is a function of the average rate over a time window, not of instantaneous rate, and it is a sigmoid function. Should the network allocate *less* than average power, figuring that later in the time window it can make up the difference? Does this depend on the auto-correlation of the channel? If late in the time window this user has experienced an average rate that places it in the convex portion of the utility function, should the network respond by increasing the transmission power? Alternatively, should the network give up on this user obtaining decent performance in this time window, and use the power to increase the utility of other users?

3 RELATED STUDY

For both elastic and inelastic applications, both non-utility based and utility based resource allocation methods have been proposed and thoroughly studied.

For inelastic applications, non-utility based resource allocation methods (such as margin adaption) generally attempt to minimize usage of wireless resources (typically power or channels)[17]. Utility based resource allocation methods generally model the utility of inelastic applications as a step function of rate, which reflects the requirement to achieve a constant bit rate if this user is active. The objective is then to maximize total user utility and bin-packing methods are often proposed to allocate resources[18].

For elastic applications, the most common non-utility based approaches attempt to maximize capacity or to minimize usage of wireless resources given a set of active users, see e.g.,[19–21]. Utility based resource allocation methods generally model the utility of elastic applications as a concave function of rate, which reflects the lack of any minimum rate requirement. Maximization of total user utility is often proposed using standard convex optimization methods[22–26].

For semi-elastic applications, there is less literature. A few articles model user utility as a sigmoid function of *instantaneous* rate or bandwidth. Lee et al.[13] consider both scheduling and user selection for CDMA systems, and propose using pricing to select which users are active and to allocate power. Hande et al.[14] similarly consider sigmoid utility functions under a single resource constraint, and give conditions under which the Nash equilibrium using marginal cost pricing is optimal. Kuo et al.[15] model utility as a sigmoid function of bandwidth, and design heuristic algorithms to choose a near optimal bandwidth allocation. However, we are unaware of any research literature on power allocation when utility is a function of the *average bit rate over a period of time*.

The optimization problem (1) could be framed as a stochastic dynamic programming problem. The base station may have knowledge of past allocations and rates, $\{\mathbf{p}_\tau, 1 \le \tau < t\}$ and $\{r_{k,n,\tau}, 1 \le \tau < t\}$, and past and current channels, $\{H_{k,n,\tau}, 1 \le \tau \le t, \forall k,n\}$. However, since the channel is assumed to be Markov, it will suffice to consider the set of current channel gains over all users and subcarriers, denoted by $\mathbf{H}_t = \{H_{k,n,t}, \forall k,n\}$, and the set of contributions as of time $t - 1$ toward the average rate of all users in the current time window, denoted by $\mathbf{S}_{t-1} = \{S_{k,t-1}, \forall k\}$.

Denote a resource allocation policy that makes decisions on the basis of the current channels and the accumulated rates by $Q(\mathbf{H}_t, \mathbf{S}_{t-1})$. A causal version of the problem (1) would require the determination of a resource allocation policy Q that maximizes total expected user utility. Denote the expected incremental utility from the current slot t to the end of the window achieved under policy Q by$V_t^{Q(\mathbf{H}_t,\mathbf{S}_{t-1})}$. The optimal resource allocation policy would allocate the power in time slot t, $\mathbf{p}_t$, to maximize the incremental utility earned in the current time slot plus the expected incremental utility earned in future time slots. This can be stated recursively as:

$$V_t^{Q(\mathbf{H}_t,\mathbf{S}_{t-1})} = \max_{\mathbf{p}_t \in \mathbf{A}_t} \left\{ \sum_{k=1}^{K} \Delta U_{k,t} + E\left(V_{t+1}^{Q(\mathbf{H}_{t+1},\mathbf{S}_{t-1})} | \mathbf{H}_t, \mathbf{S}_{t-1} \right) \right\},$$

$$\forall 1 \le t < T$$

$$V_T^{Q(\mathbf{H}_T,\mathbf{S}_{T-1})} = \max_{\mathbf{p}_T \in \mathbf{A}_T} \sum_{k=1}^{K} \Delta U_{k,T} \tag{2}$$

To complete the framing as a stochastic dynamic programming problem, we need the channel gains, power levels, and achieved rates to be discrete variables. Suppose we quantize the channel gain on each subcarrier into Z_H levels and the achieved rates for each users into Z_R levels. Then the state $(\mathbf{H}_t, \mathbf{S}_{t-1})$ has $Z_H^{NK} Z_R^K$ levels. The complexity of stochastic dynamic programming is well known to suffer a curse of dimensionality in the number of levels of the state[27]. It follows that such an approach is computationally prohibitive for any reasonably sized systems. We are thus motivated to search for less computationally complex approaches to resource allocation.

4 NON-CAUSAL VERSION

We first consider a non-causal system in which at the beginning of each time window the network knows the channel gains of each user in each time slot of the window. Such knowledge removes (for the time being) the challenge of consideration of how current resource allocation affects future utility. While knowing all the future information is unrealistic, this will serve as an upper bound to causal systems.

Because all of the channel gains are known at the beginning of each time window in this version, the resource allocation decisions for power and subcarriers can be made jointly for all users and all time slots within the window. The multiple time slot decisions are thus transformed into a single decision at the beginning of each time window. This will allow full consideration of the variation of channel for each user from time slot to slot and of the average bit rate during the time window achieved as a result of allocations made in each slot.

However, the direct solution of problem (1) requires solving *KNT* fixed point equations. We are thus motivated to solve a dual problem[28]. The idea, used previously for strictly concave utility functions[29], is to separate the determination of each user's rate (here $S_{k,T}$) and the allocation of power (here **p**) using a set of intermediate variables $\mathbf{d} = \{d_k\}$ as bounds on the achieved rates. A similar decomposition can be applied to sigmoid utility functions, transforming problem (1) into:

$$U_{tot}^{*} = \max_{\mathbf{p}\in\mathbf{A},\mathbf{d}} \sum_{k=1}^{K} U_k(d_k) \text{ s.t.} S_{k,T} \geq d_k$$

$$\sum_{k=1}^{K}\sum_{n=1}^{N} p_{k,n,t} \leq P,\ p_{k,n,t} \geq 0,\ \forall k,n,t \qquad (3)$$

The Lagrange of (3) is given by:

$$\begin{aligned}\bar{J}(\boldsymbol{\lambda},\boldsymbol{\mu}) &= \max_{\mathbf{p}\in\mathbf{A},\mathbf{d}} J(\mathbf{d},\mathbf{p},\boldsymbol{\lambda},\boldsymbol{\mu}) \\ &= \max_{\mathbf{d}} \sum_{k=1}^{K} (U_k(d_k) - \lambda_k d_k) \\ &+ \max_{\mathbf{p}\in\mathbf{A}} \left[\sum_{k=1}^{K} \lambda_k S_{k,T} + \sum_{t=1}^{T} \mu_t \left(P - \sum_{k=1}^{K}\sum_{n=1}^{N} p_{k,n,t}\right)\right]\end{aligned}$$

where λ= $\{\lambda_k\}$ are the Lagrangian multipliers associated with the rate constraints and μ= $\{\mu_t, 1 \le t \le T\}$ are the Lagrangian multipliers associated with the power constraints. The dual function is then given by:

$$\bar{J}(\boldsymbol{\lambda},\boldsymbol{\mu}) = \max_{\mathbf{p}\in\mathbf{A},\mathbf{d}} J(\mathbf{d},\mathbf{p},\boldsymbol{\lambda},\boldsymbol{\mu})$$

and the dual problem is to optimally choose the Lagrangian multipliers:

$$\bar{J}^* = \min_{\boldsymbol{\lambda},\boldsymbol{\mu}} \bar{J}(\boldsymbol{\lambda},\boldsymbol{\mu}) \text{ s.t. } \boldsymbol{\lambda} \succeq 0, \boldsymbol{\mu} \succeq 0 \tag{4}$$

For strictly concave utility functions, the dual problem always gives the same solution as the primal problem. However, for non-concave utility functions, the dual problem may result in a sub-optimal solution. The amount of the sub-optimality is measured by the duality gap$\bar{J}^*-U^*tot$. If all downlink power is allocated in all timeslots, i.e$\sum_{k=1}^{K}\sum_{n=1}^{N} p_{k,n,t} = P, \forall t$, and the target rate is achieved for all users, i.e., $R_k = d_k, \forall k$, then the duality gap is zero. Otherwise, the duality gap is positive. It can be shown that the gap is bounded[30]. We further investigate the amount of sub-optimality in simulation results in Section 7.

Using the same method in[30], the dual problem (4) can be represented as:

$$\bar{J}^* = \min_{\lambda,\mu}\left[f_1(\boldsymbol{\lambda}) + \sum_{t=1}^{T}\sum_{n=1}^{N} f_{2,n}(\boldsymbol{\lambda},\boldsymbol{\mu}) + \sum_{t=1}^{T}\mu_t P\right] \tag{5}$$

Where

$$f_1(\boldsymbol{\lambda}) = \max_{\mathbf{d}} \sum_{k=1}^{K} (U_k(d_k) - \lambda_k d_k) \tag{6}$$

$$f_{2,n,t}(\boldsymbol{\lambda},\boldsymbol{\mu}) = \max_{\mathbf{p}\in\mathbf{A}} \left(\sum_{k=1}^{K} \frac{1}{T}\lambda_k r_{k,n,t} - \mu_t \sum_{k=1}^{K} p_{k,n,t}\right) \tag{7}$$

According to the first-order condition $\partial f_{2,n,t}/\partial p_{k,n,t} = 0$, the solution of (7) is

$$p_{k,n,t} = \left(\frac{B\lambda_k}{T\mu_t \ln 2} - \frac{\delta^2 + I}{|H_{k,n,t}|^2} \right)^+ \tag{8}$$

Substituting (8) into (7) and simplifying we obtain

$$f_{2,n,t}(\boldsymbol{\lambda}, \boldsymbol{\mu}) = \max_k \Phi_{k,n,t}, \tag{9}$$

where

$$\Phi_{k,n,t} = \frac{\lambda_k}{T} B \left[\log_2 \left(\frac{B\lambda_k}{T\mu_t \ln 2} \frac{|H_{k,n,t}|^2}{\delta^2 + I} \right) \right]^+ \\ -\mu_t \left(\frac{B\lambda_k}{T\mu_t \ln 2} - \frac{\delta^2 + I}{|H_{k,n,t}|^2} \right)^+$$

Denote the relationship between $\Phi_{k,n,t}$ and $|H_{k,n,t}|^2$ as $\Phi_{k,n,t}(|H_{k,n,t}|^2)$. If the user can obtain the subcarrier, $\Phi_{k,n,t}$ should be greater than 0. Using (8), it follows that $\Phi_{k,n,t} = 0$ if and only if $|H_{k,n,t}|^2 < T\mu_t \ln 2(\delta^2 + I)/(B\lambda_k)$. Thus we define the inverse of $\Phi_{k,n,t}$ at $|H_{k,n,t}|^2 = 0$ as $\Phi^{-1}_{k,n,t}(0) = T\mu_t \ln 2(\delta^2 + I)/(B\lambda_k)$. When $\Phi_{k,n,t} > 0$, $\Phi_{k,n,t}$ is a monotonically increasing function of $|H_{k,n,t}|^2$, and thus if $|H_{k,n,t}|^2 < \Phi^{-1}_{k,n,t}(0)$, the user cannot obtain subcarrier n.

5 RESOURCE ALLOCATION WITH EXPECTATION OF FUTURE RATE

Having established the form of the optimal non-causal resource allocation, we now turn to formulation of causal resource allocation methods. Causality, combined with Markov channels, implies that resource allocation should be based on the set of current channels and accumulated rates $(\mathbf{H}_t, \mathbf{S}_{t-1})$. In addition, the system may have knowledge of the joint distribution of channels in current and future time slots, e.g., the auto-correlation of channels. The principal challenge is thus allocate resources in the current time slot given the average rate achieved so far in the time window and the likely achievable rate during the remainder of the time window.

In this section, we design a resource allocation policy that considers both the average rate achieved so far and the *expected* future rate. To estimate allocation in the future time slots in the current time window, we assume that the base station knows the channel gain $|H_{k,n,t}|$ for all users and subcarriers in the current time slot t and the conditional density $f(|H_{k,n,\tau}|^2 \,||\, |H_{k,n,t}|^2)$

for future slots τ in the current time window.

We start with the causal problem formulation in (2). Since direct solution via stochastic dynamic programming is computationally prohibitive, we propose interchanging the utility and the expectation, i.e., maximizing the utility of the expected average rate over the time window instead of maximizing the expected utility. Denote the expected user k rate in time slot τ under scheduling policy Q by $E^{Q(\mathbf{H}_t,S_{t-1})}(R_{k,\tau}|\mathbf{H}_t)$. Then the expected user k rate from time slot t through the end of the time window is given $R_{k,F} = R_{k,t} + \sum_{\tau=t+1}^{T} E^{Q(\mathbf{H}_t,\mathbf{S}_{t-1})}(R_{k,\tau}|\mathbf{H}_t)$ Denote the incremental utility of the expected average rate over the time window under scheduling policy Q byWQ($\mathbf{H}t$,$\mathbf{S}t$−1)t. This interchange transforms (2) into:

$$\begin{aligned} W_t^{Q(\mathbf{H}_t,\mathbf{S}_{t-1})} &= \max_{\mathbf{p}_t\in\mathbf{A}_t} \sum_{k=1}^{K} U_k(S_{k,t-1} + R_{k,F}/T) \\ \text{s.t.} \quad & \sum_{k=1}^{K}\sum_{n=1}^{N} p_{k,n,t} \le P;\ p_{k,n,t} \ge 0\ \forall k,n \\ & E\left(\sum_{k=1}^{K}\sum_{n=1}^{N} p_{k,n,\tau}|\mathbf{H}_t\right) \le P,\ \forall\tau \in [t+1,T], \end{aligned} \tag{10}$$

where the latter constraint places limits on the expected power allocated in future time slots.

The expected future user k rate is the sum of the expected future user k rate on each subcarrier:

$$E(R_{k,\tau}|\mathbf{H}_t) = \sum_{n=1}^{N} E(r_{k,n,\tau}|\mathbf{H}_t) \tag{11}$$

Denote the probability that under policy Q subcarrier n will be assigned to user k at time slot τ by $P(p_{k,n,\tau} > 0)$. If this probability is known, then the expected future user k rate on subcarrier n could be found by:

$$\begin{aligned} & E(r_{k,n,\tau}|\mathbf{H}_t) \\ &= \int_{\Phi_{k,n,t}^{-1}(0)}^{+\infty} r_{k,n,\tau} f\left(|H_{k,n,\tau}|^2 \,\middle|\, |H_{k,n,t}|^2\right) P\left(p_{k,n,\tau} > 0\right) d|H_{k,n,\tau}|^2 \end{aligned} \tag{12}$$

since the channel fading is assumed to be a Markov process and the fading on different subcarriers are assumed to be independent to each other. The required probability estimate is given by the following property:

Property 1 Let policy Q be dictated by λ_k and μ_τ.

$$P\left(p_{k,n,\tau} > 0\right) = \prod_{\hat{k}=1,\hat{k}\neq k}^{K} P\left\{|H_{k,n,\tau}|^2 > g\left(|H_{\hat{k},n,\tau}|^2\right)\right\}$$

where

$$g\left(|H_{\hat{k},n,\tau}|^2\right) = (\delta^2 + I)\frac{T\mu_\tau \ln 2}{B\lambda_k}\left(\frac{B\lambda_{\hat{k}}}{T\mu_\tau \ln 2}\frac{|H_{\hat{k},n,\tau}|^2}{\delta^2 + I}\right)^{\left(\frac{\lambda_{\hat{k}}}{\lambda_k}\right)^{1/2}}$$

Proof 1 ***See*** **Appendix 1.** □

This probability can also be used to estimate the expected power allocated in future time slots:

$$E\left(\sum_{k=1}^{K}\sum_{n=1}^{N} p_{k,n,\tau}|\mathbf{H}_t\right)$$
$$= \sum_{k=1}^{K}\sum_{n=1}^{N} E(p_{k,n,\tau}|\mathbf{H}_t)$$
$$= \sum_{k=1}^{K}\sum_{n=1}^{N}\int_{\Phi_{k,n,t}^{-1}(0)}^{+\infty} p_{k,n,\tau}(|H_{k,n,\tau}|^2)P(p_{k,n,\tau} > 0)$$
$$\times f\left(|H_{k,n,\tau}|^2 \,\middle|\, |H_{k,n,t}|^2\right) d|H_{k,n,\tau}|^2 \tag{13}$$

A similar decomposition approach to that used in the previous section for a non-causal system can be used here. Introduce d_k as a lower bound on the average rate $R_{k,F}/T$. Then Equation (6), which determines the target rate for each user in the non-causal system, will become:

$$f_3(\boldsymbol{\lambda}) = \max_{\mathbf{d}} \sum_{k=1}^{K}\left(U_k(S_{k,t-1} + d_k) - \lambda_k d_k\right) \tag{14}$$

in the causal system.

The decomposition indicates a method of distributing the optimization. The determination of rate in (14) indicates a role for each user, while the determination of Lagrangian multipliers μ indicates a role for the network. These two roles must be done in coordination. The decomposition suggests to us that there should be an intermediate power allocation module which determines the Lagrangian multipliers λ in (14) and determines the powers **p** in (8). The communication between the users, power allocation module, and network is illustrated in Figure2. The allocations can be determined iteratively as follows, where the iteration number is denoted by a superscript *i*:

User algorithm: Given λ_k^i,

$$d_k^{i+1} = \arg\max_{d_k}[U_k(S_{k,t-1} + d_k) - \lambda_k^i d_k].$$

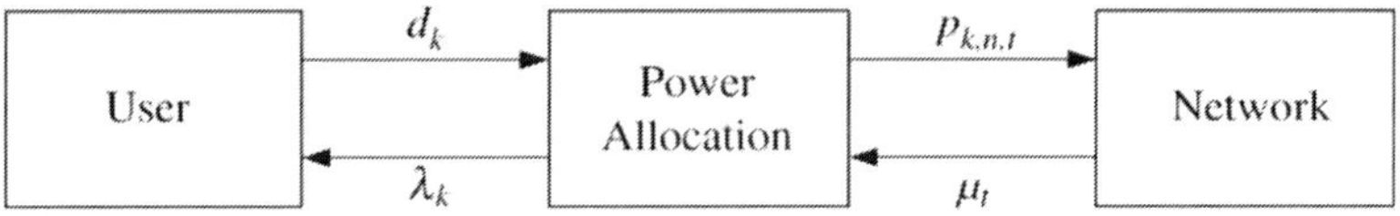

Figure 2 Communication between users, power allocation module, and network.

Network algorithm: Given tentative power and subcarrier allocations $\mathbf{p}_t \in \mathbf{A}_t$ $\forall \tau \in [t, T]$, $\mu_\tau^{i+1} = [\mu_\tau^i + s_B^i z_B^i]^+$, where*siB* is a positive scalar stepsize, and $z_B^i = \mathrm{sgn}\left(\sum_{k=1}^{K}\sum_{n=1}^{N} p_{k,n,t}^i - P\right)$ when $\tau = t$ and $z_B^i = \mathrm{sgn}\left(E\left(\sum_{k=1}^{K}\sum_{n=1}^{N} p_{k,n,\tau}^i|\mathbf{H}_t\right) - P\right)$ when $\tau > t$.

Power allocation algorithm: Given target rates $\mathbf{d}^i$ and Lagrangian multipliers*μiτ*∀τ∈[*t*,*T*] , allocate $\mathbf{p}_t$ using (8) and (9). Calculate $E(R_{k,\tau}|\mathbf{H}_t)$ $\forall \tau \in [t + 1, T]$ using (11). Update $\boldsymbol{\lambda}^{i+1} = [\boldsymbol{\lambda}^i + s_P^i \mathbf{z}_P^i]^+$, where*siP* is a positive scalar stepsize, and**z***iP* is any feasible direction that satisfiessgn$\left(\lambda_k^{i+1} - \lambda_k^i\right) = \mathrm{sgn}\left(d_k^i - R_{k,F}/T\right)$.

For the power allocation algorithm, we propose a subgradient method with bounds to update λ:

$$\lambda_k^{i+1} = \max\left[\min\left(\lambda_k^i + s_P^i(d_k^i - S_{k,T}^i), \overline{\lambda}_k\right), \underline{\lambda}\right] \tag{15}$$

with a suitable choice of step size. The lower bound $\underline{\lambda}$ can be set to a small suitable constant. The upper bound $\overline{\lambda}_k$ is given by

$$\overline{\lambda}_k = dU_k(S_{k,t-1} + R_{k,F}/T)/d(R_{k,F}/T)|(R_{k,F}/T = \Delta R_k')$$

Where $\Delta R_k'$ =argmax $R_{k,F}[U_k(S_{k,t-1} + R_{k,F}/T)U_k(S_{k,t-1})]/(R_{k,F}/T)$. The iteration for λ is terminated when:

$$\lambda_k^{i+1} - \lambda_k^i < \epsilon_1 \; \forall k \text{ or } S_{k,T}^{i+1} = S_{k,T}^i \; \forall k, \tag{16}$$

where ε_1 is a small constant.

For the update of μ in the network algorithm, we propose a bisection algorithm:

$$\text{If } z_B^i > 0, \text{ then } \mu_t^{i+1} = (\mu_t^i + \overline{\mu}_t^i)/2, \underline{\mu}_t^{i+1} = \mu_t^i, \overline{\mu}_t^{i+1} = \overline{\mu}_t^i$$

$$\text{else } \mu_t^{i+1} = (\mu_t^i + \underline{\mu}_t^i)/2, \underline{\mu}_t^{i+1} = \underline{\mu}_t^i, \overline{\mu}_t^{i+1} = \mu_t^i \tag{17}$$

where the initial lower boundμ–0t

can be set to a small suitable constant and the initial upper bound $\underline{\mu}_t^0$ can be derived from (8) as $\overline{\mu}_t^0 = \max_{k,n} (B\overline{\lambda}_k|H_{k,n,t}|^2)/[\,T(\delta^2 + I)\ln 2]$. The iteration for μ_t is terminated when:

$$\mu_t^{i+1} - \mu_t^i < \epsilon_2, \quad (18)$$

where ε_2 is a small constant.

Resource allocation that considers both the average rate achieved so far and the future expected rate can thus be logically distributed by an exchange of price and demand amongst users, the network, and an intermediate power allocation module. The network algorithm and the power allocation algorithm are both run at the base station, and the user algorithm is run in the user's wireless device. Messages must thus be exchanged between the base station and the user's device communicating the rate prices λ^i and the target rates d^i.

This solves the challenge that power allocation in the current time slot affects not only the current rate but also the utility that may be earned at the end of the time window. We call the resulting algorithm, outlined as follows, dual iteration search with prediction (DIS Prediction). The complexity of subgradient updates is polynomial in the dimension of the dual problem, and thus the complexity of the DIS algorithm is polynomial in the number of users *K*.

Dual iteration search with prediction

Every slot, initialize $\mu_t^0 = \underline{\mu}_t^0$,'t and λk=λ–$\forall k$

For current slot *t*, allocate subcarrier and power by (9) and (8)

For future slot τ, calculate

$\left(\sum_{k=1}^{K}\sum_{n=1}^{N} p_{k,n,\tau}|\mathbf{H}_t\right)$ by (13)

Update μ using (17)

Until (18)

If $\lambda_k^{i+1} = \overline{\lambda}_k$

If $R_{k,F}$= 0 then $d_k^{i+1} = 0$ Else $d_k^{i+1} = \Delta R_k'$

Else calculate d_k^{i+1} =arg max $_{d_k}[\,U_k(S_{k,t-1} + d_k) - \lambda_k^i d_k]$

Update λ using (15)

Until (16)

However, our approach required an approximation by interchanging utility and expectation, which may cause some sub-optimality. This will be examined via numerical examples in Section 7.

6 RESOURCE ALLOCATION WITHOUT EXPECTATION OF FUTURE RATE

In this section, we design a simpler resource allocation policy that considers the average rate achieved so far but ignores the expected future rate.

The contribution toward the average rate in the current time window achieved by user k as of slot $t-1$ is given by $S_{k,t-1}$. This can be take into consideration by focusing on the *incremental utility* of user k in slot t, denoted by $\Delta U_{k,t} = U_k(S_{k,t-1} + R_{k,t}/T) - U_k(S_{k,t-1})$, that would result from an allocation of power corresponding to a rate $R_{k,t}$.

A greedy version of the problem (1) would be to maximize the total user incremental utility in each time slot t:

$$\begin{aligned} &\max_{\mathbf{p}_t \in \mathbf{A}_t} \sum_{k=1}^{K} \Delta U_{k,t} \quad \forall 1 \le t < T \\ \text{s.t.} \quad &\sum_{k=1}^{K} \sum_{n=1}^{N} p_{k,n,t} \le P; \\ &p_{k,n,t} \ge 0 \; \forall k, n \end{aligned} \tag{19}$$

A similar approach to that used in the previous section can be used. The dual problem for (19) leads to a set of equations that describe the set of actions for users, the power allocation module, and the network in each time slot. From the point of view of users, the task remains the same; the utility minus cost, $U_k(S_{k,t-1} + d_k) - \lambda_k d_k$. The network algorithm is simpler than that used in DIS Prediction; it now only need determine the Lagrangian multiplier μ_t for power in the current time slot, whereas in DIS Prediction the network algorithm had to determine not only μ_t but also $\{\mu_\tau, t < \tau \le T\}$. The task of the power allocation module is also simpler; it must determine the Lagrangian multipliers λ for rate as before but it now only considers the difference between the actual and desired rate in the current time slot rather than estimating future rate.

The communication between the users, power allocation module, and network is as illustrated in Figure2, and the allocations can be determined iteratively as follows, where the iteration number is denoted by a superscript i:

User algorithm: Given λ_k^i,

$$d_k^{i+1} = \arg\max_{d_k} [\, U_k(S_{k,t-1} + d_k) - \lambda_k^i d_k].$$

Network algorithm: Given tentative power and subcarrier

$$\mathbf{p}_t \in \mathbf{A}_t, \mu_t^{i+1} = [\, \mu_t^i + s_B^i z_B^i]^+,$$

, wheres_B^i is a positive scalar stepsize, and

$$z_B^i = \operatorname{sgn}\left(\sum_{k=1}^{K}\sum_{n=1}^{N} p_{k,n,t}^i - P\right).$$

Power allocation algorithm: Given target rates $\mathbf{d}^i$ and Lagrangian multipliersμ_t^i , allocate $\mathbf{p}_t$ using (8) and (9) and update $\lambda^{i+1} = [\, \lambda^i + s_P^i \mathbf{z}_P^i]^+$ 'wheres_P^i is a positive scalar stepsize, and$\mathbf{z}_P^i$ is any feasible direction that satisfiessgn

$$\left(\lambda_k^{i+1} - \lambda_k^i\right) = \operatorname{sgn}\left(d_k^i - R_{k,t}^i/T\right).$$

We again propose to use subgradient methods for the iteration of μ_t and bisection methods for the iteration of λ_k. As above, we place bounds on each shadow cost, except that the upper bound $\overline{\lambda}_k$ is now given by

$$\overline{\lambda}_k = dU_k(S_{k,t-1} + R_{k,t}/T)/d(R_{k,t}/T)|(R_{k,t}/T = \Delta R_k') \text{ where } \Delta R_k' = \arg\max_{R_{k,t}} [\, U_k(S_{k,t-1} + R_{k,t}/T) \; U_k(S_{k,t-1})] \,/(R_{k,t}/T).$$

Greedy allocation to maximize incremental utility in the current time slot can thus be implemented in a distributed fashion by an exchange of price and demand amongst users, the network, and an intermediate power allocation module. We call the resulting algorithm, outlined as follows, Dual iteration search greedy (DIS Greedy). Its performance will be compared to that of DIS Prediction in the following section.

Greedy dual iteration search

Every slot, initialize $\mu_t^0 = \underline{\mu}_t^0, \forall t \text{ and } \lambda_k = \underline{\lambda} \;\forall k$

For current slot *t*, allocate subcarrier and power by (9) and (8)

Update μ_t using (17)

Until (18)

If $\lambda_k^i = \overline{\lambda}_k$

If $R_{k,t} = 0$ then $d_k^{i+1} = 0$ Else $d_k^{i+1} = \Delta R_k^{'}$

Else calculate $d_k^{i+1} = \arg\max_{d_k} [\, U_k(S_{k,t-1} + d_k) - \lambda_k^i d_k]$

Update λ using (15)

Until (16)

7 SIMULATION RESULTS

In this section, we compare the performance of DIS Prediction and DIS Greedy to an upper bound determined by solution of the non-causal problem. For purposes of the simulation, we set the following parameters: B = 200 KHz, N = 100, I = 0.5, δ^2 = 0.5, T = 10, and P varies from 0.01 to 0.1. We simulate K = 10 users, with identical sigmoid utility functions given by:

$$U_k(S_{k,T}) = \begin{cases} a(S_{k,T}/40)^2, & \text{if } S_{k,T} < 200\,\text{kbps} \\ (S_{k,T}/40 + b)^{1/3}, & \text{else} \end{cases}$$

where a = $(5/6)^{1/3}/25$ and b = −25/6. The utility function has a rate at the maximum average utility at average rate$S'k,T$=250kbps.

The Markov channel is given by:

$$|H_{k,n,t+1}|^2 = \rho|H_{k,n,t}|^2 + \sqrt{1-\rho^2}v,$$

where the correlation coefficient in consecutive time slots ρ = 0.5 and $v \sim Exp(1)$ is independent of $|H_{k,n,t}|^2$ so that $|H_{k,n,t}|^2$ has a Rayleigh distribution[31].

In this scenario, we set the convergence bounds for λ and μ to be very small so as to obtain accurate results. Correspondingly, convergence of μ required 20–50 iterations and convergence of λ required 200–1000 iterations. Iterations of μ^i are the domain of the network and power allocation algorithms, both of which are run at the base station, and thus these iterations may take minimal time. However, iterations of λ^i require communication between the base station and the user's device. This number of iterations is unlikely to be feasible due to the signalling time required, and thus a larger convergence bound ε_2 would be used in a real system. If channels change slowly enough, then acceptable convergence will occur over multiple time slots, and thus the proposed communication will be feasible. However, if channels change too quickly, e.g., due to high mobility, then we suggest that the time window T be lengthened to correspond to several group-of-pictures, and that the rate prices λ must be modified more slowly.

The total utility of all users per time window under DIS Prediction and under DIS Greedy is illustrated in Figure3, as a function of the total downlink power *P*. We also compare these algorithms to the utility that could be achieved in a non-causal scenario; the curve labeled DIS NC is generated by using a similar dual iteration search algorithm applied to the non-causal problem. As should be expected from any reasonable resource allocation strategy, the total utility is an increasing concave function of the total power *P* for all three algorithms.

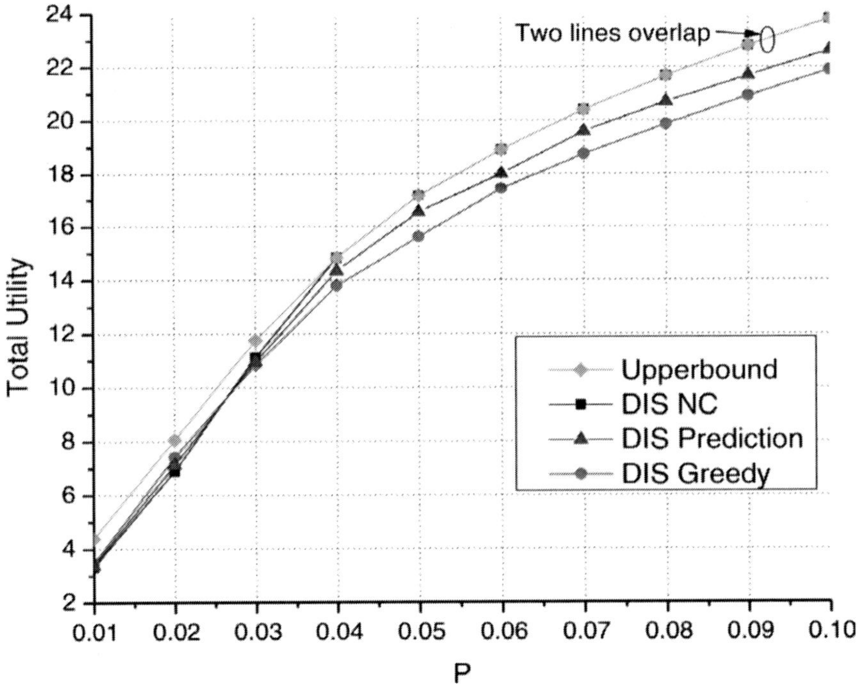

Figure 3 Simulated utility of DIS NC, DIS prediction, and DIS prediction algorithms.

We first consider the performance of the non-causal algorithm. Recall that at the beginning of each time window, this algorithm has knowledge of the channel gains of each user in each time slot of the window. However, DIS NC is not optimal, since it solves the dual problem (3) instead of the primal problem (1). Sub-optimality will occur when there is a duality gap. Because direct solution of (1) is computationally intractable, to quantify the amount of sub-optimality, we calculate an upper bound (denoted in Figure3 as Upperbound) by substituting the solution of DIS NC into (4). We see that at low powers there is a significant performance gap between DIS NC and the upper bound. This gap is caused by users who under DIS NC are allocated average rates lower than$S'k,T$, the rate at the maximum average utility. In general, it is inefficient for users to have average rates below$S'k,T$, since the required resources could often be assigned in a manner that raises many of these users above the inefficient convex portion of the utility curve. As the base station power increases, the performance gap between DIS NC and the upper bound decreases, reflecting that all users now obtain rates above$S'k,T$resulting in a duality gap of zero.

We turn next to the causal algorithm that includes future rate prediction, denoted DIS Prediction. Recall that this algorithm takes into account not only the current user channels on each subcarrier and each user's average rate achieved so far in the time window, but also the expected future rate based on the conditional distribution of the channel in future time slots. As expected, the performance of DIS Prediction falls short of that of DIS NC.

Finally we turn to the causal greedy algorithm denoted DIS Greedy. Recall that this algorithm only takes into account the current user channels on each subcarrier and each user's average rate achieved so far in the time window. As expected, the performance of DIS Greedy is lower than other two algorithms.

When the base station power is low, all three algorithms achieve similar performance. In this situation because power is limited and even DIS NC algorithm cannot achieve the optimal solution. Thus when P = 0.02, the performance of DIS NC even lower than other two algorithms. This indicates that a greedy approach that maximizes only the incremental utility in the current time slot is sufficient. Consideration of future achievable rate, even if known at the beginning of the time window, does not help. At higher base station powers, however, the performance gap becomes significant. Knowledge of future channels is now helpful. With increase of station powers, DIS NC provides optimal performance and DIS Prediction starts to outpace DIS Greedy by using its prediction of future rates.

To understand how such knowledge helps, we focus on the cumulative rate of a single user normalized by T when P = 0.06, as illustrated in Figure4 for each of 10 time slots. In slots 2,5,6,8, and 10, this user's channel is relatively poor on most subcarriers. As a consequence, both DIS NC and DIS Greedy assign no system resources to this user in any of these slots. DIS NC knows at the beginning what the user's channel will be for all 10 time slots, and it assigns substantial resources to this user during slots 1, 3, 7, and 9. In contrast, DIS Greedy must make decisions one slot at a time. It allocates fewer resources than DIS NC during slot 1, not aware of the number of time slots to come in which this user will have a poor channel. It attempts to catch up in slot 3, and further allocates a small amount of resources in slot 4. DIS Greedy acts similarly to DIS NC during slots 5 through 8. However, in slot 9, despite this user's good channel it obtains no additional resources from DIS Greedy due to a combination of the rate achieved so far and the competition with other users. DIS Prediction starts by allocating similar resources in slots 1–2 as DIS Greedy.

However, during slot 3 it allocates fewer resources than DIS Greedy, based on a prediction that the user's channel will get better in future slots. During slots 4 through 6, it allocates no resources due to poor channels, and in slot 7 it allocates similar resources to DIS NC. The biggest difference occurs in slot 9, when DIS Prediction allocates a high level of resources due to a combination of the low rate achieved so far and a good channel.

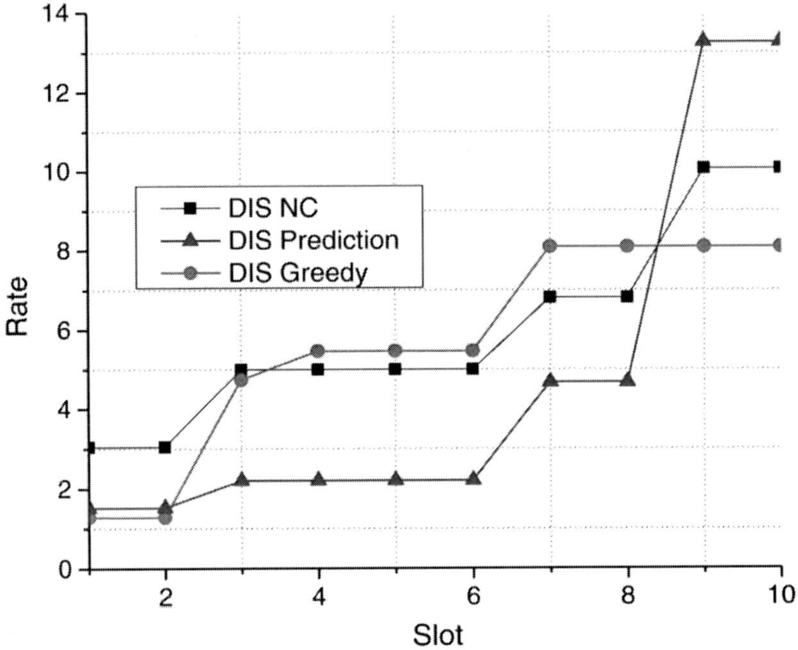

Figure 4 Cumulative rate of a user with a poor channel in slots 2, 5, 6, 8, and 10.

Even though prediction of future expected rates results in increased performance, it still does not achieve the performance resulting from exact knowledge of future channels. The major drawback of DIS Prediction and DIS Greedy is if they allocate too few resources to a user in the first several slots, they may allocate too much resources to the user in the last several slots in an attempt to ensure that the user obtains a rate higher than$S'k,T$

, as we observed in Figure4. To further illustrate this effect, we decrease the system power to $P = 0.05$. In Figure5, we illustrate the cumulative rate of a user who has poor channels in slots 5 through 9. The optimal noncausal strategy is to allocate substantial resources in slots 1, 3, and 4, due to the poor channels that will be suffered in future slots. However, not knowing this, both DIS Greedy and DIS Prediction allocate too few resources in slot 4, and then overcompensate during slot 10.

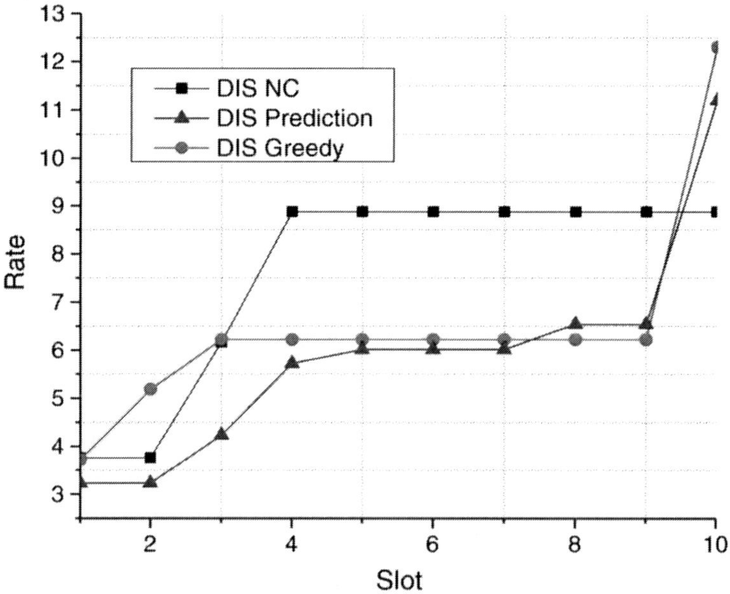

Figure 5 Cumulative rate of a user with a poor channel in slots 5,6,7, and 9.

All of these results pertained to a system with users with identical utility functions. However, it can be envisioned that in some systems not all users will be equally valuable, and utility functions can be used to express such differences. To illustrate this, we briefly examine a system with $K = 2$ users with non-identical utility functions. User 1's utility is as given above. User 2's utility function given by C times that given above, with C varying from 0.1 to 2. We focus on the situation in which both user 1 and user 2 are not allocated enough resources to obtain a rate above$S'k,T$

. The total available power is fixed at $P = 1$ and the value of μ is set to ensure that the power constraint is satisfied.

We first examine the probability that each user obtains a randomly chosen subcarrier in a given slot, given by

$$E\left[P\left(p_{k,n} > 0\right)\right] = \int_{\Phi_{k,n}^{-1}(0)}^{+\infty} \prod_{\hat{k}=1,\hat{k}\neq k}^{K} P\left\{|H_{k,n}|^2 > g\left(|H_{\hat{k},n}|^2\right)\right\} d|H_{k,n}|^2 \quad (20)$$

These expected values are plotted in Figure6. As user 2's utility function increases, its expected subcarrier allocation increases and user 1's expected subcarrier allocation decreases. The sum of the expected subcarrier allocation is less than 1, because when both users have channel gains$\|H_{k,n,t}\|^2<\Phi^{-1}_{k,n,t}(0)$

no power is allocated to either. The expected rate achieved by each users is shown in Figure7. The trend is similar. It follows that different utility functions can be used to accomplish differentiated rate allocation.

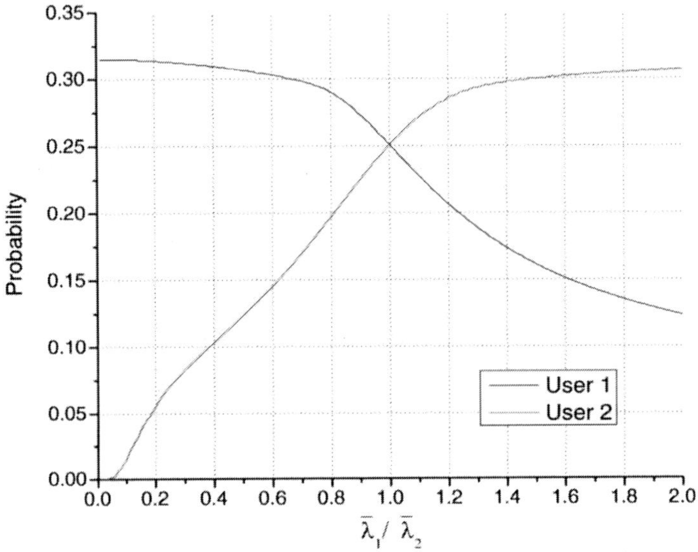

Figure 6 Expected subcarrier allocation versus the ratio of utilities **C** .

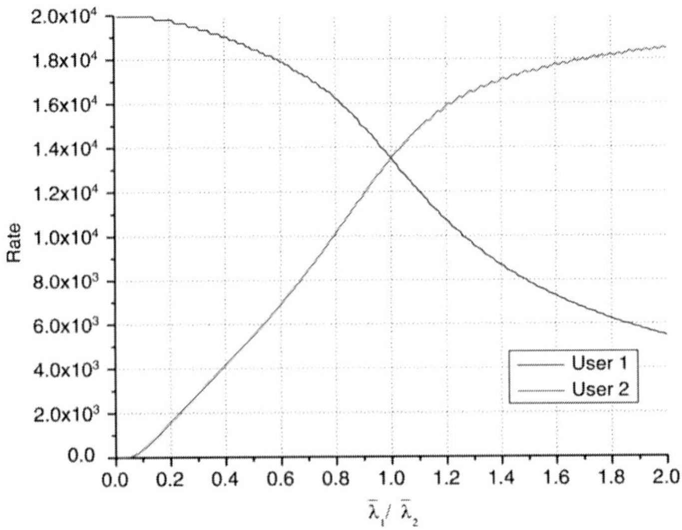

Figure 7 Expected rates versus the ratio of two utilities **C** .

8 CONCLUSION

Resource allocation for video conferencing can benefit from modeling user satisfaction as a sigmoid function of the average bit rate over a time window of a group of pictures. Some articles proposed resource allocation based on the PSNR over groups of pictures subject to average rate constraints, and others proposed resource allocation based on sigmoid utility of the instantaneous rate. None have considered sigmoid utility over groups of pictures.

We considered power and subcarrier allocation in OFDM cellular systems for semi-elastic applications whose utility is a sigmoid function of the average bit rate over multiple time slots. We proposed a resource allocation policy that considers both the average rate achieved so far and the future expected rate. We show how future expected rate can be estimated by modeling the probability that a user will be allocated a subcarrier in a future time slot. Users are unaware of this prediction, but it requires additional work by the network and by the power allocation module. The network must estimate future prices for power. The power allocation module must use these price estimates to estimate future rate.

Then a greedy allocation algorithm is proposed. The network prices power in each time slot so that the demand for power equals the base station supply. A power allocation module assigns power and subcarriers in a manner that maximizes total user utility, and transforms the price for power into a price per unit rate for each user so that users modify their desired rate to match available resources.

The performance of each algorithm is illustrated using numerical results. When the base station power is low, both algorithms have similar performance, although neither makes optimal decisions for users who cannot achieve good average rates. When the base station power is moderate or high, the algorithm that uses the prediction of future rates outperforms the greedy algorithm by taking into account both expected future channels and expected final average rate.

We have focused in this article on the downlink. Any complete design will also require power and rate allocation on the uplink. The major difference in the uplink system model is that there are power constraints on each user rather than a sum power constraint. Thus, if a similar approach to that proposed here were adopted for the uplink, then the shadow price for base station downlink power would be replaced by a set of shadow prices for each active device's uplink power. The determination of optimal power prices would therefore become more complex, perhaps requiring a subgradient algorithm instead of a bisection algorithm. Future research would be required to evaluate the resulting complexity.

DECLARATIONS

Acknowledgements

This study had been supported by the National Science Foundation. Portions of this study appeared in C Yang and S Jordan, Power and Rate Allocation for Video Conferencing in Cellular Networks, Allerton Conference on Communication, Control and Computing, Monticello, Illinois, September 2011, pp. 127–134.

Competing interests

The authors declare that they have no competing interests.

REFERENCES

1. Sandvine: Global Internet phenomena report. Tech. rep. (2011) http://www.sandvine.com
2. Shenker S: Fundamental design issues for the future Internet. IEEE J. Sel. Areas Commun 1995, 13(7):1176-1188.
3. Trestian R, Moldovan A, Muntean C, Ormond O, Muntean GM: Quality utility modelling for multimedia applications for android mobile devices. In IEEE International Symposium on Broadband Multimedia Systems and Broadcasting (BMSB). Seoul, Korea; 2012:1-6.
4. Du Q, Zhang X: Statistical QoS provisionings for wireless unicast/multicast of multi-layer video streams. IEEE J. Sel. Areas Commun 2010, 28(3):420-433.
5. Huang J, Li Z, Chiang M, Katsaggelos A: Joint source adaptation and resource allocation for multi-user wireless video streaming. IEEE Trans. Circ. Syst. Video Technol 2008, 18(5):582-595.
6. Wu D, Ci S, Wang H, Katsaggelos A: Application-centric routing for video streaming over multihop wireless networks. IEEE Trans. Circ. Syst. Video Technol 2010, 20(12):1721-1734.
7. Chen CM, Lin CW, Chen YC: Cross-layer packet retry limit adaptation for video transport over wireless LANs. IEEE Trans. Circ. Syst. Video Technol 2010, 20(11):1448-1461.
8. Mansour H, Fallah Y, Nasiopoulos P, Krishnamurthy V: Dynamic resource allocation for MGS H.264/AVC video transmission over link-adaptive networks. IEEE Trans. Multimedia 2009, 11(8):1478-1491.

9. Li P, Zhang H, Zhao B, Rangarajan S: Scalable video multicast with adaptive modulation and coding in broadband wireless data systems. IEEE/ACM Trans. Network 2012, 20(1):57-68.

10. Zhang H, Zheng Y, Khojastepour M, Rangarajan S: Cross-layer optimization for streaming scalable video over fading wireless networks. IEEE J. Sel. Areas Commun 2010, 28(3):344-353.

11. Kim IM, Kim HM: A new resource allocation scheme based on a PSNR criterion for wireless video transmission to stationary receivers over Gaussian channels. IEEE Trans. Wirel. Commun 2002, 1(3):393-401. 10.1109/TWC.2002.800538

12. ITU-T recommendation P.800: Methods for subjective determination of transmission quality.

13. Lee J, Mazumdar R, Shroff N: Downlink power allocation for multi-class wireless systems. IEEE/ACM Trans. Netw 2005, 13(4):854-867.

14. Hande P, Shengyu Z, Mung C: Distributed rate allocation for inelastic flows. IEEE/ACM Trans. Netw 2007, 15(6):1240-1253.

15. Kuo WH, Liao W: Utility-based radio resource allocation for QoS traffic in wireless networks. IEEE Trans. Wirel. Commun 2008, 7(7):2714-2722.

16. Cheung MH, Mohsenian-Rad AH, Wong V, Schober R: Random access for elastic and inelastic traffic in WLANs. IEEE Trans. Wirel. Commun 2010, 9(6):1861-1866.

17. Cioffi JM: Digital communications. Course Reader. 2003.http://www.stanford.edu/group/cioffi/

18. Liu P, Zhang P, Jordan S, Honig M: Single-cell forward link power allocation using pricing in wireless networks. IEEE Trans. Wirel. Commun 2004, 3(2):533-543. 10.1109/TWC.2004.825363

19. Jang J, Lee KB: Transmit power adaptation for multiuser OFDM systems. IEEE J. Sel. Areas Commun 2003, 21(2):171-178. 10.1109/JSAC.2002.807348

20. Shen Z, Andrews J, Evans B: Adaptive resource allocation in multiuser OFDM systems with proportional rate constraints. IEEE Trans. Wirel. Commun 2005, 4(6):2726-2737.

21. Wong CY, Cheng R, Lataief K, Murch R: Multiuser OFDM with adaptive subcarrier, bit, and power allocation. IEEE J. Sel. Areas Commun 1999, 17(10):1747-1758. 10.1109/49.793310

22. Gao X, Nandagopal T, Bharghavan V: Achieving application level fairness through utility-based wireless fair scheduling. In Proc. GlobeCom. San Antonio, Texas, USA; 2001:3257-3261.

23. Feng N, Mau SC, Mandayam N: Pricing and power control for joint network-centric and user-centric radio resource management. IEEE Trans. Commun 2004, 52: 1547-1557. 10.1109/TCOMM.2004.833191

24. Song G, Li Y: Utility-based resource allocation and scheduling in OFDM-based wireless broadband networks. IEEE Commun. Mag 2005, 43(12):127-134.

25. Zhou C, Honig M, Jordan S: Utility-based power control for a two-cell CDMA data network. IEEE Trans. Wirel. Commun 2005, 4(6):2764-2776.

26. Yang C, Wang W, Zhang X: Multi-service transmission in multiuser cooperative networks. In Proc. WCNC. Budapest, Hungary; 2009:1-5.

27. Berry R: Some dynamic resource allocation problems in wireless networks. In Proc. SPIE. Denver, Colorado, USA; 2001:37-48.

28. Boyd S, Vandenberghe L: Convex Optimization. Cambridge Univ. Press, Cambridge; 2004.

29. Ng TCY, Yu W: Joint optimization of relay strategies and resource allocations in cooperative cellular networks. IEEE J. Sel. Areas Commun 2007, 25(2):328-339.

30. Yang C, Jordan S: Resource allocation for semi-elastic applications in wireless networks. In IEEE Global Telecommunications Conference (GLOBECOM). Houston, Texas, USA; 2011:1-6.

31. Goldsmith A: Wireless Communications. Cambridge Univ. Press, Cambridge; 2005.

Index